Graph Theory:
Flows, Matrices

Graph Theory: Flows, Matrices

Béla Andrásfai
*Technical University of Budapest,
Hungary*

Adam Hilger, Bristol and New York

This book is the revised version of the Hungarian
Gráfelmélet. Folyamok, mátrixok published by Akadémiai
Kiadó, Budapest

Copyright © B. Andrásfai 1991

Copyright © English translation — O. Bíró 1991

All rights reserved. No part of this publication may be reproduced,
stored in a retrieval system or transmitted in any form or by any
means, electronic, mechanical, photocopying, recording or otherwise,
without the prior permission of the publisher. Multiple copying is
only permitted under the terms of the agreement between the Committee
of Vice-Chancellors and Principals and the Copyright Licensing
Agency.

British Library Cataloguing in Publication Data

Andrásfai, Béla
   Graph theory: flows, matrices.
   1. Graph theory
   I. Title II. Gráfelmélet. *English*
   511'.5
   ISBN 0-85274-222-3

Library of Congress Cataloging in Publication Data

Andrásfai, Béla
   Graph theory: flows, matrices.
   Translation of: Gráfelmélet. Folyamok, mátrixok.
   Bibliography: p.
   Includes index.
   1. Graph theory. 2. Matrices. I. Title.
   Q166.A532713 1991   511'.5   88-32072
   ISBN 0-85274-222-3

Published under the Adam Hilger imprint by IOP Publishing Ltd
Techno House, Redcliffe Way, Bristol BS1 6NX, England
335 East 45th Street, New York NY 10017-3483, USA.
Published in co-edition with Akadémiai Kiadó, Budapest.
Printed in Hungary
by Akadémiai Kiadó és Nyomda Vállalat, Budapest

Dedicated to my masters P. Erdős and T. Gallai

Contents

Preface ix

1 Structure of the graph model 1
   1.1 The abstract graph 1
   1.2 Geometrical realisation of graphs 2
   1.3 Components 3
   1.4 Leaves 4
   1.5 Blocks 12
   1.6 The strongly connected components of directed graphs 19
   1.7 Problems 24

2 Optimal flows 27
   2.1 Two basic problems 27
   2.2 Maximal set of independent paths 30
   2.3 The optimal assignment problem. The Hungarian method 44
   2.4 Max flow — min cut 55
   2.5 Dynamic flow. The mobilisation problem 75
   2.6 The synthesis of flow problems 77
   2.7 Optimal planning. The role of the critical path 98
   2.8 Minimal cost transportation 108
   2.9 Minimal cost flows 117
   2.10 Problems 126

3 Graphs and matrices 133
   3.1 The adjacency matrix 133
   3.2 The incidence matrix 139
   3.3 The circuit matrix 146
   3.4 Cutsets and the cutset matrix 152
   3.5 Interrelations between the matrices of graphs 161
   3.6 The spectrum of graphs. The complexity 168
   3.7 Linear electrical networks 190
   3.8 Further matrices associated with graphs 216

3.9 Problems 220

4 Solution of problems 227
   Chapter 1 227
   Chapter 2 230
   Chapter 3 246

References 273

Subject index 277

Preface

My book *Ismerkedés a Gráfelmélettel* was published in 1971 and since 1977 it has also been available in English, entitled *Introductory Graph Theory*. The present book is the continuation of the former at a higher level. It is assumed that the reader has encountered graphs before and is acquainted with some simple results, knows, for example, about trees and routes, and that she or he is able to formulate simple problems in the language of graphs. In short, some inclination and ability is expected of the reader, and even more an intention to gain insight into what is 'behind graphs'.

The first chapter is of a preparatory character. Here I show how the structure of graphs is usually characterised from the point of view of various connectivity properties.

The second chapter treats flows, transportation and planning problems. Both the possible practical applications and some feasibility considerations are presented.

I have devoted the third chapter to the relationship between graphs and matrices. I have made an effort to cover as broad a range of applications as possible. Comprehension of this part requires knowledge of the concept of linear spaces, the basic properties of bases and the operations on matrices. The solution of integro-differential equations is also touched upon in connection with the investigation of electrical networks, but only a glance at the theory of electrical networks is offered; no detailed investigations are carried out.

I have endeavoured to write this book with didactical considerations taken into account, and to organise its material to enable the reader aiming at self-reliance to progress on her or his own. Problems have been set at the end of each chapter; their solution is found in the fourth chapter.

I promised a continuation in the preface of *Introductory Graph Theory*; I even outlined some topics to be covered in this book. A glance at the contents reveals that a mere fraction of this plan has been carried out. An important reason is that since the publication of my previous book the internal development of graph theory has witnessed a substantial shift of emphasis in favour of algebraisation. Besides the internal development, the rapid expansion of the range of applications has considerably extended the literature. The applications are primarily from the fields of operations research and of computer science. It is the second chapter of my book that provides the connection with the former. In treating the problems, I have here favoured the somewhat more suggestive graph theory model as opposed to the methods of linear programming. The algorithmic implications of the problems have been stressed where appropriate. At the first occurrence of an algorithm, I have stated how the algorithms in the book should be read to integrate them into the text. But observe that these algorithms are in fact programs in the sense that, based on them, computer codes can be directly written in some high level programming language without a thorough knowledge of their mathematical significance. At the same time, if someone has to prepare a program to satisfy actual (industrial, economic, etc.) requirements in economically solving large problems, then it is indispensable to weigh further circumstances carefully and to study other literature, too, especially as regards data structures. My book does not cover these, but the book by Christofides [1] is warmly recommended to the interested reader.

I recommend my book to students of mathematics, to undergraduates in engineering and economics, to teachers of mathematics, to economists, to engineers and to researchers starting or working in scientific fields in connection with graph theory who have an interest in graph theory beyond the knowledge obtainable from textbooks.

I am fulfilling a duty dear to my mind as I express my heartfelt thanks to my master, Professor Tibor Gallai, for the pupil's inheritance, and to my referees, academician László Lovász and András Recski, D. Sc. (Math.), who helped me by their work which exceeded by far the duties of a referee; I have utilised many of their suggestions and ideas. I should like to thank András Frank, Cand. Sci., for information concerning the number of steps required by certain algorithms. My thanks are due for the careful work of the staff of Akadémiai Kiadó. Finally, I affectionately thank my wife, Zsuzsanna, without whose encouragement and devoted helpfulness this book could never have been written.

Béla Andrásfai

1

Structure of the Graph Model

In the book *Introductory Graph Theory* ([2] in the following), several seemingly different problems have been treated which, owing to their common features, can be described by the same mathematical model, by graphs. In Chapter 1 of [2], this model has been described intuitively, and now it will be defined in an abstract way.

1.1 The abstract graph

Let P and E be two disjoint sets (i.e. ones with no common element) and $\mathcal{G}$ an instruction (function) which assigns to each element of E a pair of not necessarily distinct elements in P. The system formed by the ordered triplet $(P, E, \mathcal{G})$ is called an *abstract graph* or, in short, a *graph*. The elements of P are called the *vertices* (or *points* or *nodes*) of the graph and the elements of E its *edges*. It is also said that the *edge e is incident to* the elements of P assigned by $\mathcal{G}$ to e, i.e. its *endpoints*. If the endpoints of e are p and q, the notation $e = \{p, q\}$ is also used with, for example, subscripts employed to distinguish edges with common endpoints. It is also said that e *connects* p and q, or that p is *adjacent with* q, or that p and q are *adjacent*. If P_1 and P_2 are subsets of P and $p \in P_1$, $q \in P_2$, then the edge e is also called a P_1P_2-*edge*. Generally, in what follows, P and E will be finite sets. If at least one of them is not finite, an *infinite graph* is obtained.

A *directed graph* is obtained if, for each edge, $\mathcal{G}$ also prescribes the order of its two endpoints; symbolically: $e = (p, q)$, with subscripts if needed. It is also said that q is *adjacent to* p. The vertex p is called the *tail* of e and the vertex q the *head* of e. If P_1 and P_2 are subsets of P and $p \in P_1$, $q \in P_2$, then the edge e is also called a $\overrightarrow{P_1P_2}$-*edge*.

The fact that the element x of a set is in some *relation* $\mathcal{R}$ with the element y of the set is denoted as

$$x\mathcal{R}y.$$

Such relations associating pairs of elements are called *binary relations*. In the following, the term relation is always used for binary relations.

The *relation* $\mathcal{R}_0$ is defined between graphs as follows: for the graphs $G = (P, E, \mathcal{G})$ and $G' = (P', E', \mathcal{G}')$

$$G \mathcal{R}_0 G'$$

if a correspondence with the following properties can be established between the elements of P and P' as well as between the elements of E and E':

1. Each element of P corresponds to one and only one element in P'. Each element of P' corresponds to one and only one element in P.
2. Each element of E corresponds to one and only one element in E'. Each element of E' corresponds to one and only one element in E.
3. The correspondents of the endpoints of any edge are the endpoints of the correspondent of the edge. In case of directed graphs, the correspondents of the tail and head of any edge are the tail and head of the correspondent of the edge, respectively.

It is easy to see that $\mathcal{R}_0$ is an equivalence relation*. Graphs or directed graphs belonging to the same equivalence class induced by $\mathcal{R}_0$ are called *isomorphic graphs* (cf. the concept given in [2]). A more customary symbol for the relation $\mathcal{R}_0$ is $\cong$. Everything stated about a graph (or directed graph) is valid for all graphs (or directed graphs) isomorphic to it.

1.2 Geometrical realisation of graphs

Any graph $G = (P, E, \mathcal{G})$ can be associated with a point set of the three-dimensional Euclidean space T as follows. Each element of P is associated with a point of T, distinct elements with distinct points. Each element of E is associated with a curve in T as follows: a loop is associated with a topological image** of the circle which includes the point corresponding to the endpoint

---

* The binary relation $\mathcal{R}$ defined on a set H is called an *equivalence relation* if it has the following three properties:

1. it is *reflexive*, i.e. $x\mathcal{R}x$ for any $x \in H$,
2. it is *symmetric*, i.e. if $x\mathcal{R}y$, then $y\mathcal{R}x$,
3. it is *transitive*, i.e. if $x\mathcal{R}y$ and $y\mathcal{R}z$ then $x\mathcal{R}z$.

It is well known, but can also be easily proved that if $\mathcal{R}$ is an equivalence relation defined by the set H, then the elements of H can be classified into classes: each class is constituted by elements related by the relation $\mathcal{R}$. These classes are called *equivalence classes*.

** Two sets of points are called *topological images* if a limit-preserving one-to-one correspondence can be established between their elements. The sets of points

of the loop but does not include any point associated with any other element in P; and each further edge is associated with a simple arc connecting the two points corresponding to the endpoints of the edge and not containing any point corresponding to other points of the graph. It is further required that curves corresponding to any two distinct edges can have common points in points associated with the elements of P only. The totality of points and curves in T associated with the elements of the sets P and E as above is called the *geometrical realisation* of the graph G in T.

The geometrical realisation of directed graphs in T is similarly defined with the additional requirement that an orientation is associated with each curve corresponding to an edge, from the point associated with the tail towards the one corresponding to the head.

This realisability of graphs makes their graphical representation possible and permits a graph to be thought of as its geometrical realisation, i.e. its *diagram in space*. If this geometrical realisation of a graph G is possible in the two-dimensional Euclidean space as well then G is a *planar graph*. Let us disregard the requirement that 'curves corresponding to any two distinct edges can have common points in points associated with the elements of P only'. This modified realisation is always possible in the plane. This allows the graphical representation of graphs in the usual way, by a *planar diagram*; and, further, if the common points of the curves corresponding to the edges, introduced by the modification, are regarded as distinct ones, the graph can be thought of as any of its planar diagrams. In [2], the term graph always meant a diagram; this graphical representation will be used in what follows.

Infinite graphs should be realised more carefully since the cardinality of the set formed by the points of the space T is the continuum. Therefore, a realisation in the above sense exists for an infinite graph $(P, E, \mathcal{G})$ only if the cardinalities of both P and E are the continuum at most.

The definitions given in [2] carry over word for word to abstract graphs, so the concepts presented there need no modification. If nonetheless some of them are reconsidered, the purpose lies in a new motivation. The reader is encouraged to directly redefine some further concepts for abstract graphs.

1.3 Components

The *relation* $\mathcal{R}_1$ on the set P of vertices of a graph $G = (P, E, \mathcal{G})$ is defined as follows: $p\mathcal{R}_1 p$ for each element p of P; for two distinct elements p and q of P, $p\mathcal{R}_1 q$ if and only if q can be attained along a path from p in G.

The relation $\mathcal{R}_1$ is obviously reflexive and symmetric and, according to

---

here be taken from the Euclidean space. The topological image of a section of a straight line is called *simple arc*.

Problem 1.24 of [2], it is also transitive. Therefore, $\mathcal{R}_1$ is an equivalence relation. Consequently, the elements of the set P can be uniquely classified into equivalence classes with respect to $\mathcal{R}_1$. Let $P_1, P_2, \ldots$ denote the subsets of P in the equivalence classes. These sets are pair-wise disjoint and their union is the set P, i.e. $\{P_1, P_2, \ldots\}$ is a *partition of the set* P. This partition has been determined by the equivalence relation $\mathcal{R}_1$; in short $\{P_1, P_2, \ldots\}$ is the *partition* of P *induced by* $\mathcal{R}_1$.

Let the subgraph of G induced by P_i be denoted by G_i ($i = 1, 2, \ldots$). Clearly, any of the graphs G_i is connected. The graph G can contain no edge with its endpoints belonging to different sets P_i, since the relation $\mathcal{R}_1$ holds for the endpoints of any edge of G. Therefore, placing the elements of E into the graphs G_i is also unique. The graphs G_i are called the *components* of G (cf. the definition given in [2]). The components are the 'largest possible' connected subgraphs of a graph. Recall that the graph consisting of a single vertex was called connected in [2]; otherwise, a graph was called connected if any two of its vertices were joined by a path of the graph. Any of the paths connecting the vertices p and q will henceforth be also called a *pq-path*.

1.4 Leaves

In order to 'sharpen' these considerations, let us find those 'largest possible' subgraphs in the graph in which, provided they have at least two vertices, any two vertices are connected by two paths without any common edge. Some paths of a graph no two of which share any edge (though they may share vertices) are called *edge-disjoint paths*.

The *relation* $\mathcal{R}_2$ on the set P of vertices of the graph G will yield these subgraphs. It is defined as follows: $p\mathcal{R}_2 p$ for each element p of P; for two distinct elements p and q of P, $p\mathcal{R}_2 q$ if and only if there are two edge-disjoint pq-paths in G.

The relation $\mathcal{R}_2$ is obviously reflexive and symmetric. It will now be proved to be transitive as well.

Let p, q and r be three distinct vertices, F_1 and F_2 be two edge-disjoint pq-paths and F_3, F_4 be two edge-disjoint qr-paths in the graph G. The existence of two edge-disjoint pr-paths in G must be proved. Let us first assume that F_3 includes the vertex p. (Figure 1.1 shows a possible case; it helps in following the reasoning. The edges of F_1, F_2, F_3 and F_4 in the diagram are fine, heavy, broken and dotted lines in this order.) Let the pr part and the pq part of F_3 be denoted by F_{31} and F_{32}, respectively. Let F_{41} be an sr-path of minimal length in F_4 with its end-vertex s in F_{32}. (Such a vertex s does exist.) Now, the path F_{31} and the path formed by the ps part of F_{32} and by F_{41} are two edge-disjoint pr-paths. Two edge-disjoint pr-paths can be similarly obtained if p is contained in F_4. Therefore, it suffices to treat the case when p is neither in F_3 nor in F_4.

Let the vertex t of F_1 and the vertex u of F_2 be the first vertices in F_3 and in F_4, respectively, when the paths F_1 and F_2 are traversed starting from p. (These vertices may also coincide.) Let F_1' and F_2' denote the pt part and pu part of F_1 and F_2, respectively. Let us first assume that both F_3 and F_4 contain at least one of t and u. (One case is shown in figure 1.2.) It can be assumed without loss of generality that t is a vertex of F_3 and u is a vertex of F_4.

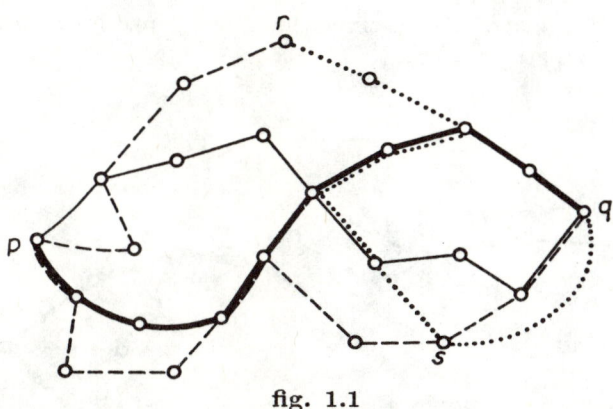

fig. 1.1

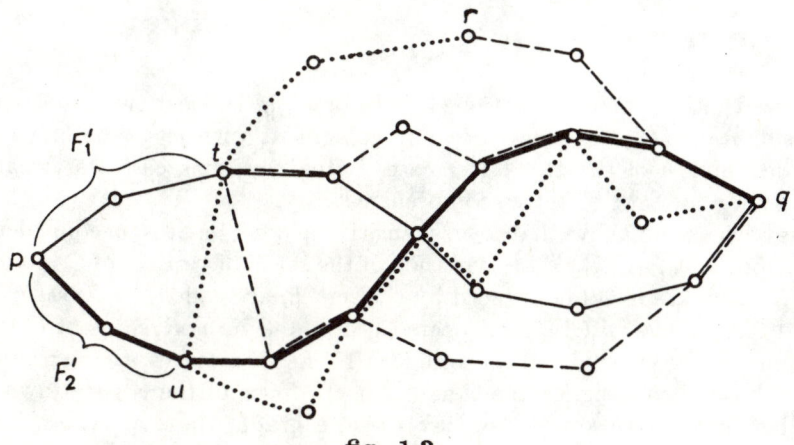

fig. 1.2

Then, the path formed by F_1' and by the tr part of F_3 as well as the path formed by F_2' and by the ur part of F_4 are two edge-disjoint pr-paths. It is therefore sufficient to consider the case when either F_3 or F_4 includes neither t nor u. Let, for example, neither t nor u be a vertex of F_4; similar reasoning can be followed in the other case.

Thus, F_3 includes both the vertices t and u (see figure 1.3). Hence, either the tr part of F_3 denoted by F_3' (its edges are indicated by 3 in figure 1.3) does not contain the vertex u as an inner vertex, or its ur part does not contain the vertex t as an inner vertex (possibly $t = u$). The former case can be assumed, otherwise similar considerations can be made. Let G_0 be the subgraph of G formed by the uq part of F_3 and by F_4. Since F_4 is a qr-path, the transitivity of the relation $\mathcal{R}_1$ ensures the existence of a ur-path in G_0; let F_4' be such a path (its edges are indicated by 4 in figure 1.3). It is now easily verified that the path formed by F_1' and F_3' as well as the path formed by F_2' and F_4' are two edge-disjoint pr-paths in G.

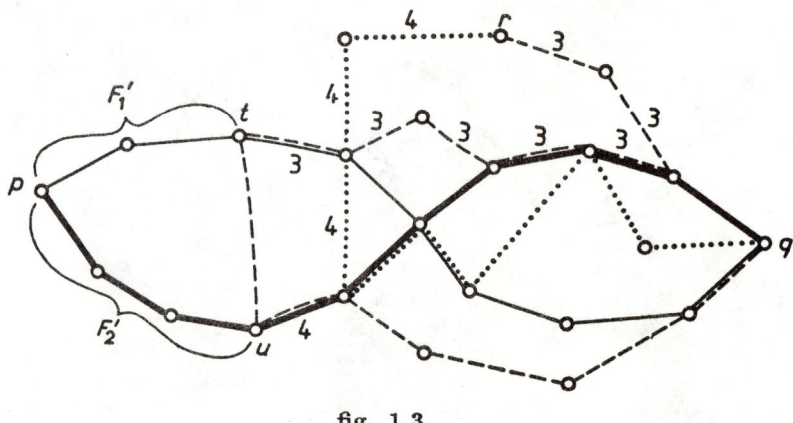

fig. 1.3

Observe that if p may be contained in F_3 or F_4 in the last two cases, then t (and simultaneously u, too) is possibly coincident with p; so the first case is included in one of the two latter cases. However, this case was treated separately for the sake of better comprehensibility.

It has been shown above that $\mathcal{R}_2$ is transitive as well, i.e. it is an equivalence relation. Let $\{P_1, P_2, \ldots\}$ be the partition of the set P of vertices of G induced by $\mathcal{R}_2$. The subgraphs of G induced by the sets P_i are called the *leaves* of G. Evidently, every leaf of G is a connected graph and no two leaves of G may have common vertices. The graph in figure 1.4 has ten leaves; the vertices of each leaf have been assigned the same number. None of the edges drawn by heavy lines belongs to any of the leaves of the graph; these are the bridges 'connecting the leaves' within the components. (See [2] for the concept of bridge.) In order to verify the general validity of our statement, the concept of bridge is extended to disconnected graphs as follows: an edge h is called a *bridge* of a graph if the graph has no circuit containing h. (This is also yielded

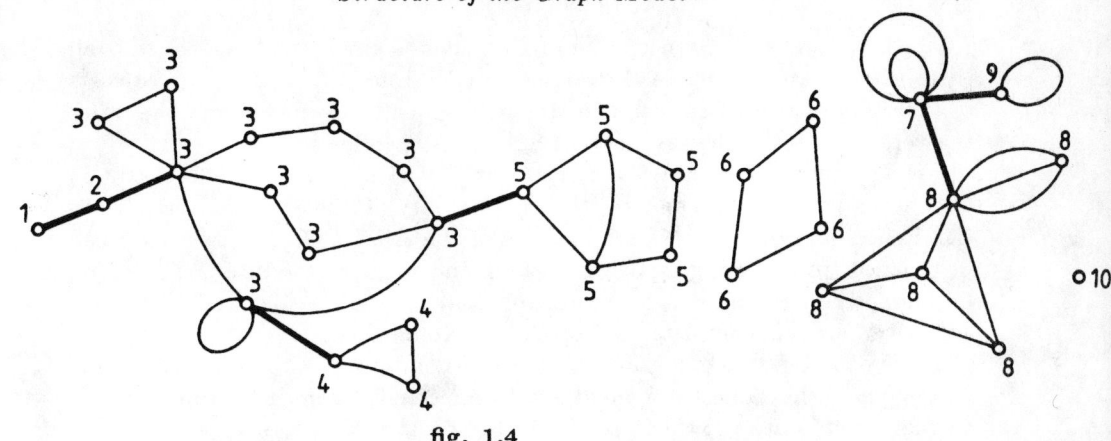

fig. 1.4

by [2] 3.12 and 3.13.) If the endpoints of a bridge h of a graph G belong to distinct leaves of G, say to L_1 and L_2, then h is said to *connect the leaves* L_1 and L_2 of the graph G. Now, the following theorem is valid:

1. *An edge of a graph G is a bridge if and only if it is contained in none of the leaves of G, i.e. it connects two leaves of G.*

Proof. Let $h = \{p, q\}$ be an edge of the graph G. Let us first assume that h is a bridge of G. The graph G cannot contain a pq-path not including h, otherwise there would be a circuit in G containing h. Therefore, there is a single pq-path: the one formed by the edge h. Consequently, p and q cannot belong to the same leaf of G. Let us now assume that the edge h is not a bridge of G. Then there is a circuit in G containing h. Hence, h is either a loop or this circuit consists of two edge-disjoint paths. Consequently, p and q belong to the same leaf of G. This completes the proof of Theorem 1.

The following statement characterising bridges will also show that our present concept of bridges coincides with the one given in [2]:

2. *Let G' be the graph obtained from the graph G by deleting the edge h of G. If h is a bridge of the graph G then the number of components of G' is exactly one more than that of G and the endpoints of the edge h belong to distinct components of G'. Conversely, if h is not a bridge of G then the number of components in G and G' is equal.*

Proof. Let $h = \{p, q\}$ be an edge of the graph G. Let us first assume that h is a bridge of G. In the graph G, q can be reached from p along the path containing h only, otherwise there would be a circuit in G containing h. Let the component of G containing its vertices p and q be denoted by K and let K' denote the graph obtained from K by deleting h. The vertices in K' attainable from p along a path not containing h belong to the same component of K'. So, K being connected, every other vertex in K' can be reached in K along a path containing h, but not along other paths, thus all these vertices are attainable along a path in K' from q. Therefore, K' consists of exactly two components: one of them contains the vertex p and the other the vertex q. Hence the first part of Statement 2 follows.

Let us now assume that h is not a bridge of G. Then there is a circuit in G containing h and this ensures the attainability of q from p along a path in G'. Hence, according to the transitivity of the relation $\mathcal{R}_1$, if a vertex b can be reached from a along a path in G then the same is true in G'. Therefore, G' cannot consist of more components than G does, evidently, nor of less components. This completes the proof of Statement 2.

Since no two leaves of a graph G can have common vertices, the leaves of G belonging to the same component are 'connected by bridges'. However, any two leaves of G are proved to be connected by at most one bridge.

Let us assume, indirectly that two distinct bridges in G connect the same two leaves of G. If the endpoints of these bridges in the same leaf are distinct, they can be connected by a path within the leaf. The two bridges, along with these possibly existing paths, form a circuit, and this circuit contains bridges, a contradiction. So, the following has been proved:

3. *No two leaves of a graph can have common vertices and at most one bridge can connect two leaves.*

In a leaf containing at least two vertices, any vertex can be attained from any other vertex along a path (even by two edge-disjoint paths), 'within the leaf', but not otherwise. To be more precise, the following holds:

4. *If both end-vertices of a path F of a graph belong to some leaf L of the graph, then all vertices of F — and hence all of its edges, too — are in L.*

Proof. Indirectly, let us assume that the end-vertices of the path F of the graph G belong to the leaf L of the graph, but a vertex p in F is not contained in L. Let F' denote a path of minimal length of F which contains the vertex p and let its end-vertices q, r be in L. Evidently, no edge or inner vertex of F' can belong to L. Accordingly, the edges of F' incident with its end-vertices are bridges of G. Now, a circuit in G containing these two bridges is obtained if a qr-path in L is joined to F'; this, however, is impossible.

Since the leaves and bridges of any component of a graph are simultaneously leaves and bridges, respectively, of the graph itself, the structure of the graph manifested by its leaves can be investigated by regarding its components only, i.e. it suffices to consider connected graphs.

Let L and L' be two distinct leaves of the connected graph G, p a vertex in L, and q a vertex in L', and let F be a pq-path in G. Let us traverse F starting from p and let us record the leaves and bridges in this order. Let the sequence thus obtained:

$$L, h_1, L_1, h_2, L_2, \ldots, L_{n-1}, h_n, L'$$

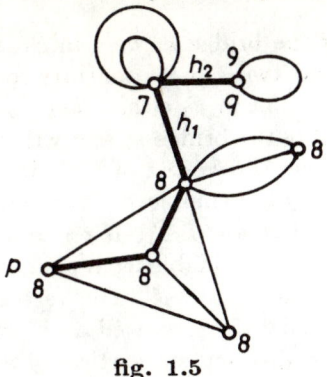

fig. 1.5

be called the *leaf-bridge sequence* corresponding to the pq-path. For example, figure 1.5 shows a component of the graph of figure 1.4. Here the following leaf–bridge sequence corresponds to the pq-path indicated by the heavy line (subscripts of the leaves refer to the number of the vertices):

$$L_8, h_1, L_7, h_2, L_9.$$

Note that no leaf appears in this leaf-bridge sequence more than once, and that the same leaf-bridge sequence corresponds to all paths connecting any vertex of L_8 with q. We prove that this observation is valid in general.

5. *The leaf-bridge sequences, corresponding to any path, have distinct leaves (and bridges, too).*

Proof. Should the leaf L_i appear several times in the leaf-bridge sequence corresponding to the pq-path then this path had a subpath between vertices of L_i and containing a bridge, a contradiction to Theorem 4.

6. *If L and L' are two leaves of a connected graph G then the same leaf-bridge sequence corresponds to any path connecting a vertex of L with one of L'.*

Proof. Let us denote by S, and by S_2 the leaf-bridge sequences

$$L = L_0, h_1, L_1, h_2, \ldots, L_{n-1}, h_n, L_n = L'$$

and

$$L = M_0, k_1, M_1, k_2, \ldots, M_{m-1}, k_m, M_m = L',$$

corresponding to the paths F_1 and F_2 in G, respectively, and assume that they are different. Without loss of generality we can assume that $m \geq n$. Obviously, if $h_1 = k_2$, then $L_1 = M_1$, if $h_1 = k_1$ and $h_2 = k_2$ then $L_1 = M_1$, $L_2 = M_2$, and so on. Consequently, there is a bridge k_i of minimal subscript in S_2 with either $i = n+1$ or $k_i \neq h_i$. In the case $i = n+1$, the leaf $M_n = L_n$ would appear twice in S_2 — which is impossible by Theorem 5. So, it can be assumed that $i \leq n$ and $k_i \neq h_i$. Owing to the minimal property of the subscript i, $M_{i-1} = L_{i-1}$. If the bridge k_i were in S_1, it could only succeed h_i; but then L_{i-1} would appear twice in S_1, contrary to Theorem 5. Thus, k_i is not included in S_1 and hence not in F_1 either. Let F_2' denote the pq part of minimal length in F_2 containing the bridge k_i and with end-vertices in leaves belonging to S_1. Obviously, one end-vertex of F_2' is in L_{i-1} and if the other end-vertex of F_2' is in L_j, then $j \geq i$ (moreover, as a consequence of Theorem 3, $j > i$). It can be assumed that $p \in L_{i-1}$. If p is not included in F_1, then there is a vertex p' of F_1 in L_{i-1} attainable from p along a path in L_{i-1} with no inner vertex in F_1; otherwise let $p = p'$. The vertex q' in F_1 is similarly assigned to q. Now, the $p'q'$ part of F_1, as well as F_2' and the paths in L_{i-1} and L_j connecting the possibly distinct end-vertices of these two paths yield a circuit in G containing the bridge k_i; which is a contradiction. This completes the proof of Theorem 6.

Let the leaf–bridge sequence

$$L_1, h_1, L_2, h_2, \ldots, L_{n-1}, h_{n-1}, L_n$$

formed by the leaves and bridges of a connected graph G be called the *leaf-chain* connecting the leaves L_1 and L_n of the graph G provided all the leaves $L_1, L_2, \ldots, L_n$ in the sequence are different and the bridge h_i connects the leaves L_i and L_{i+1} of G ($i = 1, 2, \ldots, n-1$). Obviously, for any leaf-chain of G connecting the leaves L_1 and L_n there exists a path in G involving the relevant leaf-chain as the corresponding leaf–bridge sequence. Consequently, the theorem below follows from Theorem 6:

7. *Any two leaves of a connected graph are joined by one and only one leaf-chain of the graph.*

This uniqueness of the leaf-chains is reminiscent of the fact that a graph without loops and of at least two vertices is a tree if and only if any two of its vertices are connected by exactly one path (the statement of Problem 2.9 of [2]). The similarity indicates that, as far as the structure of its leaves are concerned, the graph has a tree-like structure. This can be seen by assigning to the graph G the so-called *leaf-graph* G^L of G as follows: a vertex is associated with each leaf of G, these are the vertices of G^L. The edges of G^L are in a one-to-one correspondence with the bridges of G so that the correspondent of the bridge h in G connects the correspondents of the leaves containing the endpoints of h. The leaf-graph of the graph G can be imagined by 'contracting' the vertices of each leaf of G into a single vertex, deleting the edges in the leaf and retaining the bridges. As an example, figure 1.6 indicates the leaf-graph of the graph shown in figure 1.4. Here, each component of figure 1.4 corresponds to a tree. On the basis of Theorem 7 and of the statement of Problem 2.9 of [2], the following theorem expresses the general validity of the last remark.

8. *The leaf-graph of a connected graph is a tree.*

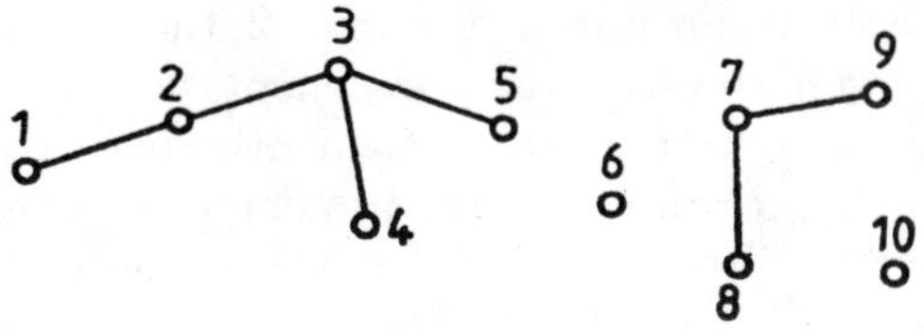

fig. 1.6

The leaves of graphs can also be obtained by the following relation. The *relation* $\mathcal{V}_2$ on the set P of vertices of a graph G is defined as follows: $p\mathcal{V}_2 p$ for each element p of P; for two distinct elements p and q of P, $p\mathcal{V}_2 q$ if and only if, for any partition of P into two parts with p and q included in distinct subsets (in short, for any *partition* $\{P_p, P_q\}$ of P), G contains at least two P_p, P_q-edges.

The relation $\mathcal{V}_2$ is obviously reflexive and symmetric. It will now be proved to be transitive as well.

Let p, q and r be three distinct vertices of G and assume $p\mathcal{V}_2 q$ and $q\mathcal{V}_2 r$. Considering a partition $\{P_p, P_r\}$ of the set of vertices P of the graph G, q is an element of exactly one of P_p and P_r; it can be assumed that $q \in P_p$, otherwise the reasoning is similar. Then this partition is simultaneously a partition of form $\{P_q, P_r\}$, too. Hence, the P_q, P_r-edges, which exist, in view of $q\mathcal{V}_2 r$, are also $P_p P_r$-edges at the same time, i.e. $p\mathcal{V}_2 r$. Therefore, $\mathcal{V}_2$ is an equivalence relation.

Let $\{P_1, P_2, \ldots\}$ be the partition of the set P of the vertices of G induced by $\mathcal{V}_2$. It will be proved that the subgraphs of G induced by the sets P_i are the leaves of G. To this end it suffices to prove the following theorem:

9. *Two vertices of any graph are related by the relation $\mathcal{R}_2$ if and only if they are related by the relation $\mathcal{V}_2$.*

Proof. Let p and q be two distinct elements of the set P of vertices of a graph G. Let us first assume that $p\mathcal{R}_2 q$. It is easily verified that then $p\mathcal{V}_2 q$, but something more is proved instead: for any partition $\{P_p, P_q\}$ of P there are at least two $P_p P_q$-edges both contained in a circuit of G. Namely, according to the assumption, there are two edge-disjoint pq-paths in G; let these be F_1 and F_2. Starting from p, let r_1 be their last common vertex in P_p and r_2 be their next common vertex. (Such vertices always exist: possibly $r_1 = p$ and possibly $r_2 = q$.) Then the $r_1 r_2$ parts of F_1 and of F_2 constitute a circuit of G and each of the two paths includes a $P_p P_q$-edge.

Let us now assume that p and q are not related by $\mathcal{R}_2$, i.e. they are not contained in the same leaf of G. If p and q are not in the same component of G, then they are evidently unrelated by $\mathcal{V}_2$. It is therefore sufficient to treat the case when p and q belong to the same component K of the graph G. Let F denote a pq-path in K. Since p and q do not belong to the same leaf of G, F includes a bridge in G, let this be $h = \{p_1, q_1\}$. The edge h is obviously a bridge of K as well. According to Theorem 2, the graph obtained from K by deleting the edge h has exactly two components K_1 and K_2 with p_1 contained in one of them and q_1 in the other. Evidently, one of p and q, too is in K_1 and the other in K_2; it can be assumed that $p \in K_1$. Now let $\{P_p, P_q\}$ be the partition of the set P in which P_p is formed by the vertices of K_1. Then the only $P_p P_q$-edge in G is the edge h. Therefore, p and q are not related by $\mathcal{V}_2$. This completes the proof of Theorem 9.

1.5 Blocks

Recall the partition of the vertex set of a graph G during the proof of Theorem 9. Two edges were related there if they were contained in one and the same circuit of G. This relation will be studied now.

The *relation* $\mathcal{T}$ on the set E of edges of a graph $G = (P, E, \mathcal{G})$ is defined as follows. $e \mathcal{T} e$ for each element e of E; for two distinct elements e and f of E, $e \mathcal{T} f$ if and only if there is a circuit in G containing both e and f.

The relation $\mathcal{T}$ is obviously reflexive and symmetric. It will now be proved to be transitive as well. In the proof, we use the concept of an arc of a circuit (see in [2]).

Let us assume that the circuit K_1 in G includes the edges $e_1 = \{p_1, q_1\}$ and $e_2 = \{p_2, q_2\}$, and the circuit K_2 includes the edges e_2 and $e_3 = \{p_3, q_3\}$. The existence of a circuit K_3 in G containing the edges e_1 and e_3 is to be verified. The proof is illustrated by figures 1.7(a) and 1.7(b) showing the circuits K_1 and K_2 in a simple and in a complex case. The edges of K_1 are drawn by dotted lines, those of K_2 by continuous lines and the common edges of the two circuits by heavy lines.

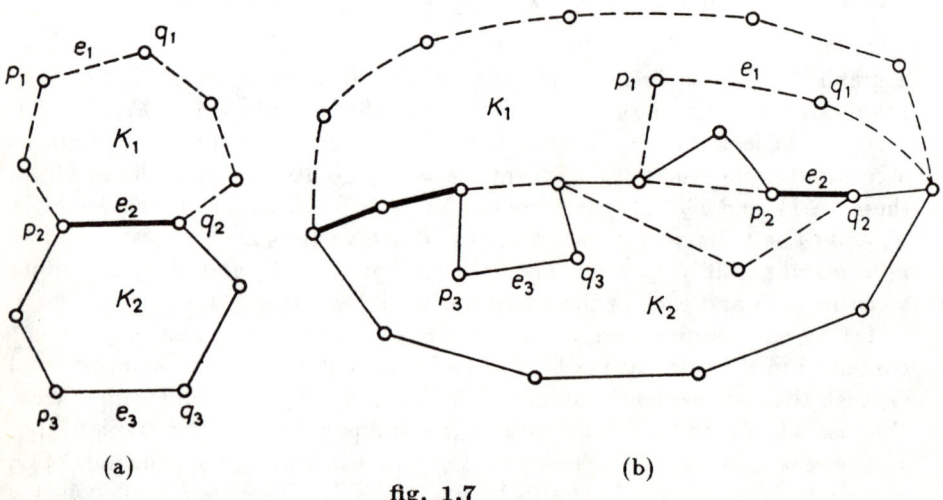

fig. 1.7

The required circuit K_3 is created by joining two paths: one of them is an arc of K_1 including e_1, and the other is an arc of K_2 containing e_3. There is certainly an arc in K_1 including e_1 with its end-vertices in K_2, for example, the one obtained from K_1 by deleting e_2. Let F denote the one of minimal length among the arcs of K_1 with this property, and let a and b be the two end-vertices of F (possibly, F contains the edge e_1 only, or e_1 may be an edge of K_2 as well). F being minimal, it has no inner vertex belonging to K_2 and it can only have an edge in K_2 if it has no edge but e_1. Accordingly, F and the arc of K_2 containing e_3 with end-vertices a and b constitute the circuit K_3 desired. This proves that T is a transitive relation.

It has been shown above that T is an equivalence relation. Let $\{E_1, E_2, \ldots\}$ be the partition of the set E of edges of G induced by T. Let T_i denote the subgraph of G with its edges being the elements of E_i and its vertices the endpoints of the elements of E_i ($i = 1, 2, \ldots$), and let each isolated vertex of G form a further graph T_i. The graphs T_i are called the *blocks* of G. Obviously, the only edge of the block containing a loop h is h itself. As an example, the blocks of the graph in figure 1.4 are shown in figure 1.8: the edges in the same blocks have been assigned the same number. This graph has 18 blocks (the isolated vertex is a block too).

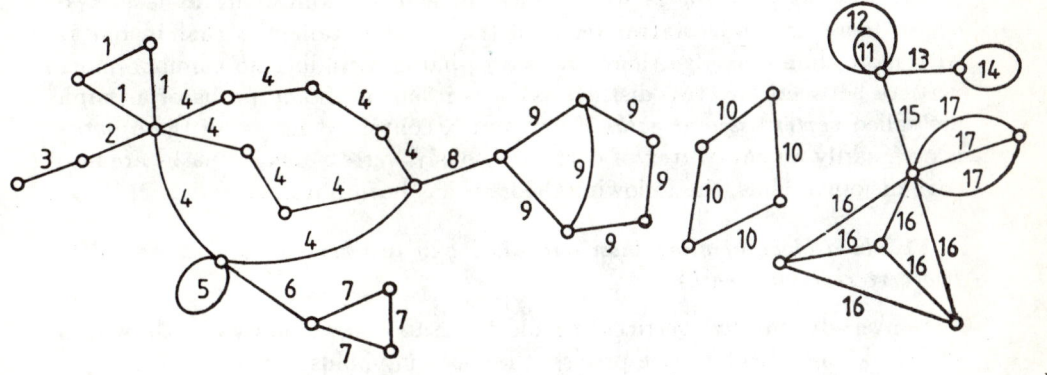

fig. 1.8

In [2], blocks have been defined in an entirely different, completely intuitive way. Both definitions can be shown to lead to the same concept. This, however, will not be carried out now, rather the structural properties of the graphs manifested in their blocks will be thoroughly investigated. To facilitate this inspection, the following definition of articulations (cut-vertices) extended to disconnected graphs will be employed: a vertex p of a graph G is an *articulation* of G if there are two edges in G incident with p with no circuit in G containing both. (In the case of connected graphs, this concept is equivalent to the one given in [2]; see Problem 33). Obviously, a graph consisting of a single block can contain no articulation, since any two edges of the same block are related by the relation T.

Observe that the blocks in figure 1.8 are all connected graphs. The following statement will now be proved which shows this observation to be valid in general:

10. *The blocks of a graph are connected graphs.*

Proof. The block consisting of a single vertex is connected. Let us now assume that two distinct vertices p and q of a graph G are the vertices of one and the same block T of G. The existence of a pq-path in T is to be verified. The graph T necessarily contains an edge e_1 incident with p as well as an edge e_2 incident with q. In the case $e_1 = e_2$, the pq-path is immediately provided, and otherwise there is a circuit in T containing both edges and thus both the vertices p and q, too. This completes the proof of Statement 10.

Carrying the reasoning further, an additional property of blocks can be discovered. It has been stated that, for any two distinct vertices p and q in a block containing at least two vertices, there is either a circuit including both or an edge $e = \{p, q\}$. If, in the latter case, there is also an edge e_0 different from e in T, then, again, there is a circuit in T containing both edges and so the vertices p and q as well. Since, in a block containing at least two edges, there are at least two vertices, the result obtained is that in blocks with more than one edge there are two pq-paths including no common inner vertices between any two distinct vertices p and q. Some paths of a graph are called *vertex-disjoint paths* if any vertex contained in two of these paths is necessarily an end-vertex of both. Obviously, vertex-disjoint paths are also edge-disjoint. Thus, the following theorem has been obtained:

11. *In a block of more than one edge, any two vertices are connected by two vertex-disjoint paths.*

Conversely, any two vertices of a block can be connected by a path 'within the block' only. To be more precise, the following holds:

12. *If both end-vertices of a path F of a graph belong to some block T of the graph then all edges of F are in T.*

Proof. Indirectly, let us assume that the end-vertices of the path F of the graph G belong to the block T of the graph, but an edge f in F is not contained in T. Let F' denote the pq part of minimal length of F which contains the edge f and its end-vertices are in T. Evidently, no edge or inner vertex of F' can belong to T. Now, any pq-path in T existing in view of Theorem 10 together with F' forms a circuit having an edge in T, but with its edge f outside T; this is, however, impossible.

If p and q are two vertices of a block T of a graph then, according to Theorem 12, all edges of each pq-path of the graph are in T. Consequently,

p and q cannot simultaneously be contained in any other block of the graph. Therefore the following statement is true:

13. *Two blocks of any graph can have at most one common vertex.*

The definition of an articulation implies that any articulation of a graph is simultaneously a vertex of at least two blocks of the graph. An articulation may be a common vertex of more than one block, as can also be seen in figure 1.8. If a vertex p is simultaneously contained in two blocks of a graph, then there is an edge incident with p in each block. Consequently, p is articulation of the graph. The following theorem has been obtained:

14. *A vertex is an articulation of a graph if and only if it is the common vertex of at least two of its blocks.*

Let us now examine a path F of a connected graph G with its two end-vertices p and q belonging to distinct blocks of G and let us record those inner vertices of F which are incident to edges of different blocks. All the marked vertices are articulations of G. Such a vertex necessarily exists, otherwise all edges, and so all vertices, of F would belong to the same block. Let the order of these vertices along F from p to q be $a_1, a_2, \ldots, a_n$. Let the sequence

$$p, T_1, a_1, T_2, a_2, \ldots, T_n, a_n, T_{n+1}, q$$

be called a *block-chain* corresponding to F; this means that, while traversing F starting from p, the edges of the block T_1 are used between p and a_1, the edges of the block T_2 are used between a_1 and a_2, etc. F being a path, no block containing a loop may appear in the block-chain. In the case of $i \neq j$, T_i and T_j denote distinct blocks in the block-chain. If $j = i + 1$, this is evident, and if T_i and T_j coincided provided $j > i + 1$, then F would have a subpath with end-vertices in T_i including edges in T_{i+1}, contrary to Theorem 12. So, the following has been proved:

15. *All blocks in a block-chain are distinct.*

In the component of figure 1.8 redrawn in figure 1.9, the block-chain corresponding to the heavy pq-path is the following:

$$p, T_3, a_1, T_2, a_2, T_4, a_3, T_8, q;$$

where T_i is the block with its edges marked by i. This also shows that F can contain further articulations of G, not among a_i, like r and q in the present case. Note that a_1, a_2 and a_3 are vertices of any pq-path in figure 1.9. It will now be proved that in general, too, any pq-path in G different from F includes the vertices $a_1, a_2, \ldots, a_n$.

Indirectly, assume that the path F_0 of G with end-vertices p and q does not contain the vertex a_i. Then neither does F_0 contain the edges e and f in F, incident with a_i, and e and f are known not to belong to the same block of G. Let us select the st-path of minimal length in F, containing the edges

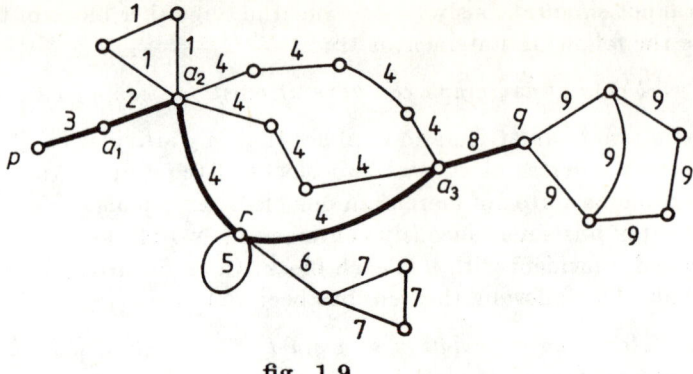

fig. 1.9

e and f and having its end-vertices in F_0. This and the st-path of F_0 yield a circuit including both e and f; this is, however, impossible, since e and f are edges of different blocks.

Let F' be a pq-path in G. F' has been shown to contain the vertices $a_1, a_2, \ldots, a_n$. Let the Pa_1 parts of F and F' be denoted by F_1 and F_1', respectively. The edges of F_1 are all in T_1. Now we prove that all edges of F_1' are also in T_1. Indeed, let us assume that an edge d of F_1' is not in T_1. Let F_1'' denote the uv part of minimal length of F_1' containing the vertex d and with end-vertices in F_1. This and the uv-part of F_1 yield a circuit simultaneously including edges in T_1 and edges not in T_1, but this is impossible. Similar reasoning leads to the conclusion that the $a_1 a_2$ part of F' contains edges in T_2 only, ..., the $a_n q$-part of F' contains edges in T_{n+1} only.

Therefore, the following theorem is true:

16. *If the vertices p and q of a connected graph G do not belong to the same block of G then the same block-chain corresponds to any pq-path of G.*

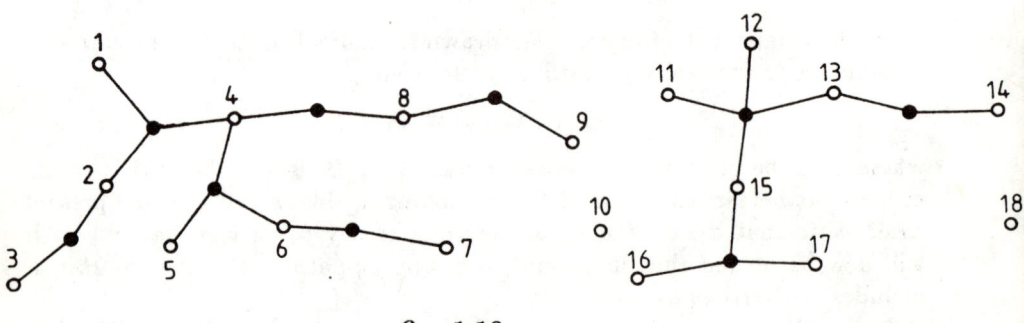

fig. 1.10

This uniqueness of block-chains indicates that the blocks of the graph also has a tree-like structure. This can be shown by assigning the so-called *block-*

graph G^T to the graph G as follows: a vertex is associated with each block and with each articulation of G, these are the vertices of G^T. The edges of G^T are obtained by connecting each vertex associated to a block with all the vertices associated with articulations in the corresponding block. The block-graph of figure 1.8 is shown in figure 1.10 with solid circles denoting the vertices which correspond to cut-vertices. This block-graph contains no circuit, it is therefore a forest, and its components are the block-graphs of the components of figure 1.8. The following theorem expressing the general validity of this observation will now be proved:

17. *The block-graph of a connected graph is a tree.*

Proof. We have to prove that the block-graph G^T of a connected graph G is connected and contains no circuit.

If a block T of G includes no articulation of G, then G, being connected, consists of the single block T only, according to Theorem 14. The block-graph of a graph consisting of a single block is formed by a single vertex and is, therefore, a tree. So, it suffices to treat the case when all blocks of G contain an articulation in G.

Let the vertices of G^T be called white or black according to their correspondence to blocks or to articulations of G, respectively. Owing to our assumption, each white vertex in G^T has a black neighbour. Hence, in view of the transitivity of the relation $\mathcal{R}_1$ introduced at the discussion of components, for the connectedness of G^T it suffices to show that any black vertex p can be reached along a path from any black vertex q in G^T. Let the correspondents of the vertices p and q in G be denoted by p_0 and q_0, respectively. If G has a block containing both of these vertices, our assertion is easily verified. Therefore, it suffices to treat the case when p_0 and q_0 do not belong to the same block of G. So, according to Theorem 16, exactly one block-chain

$$p_0, T_1, a_1, T_2, \ldots, q_0$$

corresponds to any p_0q_0-path in G. Let

$$p, t_1, b_1, t_2, \ldots, q$$

be the corresponding vertices of G^T, respectively. Here the adjacent ones are neighbours in G^T, which proves the connectedness of G^T.

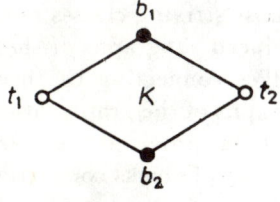

fig. 1.11

The fact that G^T contains no circuit is proved indirectly. Let us assume that K is a circuit in G^T of minimal length. Since one endpoint of each edge of G^T is black and the other white, the length of K is even ([2] 5.15). The circuit K cannot be of length 4, as on figure 1.11, since otherwise the correspondents of both b_1 and b_2 would be common vertices of the blocks of G corresponding to t_1 and t_2, contrary to Statement 13. Let the vertices of K be denoted in the order of its traversal by

$$t_1, b_1, t_2, b_2, \ldots, t_k, b_k, t_1$$

and let the corresponding blocks and articulations in G be

$$T_1, a_1, T_2, a_2, \ldots, T_k, a_k, T_1.$$

Any block appearing here can have only one common vertex with the preceding one according to Theorem 13, namely, the vertex between them. Moreover, in case $i + 1 = j$, if c were a common vertex of the blocks T_i and T_j, then, in view of Theorem 14, c would be an articulation of G and so K could be shortened, contrary to its minimality. Consequently, according to Theorem 10, there exist an $a_1 a_2$-path in G with edges in T_2 only, and an $a_2 a_3$-path with edges in T_3 only, ..., an $a_k a_1$-path with edges in T_1 only, and these paths together constitute a circuit in G with its edges not in the same block of G, which is impossible.

This completes the proof of Theorem 17.

Figures 1.4 and 1.8 show the same graph divided into leaves and blocks, respectively. It can be observed that a block either consists of a bridge and its two endpoints or it is a subgraph of some leaf of the graph. The general validity of this observation is expressed by the following theorem:

18. *If there is a vertex of degree at least two in a block of a graph then the block is the subgraph of some leaf of the graph.*

Proof. Let us assume that there is a vertex of degree at least two in some block T of a graph. If T has a single edge only, then it is a loop. In this case T contains a single vertex and the assertion is evident. If T has at least two edges then it has at least two vertices as well. Then, according to Theorem 11, any two vertices are connected by two vertex-disjoint paths, too. Since two vertex-disjoint paths are simultaneously edge-disjoint, the assertion follows in view of the transitivity of the relation $\mathcal{R}_2$.

The discussion so far, about the structure of the graph model, can be summarised as follows: the most striking classes are the components, i.e. the largest possible connected induced subgraphs of the graph. The components consist of leaves and of bridges connecting the leaves. The leaves are the largest possible induced subgraphs of the graph with at least two edge-disjoint paths connecting any two vertices, provided there are at least two vertices in them. Finally, leaves are made up of blocks connecting at cut-vertices, and, in view of Theorem 11, the blocks are the largest possible subgraphs of the graph

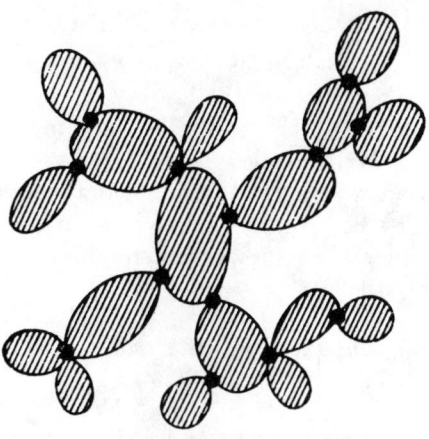

fig. 1.12

with at least two vertex-disjoint paths connecting any two vertices, provided there are at least two edges in them. Stressing also the tree-character hidden in the connection of the 'constituents', the following intuitive picture emerges.

In loose terms, all graphs are forest-like. At certain points on the trees of the forest, leaves have grown, namely the cactus-like connected leaves sketched in figure 1.12 (the 'leaves' of the cactus correspond to blocks).

1.6 The strongly connected components of directed graphs

If, in a directed graph, any vertex can be attained from any other along a directed path, the directed graph has been called *strongly connected* ([2], Chapter 3). It has been shown that a connected graph containing no bridge can be directed to make it strongly connected ([2] 3.18). Consequently, the edges of any leaf of a graph can always be directed to make the leaf strongly connected. A leaf containing at least two vertices can, however, always be directed to make it not strongly connected: e.g. by making a vertex the head of all edges incident to it. Our aim is to characterise the 'largest possible induced subgraphs' of directed graphs which are also strongly connected. First, however, some concepts are extended to directed graphs.

A graph $\vec{G}$ is said to be *connected* if the graph G obtained from $\vec{G}$ by disregarding the orientations is connected. A graph $\vec{G}_0$ is said to be a *subgraph* of the graph $\vec{G}$ or $\vec{G}_0$ is a *subgraph* of $\vec{G}$ *induced by* P_0, if the graph G_0 is a subgraph of the graph G or G_0 is a subgraph of G induced by P_0 provided the orientation of each edge of $\vec{G}_0$ is the same as in $\vec{G}$. Similarly, the set H of certain vertices of a graph $\vec{G}$ is called a *set of independent vertices* of $\vec{G}$ if H is a set of independent vertices of the graph G. The *length of a directed path* or *circuit* is the number of its edges. The directed path with starting vertex p and end-vertex q is also called $\vec{pq}$-*path*. *Directed paths* are called *vertex-disjoint* or *edge-disjoint* if no two different paths among them contain a common inner vertex or a common edge, respectively.

The following statement plays an important role in our investigations (for non-directed graphs, see [2] 1.24):

19. *If a directed graph contains a $\vec{pq}$-path and a $\vec{qr}$-path $(p \neq r)$, then it contains a $\vec{pr}$-path, too.*

Proof. Let $\vec{F}_1$ and $\vec{F}_2$ be a $\vec{pq}$-path and a $\vec{qr}$-path in $\vec{G}$, respectively. $\vec{F}_1$ can be assumed not to include r and $\vec{F}_2$ not to include p, otherwise our statement would immediately follow. Let $\vec{F}'_2$ denote the $\vec{sr}$-part of $\vec{F}_2$ of minimal length with its starting vertex in $\vec{F}_1$. (Possibly, $\vec{F}'_2 = \vec{F}_2$). Now, the $\vec{ps}$-part of $\vec{F}_1$ together with $\vec{F}'_2$ constitutes a $\vec{pr}$-path in $\vec{G}$. This completes the proof of Statement 19.

The following theorem serves to characterise strongly connected directed graphs:

20. *A graph $\vec{G}$ is strongly connected if and only if any of its edges is contained in some directed circuit of $\vec{G}$ and G is connected.*

Proof. Assume that the graph $\vec{G}$ is strongly connected. Then G is obviously connected. Let (p, q) be an edge in $\vec{G}$. $\vec{G}$ being strongly connected, it includes a $\vec{qp}$-path which, together with (p, q), constitutes a directed circuit containing the latter. This verifies the necessity of the condition.

Now let p and q be two vertices and $p = p_0, p_1, p_2, \ldots, p_n = q$ be the vertices of a pq-path in G in the order of the traversal. Then, there is a directed path in $\vec{G}$ from p_i to p_{i+1}: either the edge (p_i, p_{i+1}) exists or the part of the directed circuit containing the edge (p_{i+1}, p_i) obtained by deleting the edge (p_{i+1}, p_i). Therefore, according to Statement 19, $\vec{G}$ includes a $\vec{pq}$-path, too. This completes the proof of Theorem 20.

Now, for the investigation of directed graphs we introduce the following relationship. The *relation* $\mathcal{E}$ on the set P of vertices of the graph $\vec{G}$ is defined as follows: $p\mathcal{E}p$ for each element p of P; for two distinct elements p and q of P, $p\mathcal{E}q$ if and only if both q can be reached from p and p can be reached from q along directed paths in $\vec{G}$. (This implies the existence of two edge-disjoint pq-paths in G. Therefore, $p\mathcal{E}q$ implies that p and q belong to the same leaf of the graph G.) Obviously, $\mathcal{E}$ is reflexive and symmetric and its transitivity follows from Statement 19. Therefore, the relation $\mathcal{E}$ is an equivalence relation. Let $\{P_1, P_2, \ldots\}$ be the partition of the set P induced by $\mathcal{E}$. The subgraphs of the graph G induced by the sets P_i are called the *strongly connected components* of $\overrightarrow{G}$ or, in other words, the *directed leaves* of $\overrightarrow{G}$, in short, the *leaves* of $\overrightarrow{G}$.

As an example, see the strongly connected components of the directed graph of figure 1.13: the vertices belonging to one and the same component

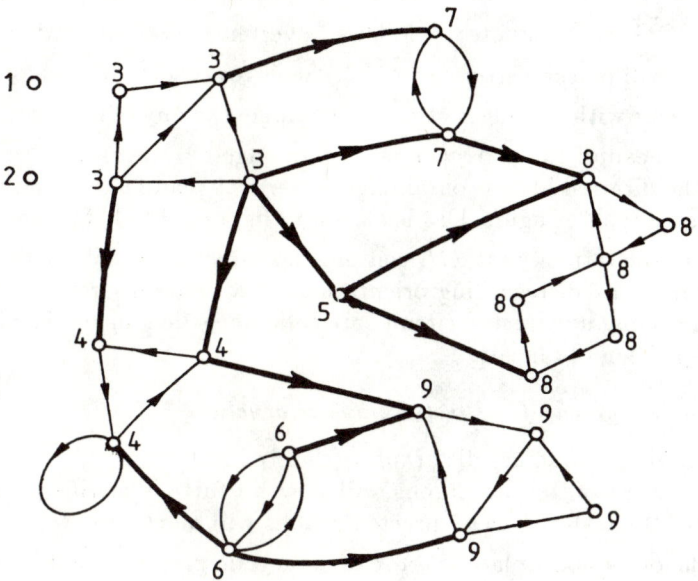

fig. 1.13

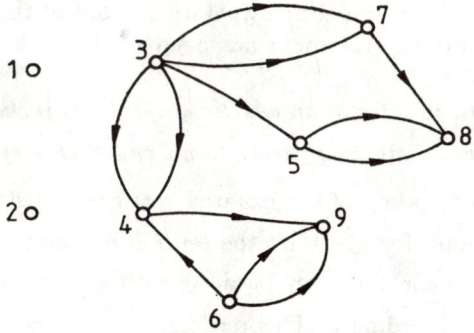

fig. 1.14

have been assigned the same number; the edges drawn by heavy lines do not belong to the directed leaves of this graph. Observe in this diagram that each edge and each vertex of any directed circuit belong to the same directed leaf. All edges of all leaves are contained in some directed circuit of the graph, an edge not included in any leaf is not contained in any directed circuit of this graph. The general validity of these observations follows from the strong connectedness of the directed leaves and from Theorem 20.

The structural feature of directed graphs expressed by their leaves is clearly shown by the graph $\vec{G}^L$ assigned to the graph $\vec{G}$. $\vec{G}^L$, the so-called *directed leaf-graph* of $\vec{G}$, is constructed as follows: a vertex is associated with each leaf of $\vec{G}$, these will be the vertices of $\vec{G}^L$. The edges of $\vec{G}^L$ are in a one-to-one correspondence with the edges of $\vec{G}$ not belonging to any of its leaves, namely the tail and head of the correspondent of an edge $e = (p, q)$ in $\vec{G}$ will correspond to the directed leaves containing the vertices p and q, respectively. The directed leaf-graph of figure 1.13 is shown in figure 1.14. It follows from the previous considerations that $\vec{G}^L$ contains no directed circuit (naturally, the graph obtained by disregarding orientations may contain circuits). Directed graphs containing no directed circuits are called *acyclic graphs*. The following theorem has been obtained:

21. *The leaf-graph of a directed graph is acyclic.*

The leaf-graph of an acyclic graph is clearly itself.

The structure of acyclic graphs will now be further clarified. Thus the structure of the leaf-graphs of directed graphs will be clearer.

Let p denote a non-isolated vertex of an acyclic graph $\vec{G}$ and let $\vec{F}$ stand for a path containing p of maximal length in $\vec{G}$. $\vec{G}$ being acyclic and $\vec{F}$ being maximal, the indegree of the starting vertex of $\vec{F}$ and the outdegree of its end-vertex are zero in $\vec{G}$. A vertex p of an acyclic graph $\vec{G}$ is called a *starting vertex* and an *end-vertex* of $\vec{G}$ if it is not an isolated vertex and $\varphi_{\text{in}}(p) = 0$ and $\varphi_{\text{out}}(p) = 0$, respectively. With the aid of the above maximal directed path, the following statement has been verified:

22. *Any non-isolated vertex of an acyclic graph $\vec{G}$ is included in a directed path of $\vec{G}$ leading from a starting vertex to an end-vertex of $\vec{G}$.*

Let the set T_0 be composed of the isolated vertices and starting vertices of an acyclic graph $\vec{G}$, and, for $i > 0$, let the set T_i be composed of the vertices in $\vec{G}$ attainable from some vertex in T_0 along a directed path of length i, but along no longer one. According to Theorem 22, all vertices of $\vec{G}$ are included in one set T_i only. We shall show that the subscript of the set T_i containing the tail of any edge of $\vec{G}$ is smaller than that for the head of that edge. Hence each set T_i constitutes a set of independent vertices of $\vec{G}$.

So, let $e = (p, q)$ be an edge in $\vec{G}$. If $p \in T_0$ then $q \notin T_0$. If $p \in T_i$ ($i \neq 0$) then there is a path $\vec{F}$ of length i in $\vec{G}$, starting from a vertex of T_0 and leading to p. The path $\vec{F}$ cannot contain the vertex q since, otherwise, the $\vec{qp}$-part of $\vec{F}$ and e together would yield a directed circuit in $\vec{G}$, contrary to the acyclic property of $\vec{G}$. Therefore, if $\vec{F}$ is extended by the edge e, a path of length $i+1$ starting from a vertex of T_0 and leading to q is obtained. Hence q is contained in a set T_j with $j \geq i+1$.

This proves the following theorem characterising the structure of acyclic graphs (cf. [2] page 158):

23. *There exists a unique partition $\{T_0, T_1, T_2, \ldots\}$ of the set of vertices of an acyclic graph $\vec{G}$ so that the set T_0 is composed by the isolated vertices and starting vertices of $\vec{G}$ and, for $i > 0$, the sets T_i contain those vertices of $\vec{G}$ which can be reached from some vertex of T_0 along a path of length i, but along no longer path. Each T_i is a set of independent vertices of $\vec{G}$ and the subscript of the set T_i containing the tail of any edge of $\vec{G}$ belong to a T_i with smaller subscript than does its head.*

Figure 1.15 shows the structure of figure 1.14 according to Theorem 23.

Remark. Note that a directed graph is acyclic if and only if there exists an order of its vertices with the tail of each edge preceding its head. It hence follows that all graphs without loops have acyclic orientations.

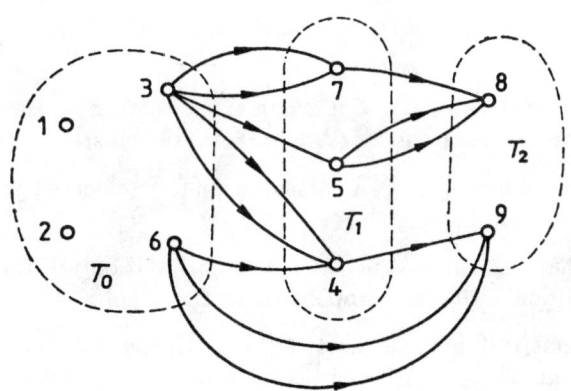

fig. 1.15

1.7 Problems

24. Prove that all simple graphs have a diagram in the space where all the edges are straight lines.

25. The *relation* $\mathcal{P}$ on the set of vertices of a connected graph G is defined as follows: $p\mathcal{P}p$ for each element p of P; for two distinct elements p and q of P, $p\mathcal{P}q$ if and only if no pq-path of odd length exists in G. Show that $\mathcal{P}$ is an equivalence relation. What can be said of the partition of P induced by $\mathcal{P}$, if G is a bipartite graph?

26. Show that all the edges of a tree are bridges of the graph.

27. Prove that if every edge of a connected graph is included in some circuit of the graph then the graph consists of a single leaf.

28. The *relation* $\mathcal{V}_1$ on the set of vertices P of a graph G is defined as follows: $p\mathcal{V}_1 p$ for each element p of P; for two distinct elements p and q of P, $p\mathcal{V}_1 q$ if and only if, for any partition $\{P_p, P_q\}$ of P (where $p \in P_p$, $q \in P_q$), G contains a $P_p P_q$-edge. Show that $\mathcal{V}_1$ is an equivalence relation. Further let $\{P_1, P_2, \ldots\}$ be the partition of P induced by $\mathcal{V}_1$. What can be said of the subgraphs of G induced by the sets P_i?

29. The *relation* $\mathcal{L}$ on the set of vertices P of a graph G is defined as follows: $p\mathcal{L}p$ for each element p of P; for two distinct elements p and q of P, $p\mathcal{L}q$ if and only if there exists a pq-path in G with all of its edges included in some circuit of G. Prove that $\mathcal{L}$ is an equivalence relation, and that if $\{P_1, P_2, \ldots\}$ is the partition of P induced by $\mathcal{L}$, then the subgraphs of G induced by the sets P_i are the leaves of G.

30. Show that if the vertex p of a graph G with a degree not equal to one is the endpoint of a bridge of G then p is an articulation of G.

31. Prove that if the degree of all vertices of a graph is at most 2 then any component and any block of the graph is either an isolated vertex or a path or a circuit.

32. Assume that each edge of a graph G containing no isolated vertex, is included in exactly one circuit of G. Characterise the structure of G.

33. Show that the concept of an articulation for connected graphs coincides with the one given in [2].

34. Show that a graph has no articulation if and only if the blocks of the graph are simultaneously the components of the graph.

35. Show that if T is a block of a graph without circuits of even length then T either consists of a single vertex or a single edge or it is a single circuit of odd length.

36. Prove that if T is a block of a graph containing at least three vertices, and p, q and r are three arbitrary vertices in T, then there exists a pq-path in T passing through r. Also show that a circuit in T containing all of p, q and r does not necessarily exist.

37. Demonstrate by an example that if p and q are two arbitrary vertices of a strongly connected graph $\vec{G}$, then a closed edge-train in $\vec{G}$ containing both p and q does not necessarily exist. (For the concept of edge-trains see in [2] page 61.)

38. The *relation* $\mathcal{V}'$ on the set of vertices P of a graph $\vec{G}$ is defined as follows: $p\mathcal{V}'p$ for each element p of P; for two distinct elements p and q of P, $p\mathcal{V}'q$ if and only if for any partition $\{P_p, P_q\}$ of P (where $p \in P_p$, $q \in P_q$), $\vec{G}$ contains both a $\overrightarrow{P_p P_q}$-edge and a $\overrightarrow{P_q P_p}$-edge. Show that $\mathcal{V}'$ is an equivalence relation, and that if $\{P_1, P_2, \ldots\}$ is the partition of P induced by $\mathcal{V}'$, then the subgraphs of $\vec{G}$ induced by the sets P_i are the leaves of $\vec{G}$.

2

Optimal Flows

2.1 Two basic problems

Before starting the discussion of the topics of this chapter, two problems are raised, one of them presented in a form befitting a thriller. These problems are the basic objectives of the chapter.

Let us assume that the police intend to collar a burglar. In the course of the investigation the burglar has been identified and he is known to stay at present in city x with the intention of getting to city y by car as soon as possible. It is therefore necessary to check up on people travelling along the network of roads connecting the two cities. This can be attained, for example, by selecting only certain points of the network of roads as checkpoints. But any route connecting x and y must contain at least one selected point. It is reasonable to accomplish this with as few points as possible. The only practical places for checkpoints are at road-junctions since, if a checkpoint on a road-section is relocated to the nearest junction encountered on travelling along the road-section in any direction then the number of road-sections already checked is not decreased. It is not impossible that there is no road-junction along some route connecting x and y; it is, however, necessary to select a checkpoint along this route, too. For each road leaving x and y, let us regard also their points at the city limits as road-junctions (these are two-way junctions only). This eliminates the above case but does not influence the minimal number of checkpoints we wish to determine.

The map of this network of roads is a graph $G = (P, E, \mathcal{G})$: the vertices of G stand for the junctions (the cities x and y themselves are called junctions in this context), and the edges of G correspond to road-sections with junctions situated at their endpoints but nowhere in between. Let x and y also denote the vertices corresponding to the cities x and y, respectively. Let the complexity of the network of roads in these two cities preclude the possibility of x and y being checkpoints. A subset of the set P excluding x and y and containing at least one vertex of all xy-paths of G is called a *vertex xy-cut* in G. (The possibility of getting to y from x can be cut along these.) Let the set of vertex xy-cuts in G be denoted by $\mathcal{V}$ and, in general, the number of elements of a set H by $|H|$. So, the above problem requires the selection of a vertex xy-cut V_0 in G with the number of its elements satisfying

$$|V_0| = \min_{V \in \mathcal{V}} |V|.$$

The set V_0 is also called a *minimal vertex xy-cut* in G and the number $|V_0|$ the *minimal number of xy-cut vertices*.

The development of graph theory has been greatly stimulated by a transportation problem which emerged during World War II in the United States. Its solution facilitated the redeployment of army units as quickly as possible. Such a transportation resulting in redeployment can be carried out on roads, railways, airways or waterways. The network of roads can be deemed to be given (by its map, for example), just like the place or places of departure and destination. Let us assume the availability of a suitable number of transport facilities. For the sake of simplicity, the model is illustrated by the transportation of some goods. The transportation or, to be more precise, the flow of goods is limited by the transmission capabilities, i.e. *capacities*, of the road-sections and nodes (junctions, cities, railway stations, airports, ports) indicating the maximal quantity of goods transferable in unit time through the 'cross sections' of certain sections or through the junctions of the network of roads. The capacities can be assessed, i.e. they can be deemed to be given. An optimally executable transportation plan is to be devised or, to be more precise, the maximal quantity of goods despatchable in unit time is to be determined and a corresponding itinerary of the transportation is to be given with the quantity of goods in flow nowhere exceeding the capacity limits. Naturally, the time consumed by the entire transportation process depends upon the lengths of the road-sections and the speeds of the transporting vehicles. Now, at the examination of the so-called *static problem*, this is disregarded, its treatment is postponed until the so-called *dynamic problem* is scrutinised. It is only sought instead, how can the flow of goods maximally load the network of roads.

Our problem is formulated in the language of graphs, too, but only in the simple case when goods are to be transported from a single fixed place to another place and, further, the transportation is only limited by the capacities of the road-sections and not by the nodes. The general case will later be seen to be easily reducible to this special case.

Let us therefore regard the map of the network of roads as a graph with its edges denoting the road-sections and its vertices the nodes in question. The vertex x called *source* stands for the place from where the goods are despatched, and the point y called *sink* for the destination. Let κ denote the non-negative, real *edge-capacity function* with its value being $\kappa(e)$ along edge e. If the capacity is not constant along a road-section represented by a single edge, then the places separating the sections with different capacities are also taken to be nodes. Such is the case, for example, if the road-section narrows, its quality changes or it climbs to an extent resulting in loss of speed. If, however, a road-section ascends in one direction, then it descends in the other and, hence, the value of its capacity varies with direction. This fact is highlighted even more sharply by one-way streets whose capacity in the disallowed directions should be taken as zero. It seems, therefore, right to orient our graph: the orientations of the edges indicate the allowed directions of transport. If, along an edge $\{p, q\}$, both directions of transport are allowed then this fact is indicated by a pair of directed edges (p, q) and (q, p) with both assigned the appropriate capacity value. If, along a section represented by the edge $\{p_1, q_1\}$, the capacity is κ in both directions, then the corresponding edges (p_1, q_1) and (q_1, p_1) are both taken to be of capacity κ. The question of possible confusion may arise if the relevant section is, for example, a one-track railway since, then, simultaneous transport is impossible in two directions. This can be resolved by saying that the quantity of goods flowing through the edge (p_1, q_1) in unit time is f_1 and through the edge (q_1, p_1) it is f_2 ($f_1 \leq f_2 \leq \kappa$) since this corresponds to a quantity $f_2 - f_1$ flowing through edge (q_1, p_1) only and the condition $f_2 - f_1 \leq \kappa$ is obviously satisfied.

Let the sets of vertices and edges of the relevant directed graph $\vec{G}$ be P and E, respectively. The value of the capacity function along edge (p, q) is $\kappa(p, q)$; it is also called the *capacity of edge* (p, q) in short. Now, a non-negative real function f defined over the edges of $\vec{G}$ is called an *xy*-flow in $\vec{G}$ provided the following three conditions are satisfied:

(1°) $0 \leq f(p, q) \leq \kappa(p, q)$ for each edge (p, q) in E in accordance with the fact that the capacities limit the flow of goods.

(2°) $\sum_{\substack{(p,q) \in E \\ p \text{ fixed}}} f(p, q) = \sum_{\substack{(q,p) \in E \\ p \text{ fixed}}} f(q, p)$ for all vertices p different from the source x and sink y, expressing the fact that the amount of goods despatched from p and reaches p in unit time.

(3°) $\sum_{(x,p) \in E} f(x, p) - \sum_{(p,x) \in E} f(p, x) = \sum_{(q,y) \in E} f(q, y) - \sum_{(y,q) \in E} f(y, q),$

expressing the fact that the amount of goods despatched from the source in unit time equals that reaching the sink in unit time (provided the flow is established). The quantity appearing on the two sides of the equality in (3°) is called the *value of the flow* f and is denoted as $\sigma(f)$.

Denoting the set of *xy*-flows in $\vec{G}$ by $\mathcal{F}$, the solution of the relevant problem is yielded by the specification of an *xy*-flow f_0 in $\vec{G}$ whose value is

$$\sigma(f_0) = \max_{f \in \mathcal{F}} \sigma(f).$$

(This maximum will be shown to exist.) The plan of the transportation can be obtained by giving the values of f_0 along the edges of $\overrightarrow{G}$.

Notice that the flow defined here may relate to any so-called stationary process (which does not depend on time). The capacity can be actually expressed, e.g. by the number of vehicles per hour on roads and railways, by cubic metres per second on pipelines carrying gas, oil or water.

Remark. The relationship (2°) for the flow f in $\overrightarrow{G}$ implies the Equality (3°). This statement which can be regarded as the law of conservation of matter can be verified as follows. Let us add the values of the function f along the edges in $\overrightarrow{G}$ in two ways: once, taking the tails of the edges into account, and once considering their heads. This yields the following:

$$\sum_{\substack{p \in P \\ p \text{ fixed}}} \sum_{(p,q) \in E} f(p,q) = \sum_{\substack{q \in P \\ q \text{ fixed}}} \sum_{(p,q) \in E} f(p,q).$$

Hence subtracting the equalities in (2°) relating to the vertices different from the source and the sink, the following relation, equivalent to (3°), is obtained:

$$\sum_{(x,p) \in E} f(x,p) + \sum_{(y,q) \in E} f(y,q) = \sum_{(p,x) \in E} f(p,x) + \sum_{(q,y) \in E} f(q,y).$$

2.2 Maximal set of independent paths

Let us return to Theorem 11 of Chapter 1, (to (1.11), in short) stating that in a block of more than one edge, any two vertices are connected by two vertex-disjoint paths. Let us also consider Theorem (1.16). According to this, if x and y are two points of a connected graph G in distinct blocks then any xy-path in G determines the same block-chain BCh. This also means that all articulations in BCh are simultaneously inner vertices of each xy-path. Since the two vertices of a block containing one edge and two vertices are adjacent by all means, the following theorem is true:

1. *If x and y are two non-adjacent vertices of a connected graph G and there are no two vertex-disjoint xy-paths in G then there is vertex in G constituting an inner vertex of each xy-path in G.*

In §1.4 of Chapter 1, the leaves of a graph have been constructed with the aid of the relationship $\mathcal{R}_2$. If two distinct vertices x and y of a connected graph $G = (P, E, \mathcal{G})$ meet $x\mathcal{R}_2 y$ then there are two edge-disjoint xy-paths. If x and y are not related by the relationship $\mathcal{R}_2$ then, according to Theorem (1.9), they are neither related by the relationship $\mathcal{V}_2$. The definition of $\mathcal{V}_2$ implies that, in the latter case, there is a partition $\{P_x, P_y\}$ of P with no two P_xP_y-edges in G. G being connected, it must contain a P_xP_y-edge. This single P_xP_y-edge is necessarily contained in all xy-paths in G. So, the following theorem has been proved:

2. *If x and y are two vertices of a connected graph G and there are no two edge-disjoint xy-paths in G then G contains an edge included in all xy-paths in G.*

This theorem will be generalised below. As a preparation, the relations $\mathcal{R}_2$ and $\mathcal{V}_2$ are generalised.

For any ordinal number i, the *relationship* $\mathcal{R}_i$ on the set P of vertices of a graph G is defined as follows: $p\mathcal{R}_i p$ for each element of P; for two distinct elements p and q of P, $p\mathcal{R}_i q$ if and only if there are i edge-disjoint pq-paths in G.

For any ordinal number i, the *relationship* $\mathcal{V}_i$ on the set of vertices P of a graph G is defined as follows: $p\mathcal{V}_i p$ for each element of P; for two distinct elements p and q of P, $p\mathcal{V}_i q$ if and only if for any partition $\{P_p P_q\}$ of P, G contains at least i P_pP_q-edges.

In cases $i = 1$ and $i = 2$, the two above relations have already been defined in Chapter 1. Both have been shown to be equivalence relations and to induce the same partition of P (§§3 and 4, Theorem 9, Problem 28 and its solution). The same will be shown to hold for the cases $i > 2$. An appropriate edge of each of the i edge-disjoint paths in case $p\mathcal{R}_i q$ also ensures the validity of the relationship $p\mathcal{V}_i q$. We shall further prove that, in the case $p\mathcal{V}_i q$, $p\mathcal{R}_i q$ also holds for all values of i.

For any partition $\{P_p, P_q\}$ of the set of vertices of a graph G, all pq-paths contain a P_pP_q-edge. Therefore, the set of P_pP_q-edges is called an *edge pq-cut* in G. If W is an edge pq-cut with a minimal number of elements then the set W is also called a *minimal edge pq-cut* in G and the number $|W|$ the *minimal number of pq-cut edges*. The highest number i for which the relationship $p\mathcal{R}_i q$ holds is called the *maximal number of edge-disjoint pq-paths*. Since $p\mathcal{R}_i q$ implies $p\mathcal{V}_i q$, the following can be stated:

3. *If x and y are two vertices of a graph G then the maximal number of edge-disjoint xy-paths is not greater than the minimal number of xy-cut edges.*

In order to fulfil our previous promise we prove that the inequality of Theorem 3 is always met by equality. Indeed, let us assume that $x\mathcal{V}_j y$ holds for two arbitrary vertices x and y of G, i.e. the minimal number of xy-cut edges in G is at least j. Let i be the maximal satisfying $x\mathcal{R}_i y$. Now, if equality holds in Theorem 3 then $i \geq j$, i.e. there exist j edge-disjoint xy-paths in G, in other words, $x\mathcal{R}_j y$ is valid.

If x and y belong to two distinct components of G, then the empty set also constitutes an edge xy-cut, and so both numbers given in Theorem 3 are zero. Therefore, in the proof of the equality in Theorem 3 we may suppose that x and y belong to the same component of G, and a similar argument will be used in the proofs of Theorems 5, 8, 11 and 12.

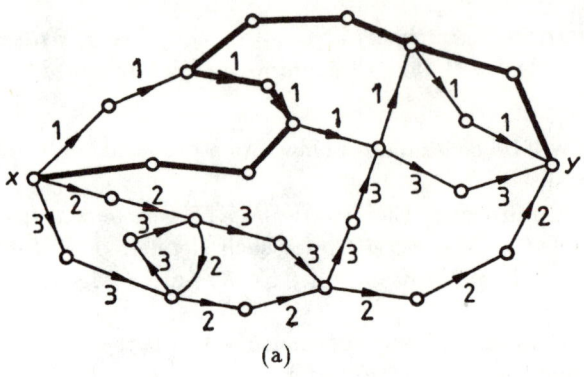

(a)

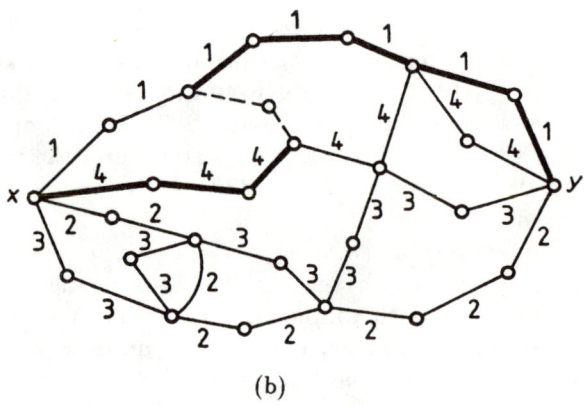

(b)

fig. 2.1

The above sharpening of Theorem 3 will have been proved if the existence of a number k is shown with the property that one can find k edge-disjoint xy-paths in G edges, since then $k \leq$ the maximal number of edge-disjoint paths $\leq$ the minimal number of xy-cut edges $\leq k$. A procedure will be given for selecting k edge-disjoint xy-paths and, further, an instruction for marking certain edges of k edge-disjoint xy-paths. Then the marked edges will be shown to constitute an edge xy-cut. During the procedure, the knowledge of h edge-disjoint xy-paths will allow finding $h+1$ edge-disjoint xy-paths as long as h has not reached the value k with the above property. The main idea of this augmentation will be illustrated by examples. In figure 2.1(a) the numbers mark three edge-disjoint xy-paths, with arrows indicating the walk from x

towards y. It is clear that no further edge-disjoint xy-paths can be found. A further xy-path has been marked by heavy line. Let us traverse this, also starting from x, and observe that some edges of the original xy-path marked by 1 are also encountered but in a direction opposite to their orientation. Therefore, the heavy path and the three previous paths do not constitute four edge-disjoint paths. It is, however, possible to obtain four xy-paths from these without using the doubly covered edges. This is indicated in figure 2.1(b) where the doubly covered edges are indicated by dotted lines. A similar augmentation is shown in figure 2.2 in a more complex case. Notice that, traversing the heavy xy-path of figure 2.2(a) starting from x, our direction along the original paths is always opposite to the indicated orientation. The dotted edges of figure 2.2(b) have either been thick or they form a circuit. If possible, the number of edge-disjoint xy-paths can also be increased by adding an xy-path without a common edge with the paths already selected. This simple method of augmentation has not been illustrated.

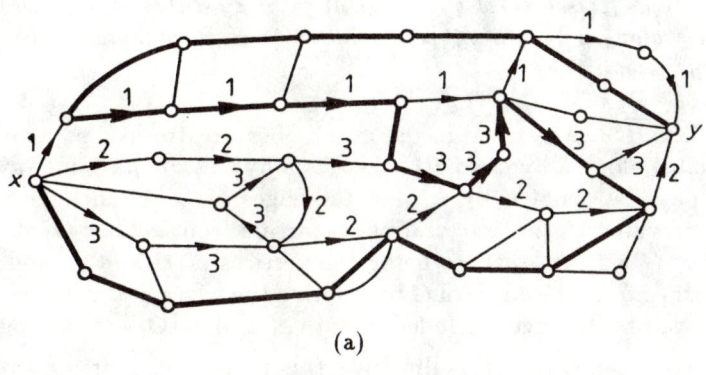

(a)

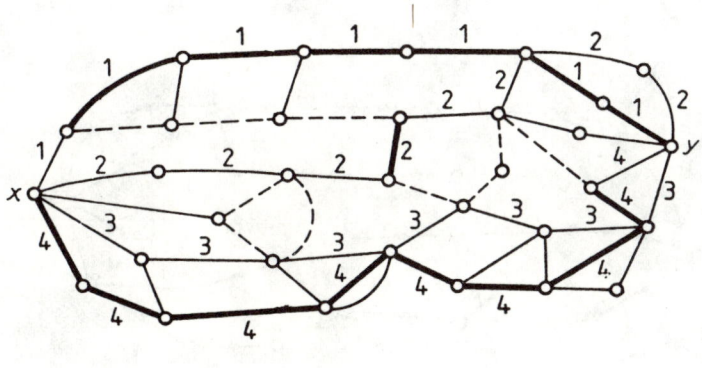

(b)

fig. 2.2

For the general description of the augmenting procedure, let us assume that h edge-disjoint xy-paths have somehow been selected in a graph G. Let H denote the subgraph of G formed by these h paths. Let us traverse all the h paths in H starting from x and let us orient the edges of H accordingly. This yields the graph $\overrightarrow{H}$. Now, an xp-path in G is called a *bypass xp-path relative to H* if the orientation of its common edges with H along a walk starting from x is opposite to their orientation in $\overrightarrow{H}$. Then the vertex P of the graph G is called *accessible along a bypass path relative to H*. The words 'relative to H' will be omitted if the subgraph is clear from the context. The point x itself is called accessible along a bypass path even if there is no bypass path in G. Since all initial segments of a bypass path are bypass paths, all vertices of any bypass path are accessible along a bypass path.

The situation indicated in figures 2.1 and 2.2 is valid in general, since the following statement is true:

4. *If a graph H is formed by h edge-disjoint xy-paths of a graph G and y is accessible along a bypass xy-path relative to H then there are $h+1$ edge-disjoint xy-paths in G.*

Proof. Let $\overrightarrow{H}$ denote the directed graph obtained from H as above. Since y is accessible along a bypass path, there is a bypass xy-path in G. Let such a bypass path be denoted by L. Let the edges of L not included in H be oriented according to the traversal of L starting from x. Let G_1 denote the subgraph of G with its vertices being the vertices of H and L, and with its edges obtained from the edges of G by deleting the edges not included in either H or L as well as the edges included in both H and L. Orient the edges of G_1 as above to obtain $\overrightarrow{G}_1$. Let us illustrate this procedure, starting from figure 2.2(a): here $h = 3$ (with the numbered paths included in H), L is the heavy path and $\overrightarrow{G}_1$ is shown in figure 2.3.

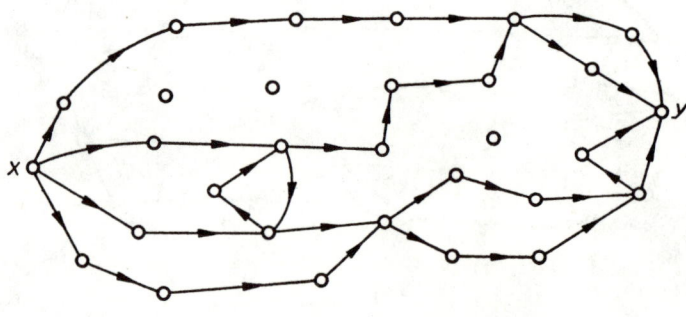

fig. 2.3

Observe that, in general

$$\varphi_{\text{out}}(x) = \varphi_{\text{in}}(y) = h + 1,$$
$$\varphi_{\text{in}}(x) = \varphi_{\text{out}}(y) = 0$$

and

$$\varphi_{\text{out}}(p) = \varphi_{\text{in}}(p) \quad \text{if} \quad p \neq x \quad \text{or} \quad y$$

hold for $\vec{G}_1$. Indeed, writing h instead of $h+1$, these equalities obviously hold in $\vec{H}$. When forming the graph $\vec{G}_1$, the indegree of x and the outdegree of y have been increased to $h+1$ and, for any other vertex, one can easily see that the changes of the indegree and the outdegree are equal.

The existence of $h+1$ edge-disjoint $\overrightarrow{xy}$-paths in $\vec{G}_1$ will now be shown. Clearly, starting from x along the edges and following the orientations, we can get stuck in y only. Disregarding the edges of the traversed circuits, if any, the traversed edges yield an $\overrightarrow{xy}$-path. Let us delete the edges of this path. Writing h instead of $h+1$ for the degrees of the vertices in the remaining graph, the above equalities hold. So, repeating this argument, we obtain

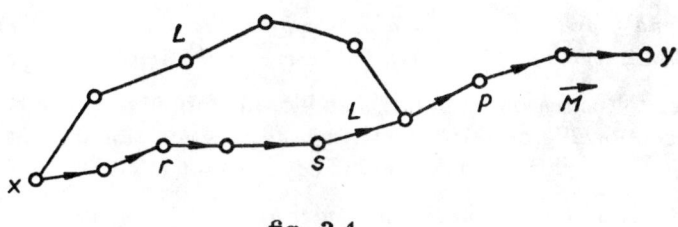

fig. 2.4

$h+1$ edge-disjoint $\overrightarrow{xy}$-paths in $\vec{G}_1$. Such a system of paths constitutes $h+1$ edge-disjoint xy-paths in G_1 and hence also in G. This completes the proof of Statement 4.

Hence, starting from an xy-path of a graph, the number of edge-disjoint xy-paths can be increased as long as y is accessible along a bypass path relative to the paths so far selected.

Therefore, consider the case when $\vec{H}$ consists of h edge-disjoint xy-paths of the graph G, with the previously explained orientations, and assume now that y is not accessible along a bypass path. Let $\{P_x, P_y\}$ be a partition of the set of vertices of the graph G, where P_x is the set of vertices accessible along a bypass path. Let $\vec{M}$ denote one of the h $\overrightarrow{xy}$-paths in $\vec{H}$ and let p be one of its vertices accessible along a bypass path. We claim that, in the partial xp-path $\vec{M}_1$ of $\vec{M}$, all vertices are accessible along a bypass path. Indirectly, assume that r is a vertex of $\overrightarrow{M}_1$ and $r \in P_y$ (see figure 2.4). Let us start from x along $\overrightarrow{M}$ and let us follow the orientation. Let s be the vertex in P_x first encountered after r (possibly, $s = p$). This means that there is a bypass xs-path L in G. Augmenting the bypass path L by the sr part of M, a bypass xr-path is obtained since only the vertex s is accessible along a bypass path in the augmenting part. This, however, contradicts $r \in P_y$. Our assertion implies that if a vertex q in $\overrightarrow{M}$ is not accessible along a bypass path then no vertex in the $\overrightarrow{qy}$ part of $\overrightarrow{M}$ is accessible along a bypass path. Hence, $\overrightarrow{M}$ has exactly one edge (p,q) with $p \in P_x$ and $q \in P_y$, and so, all vertices in the $\overrightarrow{xp}$ part of $\overrightarrow{M}$ are accessible along a bypass path and no vertex in its $\overrightarrow{qp}$ part is accessible along a bypass path. It will now be shown that if $\{a,b\}$ is an edge of the graph G and $a \in P_x$, $b \in P_y$ then the edge $\{a,b\}$ is simultaneously an edge of H. Indeed, if $\{a,b\}$ were not an edge of H, then, as a consequence of $a \in P_x$, there would be a bypass xa-path in G which, augmented by $\{a,b\}$, would still be a bypass path, contrary to the fact that $b \in P_y$.

Thus, exactly these distinguished edges of each of the h edge-disjoint xy-path in H are the $P_x P_y$-edges in G, and so this edge xy-cut is formed by h edges. Consequently, taking Theorem 3 also into account, we obtained the following theorem, also expressing the generalisation of Theorem 2.

5. *If x and y are two vertices of a graph G then the maximal number of edge-disjoint xy-paths in G is equal to the minimal number of xy-cut edges.*

According to the reasoning so far, the following formulation can be given to the above-mentioned procedure permitting the selection of a maximal number of edge-disjoint xy-paths and a minimal edge xy-cut in G:

6. Algorithm* *for searching a maximal number of edge-disjoint xy-paths and a minimal edge xy-cut:*

(1) If there is no xy-path in G,

   then $\begin{cases} \text{let } H \text{ be the empty graph with no vertices and} \\ \text{jump to (5).} \end{cases}$

(2) Select an arbitrary xy-path in G and orient its edges in accordance with its traversal starting from x; let this be $\overrightarrow{H}$. (The vertices of G accessible along a bypass path relative to H are now marked.)

---

* An *algorithm* is a finite sequence of instructions which are to be executed sequentially except that if the current instruction requires a 'jump' to some instruction different from the following one then the instructions are to be executed sequentially starting from that instruction. If the conditions of the current instruction are not satisfied then the following one (or the one to be jumped to) is to be executed next. In our algorithms, the end of each instruction is denoted by a period, texts in parentheses merely serve as comments.

Mark the vertex x.

(3) If y is marked, then jump to (4).

If there is a yet unselected edge $e = \{p, q\}$ with its head marked but its tail not marked

then $\begin{cases} \text{select the edge } e \text{ and, furthermore,} \\ \text{if the edge } e \text{ is not in } H \text{ or if } q \text{ is the tail of the edge in } \overrightarrow{H} \\ \text{obtained from } e \text{ by orientation, then mark the vertex } q \text{ and,} \\ \text{furthermore} \\ \text{return to (3).} \end{cases}$

Jump to (5).

(4) Mark the edge $\{r, y\}$ in G whose selection resulted in marking y.

If $r \neq x$, then $\begin{cases} \text{let } r \text{ take over the role of } y \text{ and} \\ \text{return to (4). (This results in marking the edges} \\ \text{of a bypass } xy\text{-path relative to } H.) \end{cases}$

Let L denote the bypass xy-path relative to H whose edges have been marked.

Let us orient the edges of L not in H according to the traversal of L starting from x.

Let G_1 denote the subgraph of G with its vertices being the vertices of H and L and its edges obtained from the edges of G by deleting the edges not included in either H or L as well as the edges included in both H and L.

Let us select, in the above way, in $\overrightarrow{G}_1$ obtained from G_1 by the given orientations, one more edge-disjoint $\overline{xy}$-paths than the number of those constituting $\overrightarrow{H}$ and let the graph formed by these take over the role of $\overrightarrow{H}$.

Consider the vertices marked and edges selected in (3) and (4) as unmarked and unselected, respectively, and return to (3).

(5) Record the result: a maximal number of edge-disjoint paths is constituted by the edge-disjoint paths forming H; and a minimal edge xy-cut is formed by those edges of H with one of their ends marked (accessible along a bypass path) and the other end not.

As an example for the final stage of the algorithm: the last application of (3) arises if, in figure 2.2(b), $\overrightarrow{H}$ is formed by the numbered edges with the appropriate directions. The edges of $\overrightarrow{H}$ are heavy, the vertices accessible along a bypass path are solid, the four edges forming the minimal edge xy-cut are indicated by a dotted line in figure 2.5.

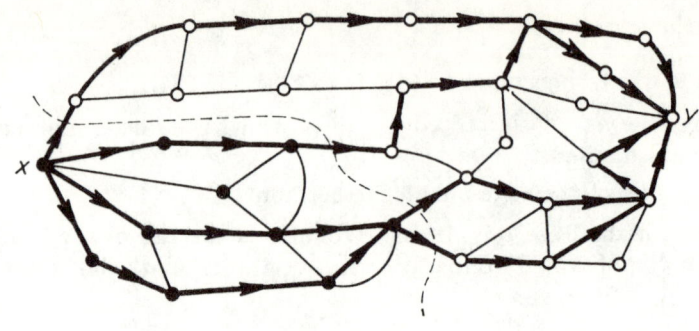

fig. 2.5

Remark. It can be verified that this algorithm terminates in a finite number of steps. But what is its advantage over the trivial algorithm of scanning all possible cases which also requires a finite number of steps? The examination of the theory of algorithms (see, for example [3]) indicates that even a simple enumeration may be theoretically problematic besides the fact that it requires a relatively large number of steps. If the graph has n vertices and e edges then the maximal number of edge-disjoint paths is at most e and, since a bypass path is found in no more than n^2 steps, the number of necessary steps in this algorithm is at most en^2.

Similarly to Theorem 2, Theorem 1 can also be generalised. Let p and q be two vertices of a graph G. The highest number of existing vertex-disjoint pq-paths in G is called the *maximal number of vertex-disjoint pq-paths* in G. The following analogue of Theorem 3 is obvious:

7. *If x and y are two non-adjacent vertices of a graph G then the maximal number of vertex-disjoint xy-paths in G is not greater than the minimal number of xy-cut vertices.*

As a generalisation of Theorem 1, we now prove that equality can always replace the inequality in Theorem 7, i.e. the following fundamental theorem of graph theory of K Menger is true ([4], *n*-Kettensatz, pages 221–228).

8. *If x and y are two non-adjacent vertices of a graph G then the maximal number of vertex-disjoint xy-paths in G equals the minimal number of xy-cut vertices.*

The proof of Menger's theorem can be carried out similarly to that of Theorem 2 by successively increasing the number of vertex-disjoint paths with bypasses. Since the structure of the bypass can be more complex (for example it may be an edge-train) and the proof itself is also more intricate, we only mention it here: [5]. Several attractive proofs are known for the theorem; the following one is due to G A Dirac [6].

The proof of Theorem 8. According to Theorem 7, in order to prove Theorem 8, it suffices to verify the following theorem:

9. *If x and y are two non-adjacent vertices of a graph G, and the minimal number of xy-cut vertices in G is k, then there are k vertex-disjoint xy-paths in G.*

Indirectly, let h denote the smallest value k when a counter example exists. In any case, $h > 1$. Let G_0 be a counter example at $k = h$ with a minimal number of vertices and, if there are several of these, let it have a minimal number of edges. Therefore, there are no h vertex-disjoint xy-paths in G_0.

The graph G_0 will be proved to have the following properties:

1. x and y have no common neighbour in G_0.

2. If V is a vertex xy-cut of h elements in G_0 then either x or y is adjacent in G_0 to all the vertices of V.

3. Let $G(-e)$ be the graph obtained from G_0 by deleting the edge $e = \{a_1, a_2\}$, and $V(e)$ be a vertex xy-cut of $h-1$ elements in $G(-e)$. Then, $a_1, a_2 \notin V(e)$, and if a_i is different from x and y, then the set $V(e) \cup \{a_i\}$ is a vertex xy-cut in G_0 ($i = 1, 2$).

Before proving these three statements we show how they lead to the verification of Theorem 9. Let L be an xy-path of minimal length in G_0. Let the vertices of L be denoted by $x, a_1, a_2, \ldots, y$ in the order of the traversal of L. Let the edge $e = \{a_1, a_2\}$ be in L and let $v_1, v_2, \ldots, v_{h-1}$ denote the elements of the set $V(e)$ defined in 3.

According to 1, $a_2 \neq y$ and a_1 is not adjacent to y in G_0. Hence, according to 3 the set $V_i = V(e) \cup \{a_i\}$ is a vertex xy-cut in G_0 for $i = 1$ and $i = 2$. Since a_1 is not adjacent to y in G_0, 2 implies with the substitution $V_1 = V$ that x is adjacent to all the vertices v_i in G_0. Hence, according to 1, no vertex v_i is adjacent to y in G_0. (Figure 2.6 is an illustration at $h = 5$. Here, the

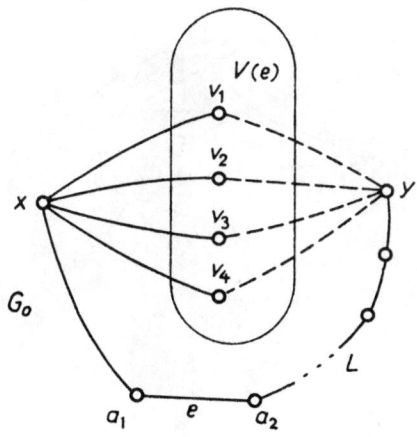

fig. 2.6

dotted lines are not edges, on the contrary, they denote non-adjacency.) Since $V_2 = V(e) \cup \{a_2\}$ is also a vertex xy-cut in G_0 and no v_i is adjacent to y, according to 2 a_2 is adjacent to x in G_0. Then, however, substituting the two edges in L incident with a_1 by $\{x, a_2\}$, the minimality of L is violated. Hence, the proof of Theorem 9 is complete provided the above three statements are verified.

In order to prove 1, assume that p is a common neighbour of x and y in G_0. Let $G(-p)$ be the graph obtained from G_0 by deleting the vertex p and all edges incident to p. The minimal number of xy-cut vertices in $G(-p)$ is $h - 1$. Owing to the minimality of G_0, there are $h - 1$ vertex-disjoint xy-paths in $G(-p)$. Such a set of paths, however, yields h vertex-disjoint xy-paths in G_0 together with the path of length 2 via p, contradicting our indirect assumption.

In order to prove 2, let $V = \{v_1, v_2, \ldots, v_h\}$ be a vertex xy-cut in G_0. Let L_x and L_y denote the sets of paths in G_0 with one of their endpoints being x and y, respectively, their other endpoint being in V and having no inner vertex in V. The common vertex q of a path in L_x and of another path in L_y can only be their endpoint in V since, otherwise, the xq part of the former and the qy part of the latter would together constitute an xy-path in G_0 containing no vertex of V, although V is a vertex xy-cut in G_0. We claim that either the length of all paths in L_x or that of all paths in L_y is 1. Assume the contrary: see figure 2.7 for a special case. The graphs G_x and G_y are obtained from G_0 by deleting the inner vertices of the paths in L_x and L_y, respectively, along with the edges incident to them and by further adding the edges $\{x, v_i\}$ and $\{v_i, y\}$, respectively ($i = 1, 2, \ldots, h$). The graphs

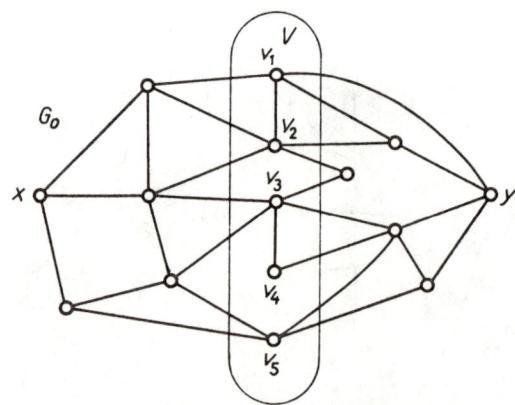

fig. 2.7

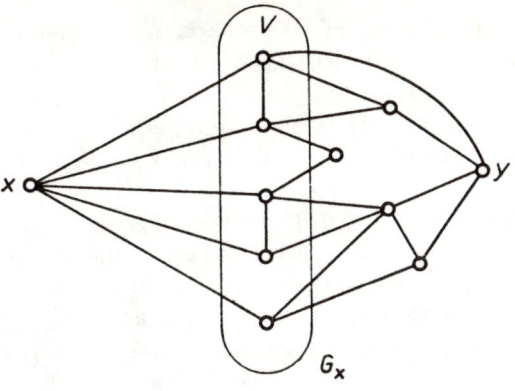

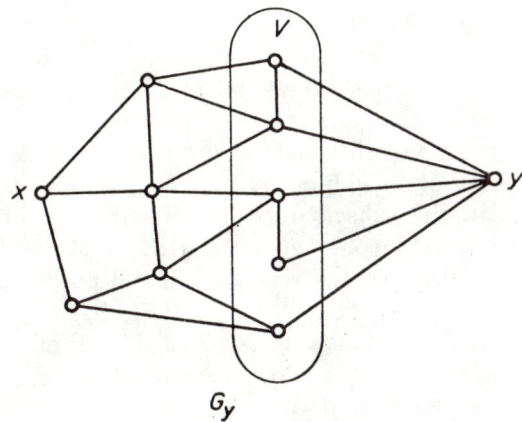

fig. 2.8

G_x and G_y obtained from figure 2.7 are shown in figure 2.8. Evidently, both G_x and G_y contain fewer vertices than G_0 and in none of them are x and y adjacent. The graph G_x cannot have a vertex xy-cut V_0 of fewer vertices than h since, otherwise, there would be an xy-path L_0 in G_0 containing no vertex of V_0 as V_0 cannot be a vertex xy-cut in G_0. Now, substituting the xv_j part of L_0 in L_x by the edge $\{x, v_j\}$, an xy-path in G_x would be obtained with no vertex in V_0. Therefore, V_0 cannot be a vertex xy-cut in G_x. The fact that there is no vertex xy-cut of fewer than h elements in G_y either, can be similarly verified. Since both G_x and G_y contain fewer vertices than G_0, both graphs have h vertex-disjoint xy-paths due to the minimality of G_0. The parts in G_x belonging to L_y and those in G_y belonging to L_x of these paths together constitute h vertex-disjoint xy-paths in G_0, contrary to our indirect assumption.

Therefore, either the length of all paths in L_x or that of all paths in L_y is 1. Without loss of generality we suppose that the length of paths in L_x is 1 in the graph G_0. Since the endpoints in V of the paths in L_x constitute a vertex xy-cut in G_0 and h is the minimal number of xy-cut vertices in G_0, all edges $\{x, v_i\}$ must be included in the paths in L_x. This completes the proof of Property 2.

In order to verify 3, let $G(-V(e))$ denote the graph obtained from G_0 by deleting the vertices of $V(e)$ along with the edges incident to them. The set $V(e)$ of $h-1$ elements is not a vertex xy-cut in G_0, therefore, there is an xy-path in $G(-V(e))$. No such xy-path can exist in $G(-e)$ since $V(e)$ is a vertex xy-cut in $G(-e)$. Hence, all xy-paths in $G(-V(e))$ must include the edge $e = \{a_1, a_2\}$. Consequently, $a_1, a_2 \notin V(e)$ and, further, each xy-path in G_0 either contains a vertex in $V(e)$ or includes the endpoints of the edge e, which yields Property 3.

This completes the proof of Theorem 8.

This theorem gives a solution to the road-covering problem raised in §2.1. In the language of graphs, a possible modification of this problem is obvious: instead of the vertices x and y of a graph G, two subsets A and B of the set of vertices of G are given and all ab-paths with $a \in A$ and $b \in B$ are intended to be 'covered', possibly at an endpoint, by a minimal number of vertices. So, let A and B be two disjoint subsets of the set of vertices P of a graph G. An AB-*path* of the graph G is one of its ab-paths with $a \in A$ and $b \in B$. A subset of the set P containing at least one vertex of all AB-paths of G (possibly an endpoint) is called a *set of vertices covering the AB-paths* in G. Consider one of these with a minimal number of vertices. The number of its elements is called the *minimal number of vertices covering the AB-paths* in G. The *independent paths* of a graph are paths with no two of them sharing a vertex (not even endpoints). The highest number of existing independent paths in the graph G is called the *maximal number of independent paths* in G. Now, applying Theorem 9, the following theorem will be proved:

10. *If A and B are two disjoint subsets of the set of vertices of a graph G then the maximal number of independent AB-paths in G is equal to the minimal number of vertices covering the AB-paths.*

Proof. Let us augment this graph $G = (P, E, \mathcal{G})$ as follows. Let us add two new vertices x and y to P and let us increase E by $|A| + |B|$ new edges to make x adjacent to all the vertices in A and y to all the vertices in B. Let the graph thus constructed from G be G'. By this augmentation, any AB-path in G is uniquely augmented to an xy-path in G', and any xy-path in G' is uniquely reduced to an AB-path in G if we return to G from G'. It is further obvious that any system H of independent AB-paths in G is uniquely augmented to a system H' of xy-paths in G', and H' is uniquely reduced to H if G is returned to from G'. Similarly, a unique correspondence can be discovered between the sets of vertices covering the AB-paths in G and the vertex xy-cuts in G'. Hence, applying Theorem 8, we obtain the assertion of Theorem 10.

However, Theorem 8 can also be deduced from Theorem 10: see Problem 57.

The theorems stated in this section are valid for directed graphs, too, clearly with a suitably modified definition of the concepts encountered. Vertex-disjoint and edge-disjoint directed paths have already been mentioned in 1.6 of Chapter 1. The following concepts for directed graphs can be obtained according to the pattern of the definitions given for non-directed graphs: *the maximal number of vertex-disjoint and edge-disjoint $\overrightarrow{xy}$-paths, the maximal number of independent $\overrightarrow{AB}$-paths, edge $\overrightarrow{xy}$-cut, vertex $\overrightarrow{xy}$-cut, minimal edge $\overrightarrow{xy}$-cut, minimal vertex $\overrightarrow{xy}$-cut, the minimal number of $\overrightarrow{xy}$-cut edges, the minimal number of $\overrightarrow{xy}$-cut vertices, the minimal number of vertices covering the $\overrightarrow{AB}$-paths.* The edge $\overrightarrow{xy}$-cut denoted by $\overrightarrow{W}$ involves now also a partition $\{P_x, P_y\}$ of the set of vertices P of the graph $\overrightarrow{G}$ with the edges of $\overrightarrow{W}$ constituted by the $\overrightarrow{P_x P_y}$-edges. $\overrightarrow{G}$ may simultaneously have $\overrightarrow{P_y P_x}$-edges, too which, clearly do not belong to $\overrightarrow{W}$. A bypass $\overrightarrow{xp}$-path in the graph $\overrightarrow{G}$ can also be defined according to the pattern of non-directed graphs. Now, however, the necessary modifications require more care. For the definition of this concept, the direction of $\overrightarrow{H}$ is now given by the direction of $\overrightarrow{G}$. The *bypass $\overrightarrow{xp}$-path* is a subgraph of $\overrightarrow{G}$ constituting an xp-path with the orientation disregarded and at its one-fold traversal starting from x, the direction of walk coincides with the orientations in $\overrightarrow{G}$ along the edges not in $\overrightarrow{H}$ and it is contrary to these along the edges in $\overrightarrow{H}$. It can be verified that, with these insignificant modifications, the previous reasonings are valid for directed graphs, too, and so the following analogue of Theorem 5 can be proved:

11. *If x and y are two vertices of a graph $\overrightarrow{G}$ then the maximal number of edge-disjoint $\overrightarrow{xy}$-paths in $\overrightarrow{G}$ is equal to the minimal number of $\overrightarrow{xy}$-cut edges.*

Furthermore, the following analogues of Theorems 8 and 10 can be proved:

12. *If x and y are two vertices of a graph $\overrightarrow{G}$ and there is no edge (x, y) in $\overrightarrow{G}$, then the maximal number of vertex-disjoint $\overrightarrow{xy}$-paths in $\overrightarrow{G}$ equals the minimal number of $\overrightarrow{xy}$-cut vertices.*

13. *If A and B are two disjoint subsets of the set of vertices of a graph $\overrightarrow{G}$, then the maximal number of independent $\overrightarrow{AB}$-paths in $\overrightarrow{G}$ is equal to the minimal number of vertices covering the $\overrightarrow{AB}$-paths.*

The substantial difference between the cases of directed and non-directed graphs is that for non-directed graphs the maximal number of vertex-disjoint and edge-disjoint xy-paths is equal to the maximal number of vertex-disjoint and edge-disjoint yx-paths, respectively, which is by far false for $\overrightarrow{xy}$-paths. This, however, in no way affects the validity of the above theorems.

I mentioned that Theorems 5, 8 and 10 are also valid for infinite graphs G provided there are edge xy-cuts, vertex xy-cuts and sets of vertices covering AB-paths, respectively with a finite number of elements. A similar assertion is true for directed graphs, too.

2.3 The optimal assignment problem. The Hungarian method

As noted before, Theorem 10 was discovered by K Menger. He published his results, phrased in a way different from the one adopted here, in 1927 [7]. His proof presented at that time, however, did not cover the special case when all edges of the graph G are AB-edges, i.e. when the graph $G(A, B)$ is a *bipartite graph*. The first proof for this case was given by Dénes König in 1931 (see footnote 1 on page 234 and footnote 2 on page 244 of [8]). This theorem is 5.22 in [2]). The main idea of König's proof can be profitably utilised in the solution of many a problem. It also constitutes the root of the above augmentation by bypass paths.

Therefore, the special case of Theorem 10 relating to bipartite graphs $G(A, B)$ is worth considering. Then the length of each AB-path is 1 and each AB-edge, along with its endpoints, yields an AB-path. Certain *edges* of a graph are called *independent*, if no two of them share an endpoint. If R is a subset of the set vertex of a graph containing at least one endpoint of any edge of the graph then R is called a *set of covering vertices* of the graph. Let $\mathcal{R}$ denote the set formed by all sets of covering vertices in the graph G. The *minimal number of covering vertices* of a graph is the number

$$\min_{R \in \mathcal{R}} |R|$$

(see the definitions of these concepts in [2], too). The pertinent special case of Theorem 10, i.e. the theorem of Dénes König can be formulated as follows:

14. *The maximal number of independent edges in a bipartite graph is equal to the minimal number of covering vertices.*

König's proof is based on the so-called *method of alternating paths* detailed below (cf. [2] pages 122–126). If, in the course of the traversal of a path L of a graph, consistently two types of edges, say heavy and thin ones, alternate then the path L is called an *alternating path*. It is also said that any vertex of L is *accessible along an alternating path* from any of its other vertices. Now, the heart of the method is that if both ends of an alternating path are formed by a thin edge, then, replacing the thin edges by heavy ones and vice versa, the number of heavy lines is increased by one. Such an exchange is illustrated in figure 2.9. Also notice that, if the heavy edges of a graph are independent and no heavy edge is incident to the end-vertices of the alternating path then the heavy lines remain independent after the exchange, too.

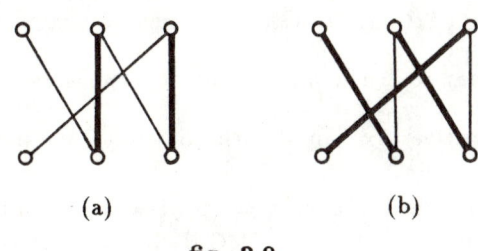

fig. 2.9

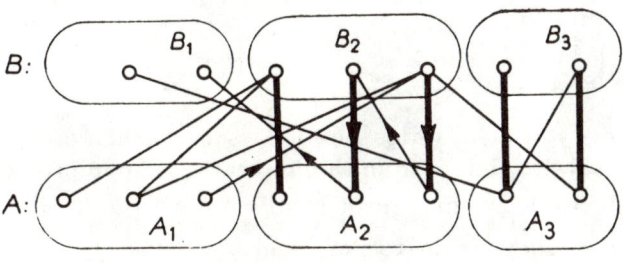

fig. 2.10

Proof of Theorem 14. Let us select in the bipartite graph $G = G(A, B)$ as many independent, i.e. heavy, edges as possible. Let A_1 and B_1 consist of those vertices of A and B, respectively which are incident to no heavy edges. Then there is obviously no $A_1 B_1$-edge in G (see figure 2.10). The endpoints of the heavy lines belong to the subsets of A and B marked by subscripts 2 and 3 depending on whether their vertices from A_1 along an alternating path are accessible or inaccessible, respectively. Then, obviously, there are neither $A_1 B_3$-edges nor $A_2 B_3$-edges in G. Now, if there is an $A_2 B_1$-edge in G then there is an alternating $A_1 B_1$-path, too (the one marked by arrows in the diagram). The number of heavy lines can be increased using the method of the alternating paths. If, however, there is no $A_2 B_1$-edge in G then $B_2 \cup A_3$ is obviously a set of covering vertices and, $|B_2 \cup A_3|$ being equal to the number of heavy lines, the latter is maximal and $B_2 \cup A_3$ is minimal. This concludes the proof of Theorem 14.

The following procedure has been formulated on the basis of the above proof.

15. Algorithm *for finding a maximal number of independent edges and a minimal number of covering vertices in the bipartite graph $G = G(A, B)$.*

(1) If there is no AB-edge in G,

then $\begin{cases} \text{there are neither heavy edges nor marked vertices, and} \\ \text{jump to (5).} \end{cases}$

Select an arbitrary AB-edge in G and let this be heavy.

(2) Mark those vertices of A which are incident to no heavy edges.

(3) If there is a marked vertex r in B with no heavy edge incident to it then jump to (4).

If there is a yet unselected edge $e = \{p, q\}$ with its endpoint p marked but its endpoint q not and either $p \in A$ and e is thin or $p \in B$ and e is heavy,

then $\begin{cases} \text{select the edge } e, \\ \text{mark the vertex } q \text{ and} \\ \text{return to (3).} \end{cases}$

Jump to (5).

(4) Mark the edge $f = \{s, r\}$ whose selection resulted in marking r and change f from heavy to thin or from thin to heavy, as appropriate.

If the number of heavy lines incident to s is not 1 (i.e. 0 or 2),

then $\begin{cases} \text{let } s \text{ take over the role of } r \text{ and} \\ \text{return to (4).} \end{cases}$

Regard the vertices marked and edges selected in (2), (3) and (4) as not marked and not selected, respectively.

Return to (2).

(5) Record the result: a maximal number of independent edges is constituted by the heavy lines. A minimal number of covering vertices is formed by the unmarked endpoints in A of the heavy lines and by their marked endpoints in B.

Remark. Just as in the remark following Algorithm 6, in the case $|A| = n \leq m = |B|$, the upper bound $n(m + n)^2$ can be obtained for the number of steps.

Let $G = G(A, B)$ be a bipartite graph, let the vertices belonging to A and B be denoted as $a_1, a_2, \ldots, a_m$ and $b_1, b_2, \ldots, b_n$, respectively and let us associate a matrix $M = M_{m,n} = [c_{ij}]$ with the graph G as follows:

$$c_{ij} = \begin{cases} 1 & \text{if } a_i \text{ and } b_j \text{ are adjacent in } G, \\ 0 & \text{otherwise.} \end{cases}$$

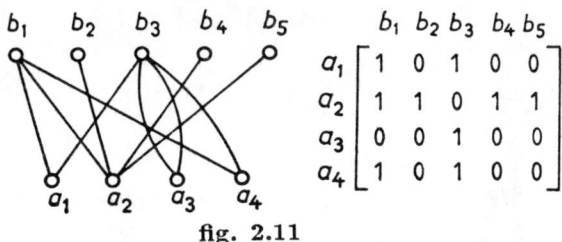

fig. 2.11

Figure 2.11 illustrates such an assignment. What bearing does Theorem 14 mean in terms of M? The independent edges in G correspond to one-entries in M with no two of them in the same row or the same column. Let us call such *one-entries in M independent*. The elements of a set of covering vertices in G correspond to rows or columns (*bands*, with a common name) in M together including all one-entries. Let us call such bands of M a set of *bands covering the one-entries in M*. Accordingly, Theorem 14 can be reformulated as follows:

16. *If all elements of a matrix M are 0 or 1 then the maximal number of independent one-entries is equal to the minimal number of the bands covering the one-entries.*

A maximal number of independent one-entries in figure 2.11 are, for example those at the positions c_{11}, c_{24}, c_{43} and the bands a_2, b_1, b_3 are covering ones.

Theorem 16 and its generalisation detailed below is due to Jenő Egerváry ([9], §3).

Let us assign the 'weights' $\alpha_1, \alpha_2, \alpha_3, \alpha_4, \beta_1, \beta_2, \beta_3, \beta_4, \beta_5$ to the bands $a_1, a_2, a_3, a_4, b_1, b_2$ of the matrix M in figure 2.11. If $c_{ij} \leq \alpha_i + \beta_j$ for each element of M then the weights 'cover the elements of M'. In this case, regarding the independent elements c_{11}, c_{24} and c_{43},

$$3 = c_{11} + c_{24} + c_{43} \leq \alpha_1 + \beta_1 + \alpha_2 + \beta_4 + \alpha_4 + \beta_3.$$

Therefore, at any covering system of weights, the sum of the weights covering the elements of M is at least 3. If a unit weight is associated with the covering bands and zero weight with the rest then this system of weights is obviously covering and, according to the above, there is no covering system of weights with a smaller sum of the weights. As a generalisation of Theorem 16 we prove that for an arbitrary matrix with non-negative elements, the maximum of the sum of independent elements is equal to the minimum of the sum of covering weights.

Let $c_{ij} \geq 0$ hold for all pairs i, j in a matrix $M = M_{m,n} = [c_{ij}]$. It can be assumed that $m \leq n$ since, otherwise, the transpose of M is considered. Let the rows and columns of M be denoted by $a_1, a_2, \ldots, a_m$ and by $b_1, b_2, \ldots, b_n$, respectively. Let f be a real function defined over the bands of M with the following property:

$$f(a_i) = \alpha_i \geq 0, \quad f(b_j) = \beta_j \geq 0 \quad \text{and} \quad \alpha_i + \beta_j \geq c_{ij}$$

$$\text{for} \quad i = 1, 2, \ldots, m, \quad j = 1, 2, \ldots, n.$$

Such functions f can also be called *covering systems of weights*. The *sum* $s(f)$ *of weights of a covering system of weights* f is:

$$s(f) = \sum_{i=1}^{m} \alpha_i + \sum_{j=1}^{n} \beta_j.$$

Now let P denote a permutation of m distinct numbers selected from among the numbers $1, 2, \ldots, n$ and let the ith number therein be $P(i)$. Let, further

$$c(P) = \sum_{i=1}^{m} c_{iP(i)}.$$

No two terms in this sum are in the same band. Therefore $c(P)$ can be called the sum of the maximal number of independent elements in M selected by P. Now, the theorem of J Egerváry generalising Theorem 16 is the following:

17. $\max_{P} c(P) = \min_{f} s(f)$.

Proof. At first we prove that, for any permutation P and covering system of weights f, the following holds:

$$c(P) \leq s(f).$$

Indeed,

$$c_{iP(i)} \leq \alpha_i + \beta_{P(i)}, \quad (i = 1, 2, \ldots, m),$$

and thus

$$c(P) = \sum_{i=1}^{m} c_{iP(i)} \leq \sum_{i=1}^{m} (\alpha_i + \beta_{P(i)}) \leq s(f).$$

Consequently,

$$\max_{P} c(P) \leq \min_{f} s(f).$$

Therefore, it suffices to prove the existence of P and f with

$$c(P) = s(f).$$

Evidently, in this case

$$\alpha_i + \beta_{P(i)} = c_{iP(i)} \quad \text{for all } i,$$

and if there is no i with $j = P(i)$, then

$$\beta_j = 0.$$

It is sufficient to treat the case $m = n$ since if $m < n$, then the matrix can be extended to a square matrix by adding rows consisting of zeros, i.e. let $c_{ij} = 0$ if $i = m+1, m+2, \ldots, n$. Then

$$\alpha_i = \beta_{P(i)} = 0$$

is necessarily valid, too, for the above P and f, in the case $i = m+1$, $m+2, \ldots, n$. Thus the equality $c(P) = s(f)$ holds with the same value $s(f)$ for the original matrix, too.

So, let M be the n-order square matrix obtained in the above way. Let a *complete bipartite graph* $G = G(A, B)$ be associated with M as follows: let $a_1, a_2, \ldots, a_n$ and $b_1, b_2, \ldots, b_n$ denote the vertices of the sets A and B, respectively and let all vertices of A be adjacent to all vertices of B. Let us further assign the *value c_{ij} to the edge* $\{a_i, b_j\}$. A permutation P selects n independent edges in G and $c(P)$ is the sum of the numbers assigned to the edges selected by P, in short, the *value of* this *system of* independent *edges*. It is also said that the *covering system of weights* f *covers the edges* of G, the value $f(a_i)$ of the function is called the *weight* of a_i and, further, the edge $\{a_i, b_j\}$ is called *covered* or *exactly covered* or *over-covered*, in this order, depending on which of the following relations is satisfied:

$$\alpha_i + \beta_j \left\{ \begin{array}{c} \geq \\ = \\ > \end{array} \right\} c_{ij}.$$

Let f_0 be a minimal covering system of weights, i.e.

$$s(f_0) = \min_f s(f)$$

(it follows from the well-known theorem of Weierstrass that this minimum exists and that the minima and maxima in the similar subsequent cases also exist). Let, further, $G_0 = G_0(A, B)$ be the subgraph of the graph G formed by those edges of G which are exactly covered by f_0. In order to verify Theorem 17, it suffices to show that an independent system E of n edges can be selected from G_0. So, if P_0 is the permutation with $\{a_i, b_{P_0(i)}\}$ included in E for all values of i then, the edges of E being exactly covered, the equality

$$c(P_0) = s(f_0)$$

holds which, according to the above considerations, will yield the proof of Theorem 17.

Our reasoning uses the method of alternating paths. Let E_0 denote a system of a maximal number of independent edges in G_0. We must show

$$|E_0| = n.$$

Indirectly, assume that $|E_0| < n$. Let us consider the partition of A and B shown in figure 2.10: the edges of E_0 are heavy. The assumption $|E_0| < n$ implies that neither A_1 nor B_1 is empty.

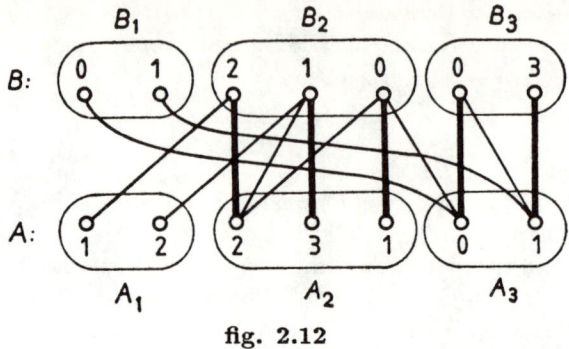

fig. 2.12

A particular case is illustrated in figure 2.12 with the weights written beside the vertices. For the sake of simplicity, integer numbers have been used, while the values c_{ij} can be determined since the edges are exactly covered.

Let us recollect the basic properties of the graph G_0 owing to the maximality of E_0. Obviously, there are neither A_1B_1-edges, nor A_1B_3-edges, nor A_2B_3-edges.

Let us first assume that $A_1 \cup A_2$ contains no vertex with weight 0. Let ε_1 be the minimum of the extent of over-covering among the over-covered edges in G, i.e.

$$\varepsilon_1 = \min_e \left(f_0(a_i) + f_0(b_j) - c_{ij} \right),$$

where $e = \{a_i, b_j\}$ is an over-covered edge in G. Let ε_2 be the least non-zero weight, i.e.

$$\varepsilon_2 = \min_a f_0(a)$$

where $a \in A \cup B$ and $f_0(a) \neq 0$. Let, further,

$$\varepsilon_0 = \min(\varepsilon_1, \varepsilon_2).$$

Let us decrease the weight of all vertices in $A_1 \cup A_2$ by ε_0 and increase the weight of all vertices in B_2 by ε_0. Since the neighbours in G_0 of the vertices in $A_1 \cup A_2$ are all in B_2, the system f_1 obtained from f_0 by this modification is also a system of weights covering the edges of G. However, $|B_2| < |A_1 \cup A_2|$ and so, $s(f_1) < s(f_0)$, contrary to the minimality of f_0.

Let us now assume that $A_1 \cup A_2$ includes a vertex with weight 0. Then, $B_1 \cup B_3$ cannot contain a vertex with weight 0 because any two vertices with zero weight are adjacent in G_0 since the edge in G implying this adjacency cannot be over-covered. However, there is no edge in G_0 with one of its endpoints in $A_1 \cup A_2$ and the other in $B_1 \cup B_3$. Let us now decrease the weight of all vertices in $B_1 \cup B_3$ by ε_0 and increase the weight of all vertices in A_3 by ε_0. Since the neighbours in G_0 of the vertices in $B_1 \cup B_3$ are all in A_3, the system f_2 obtained from f_0 by this modification is also a system of weights covering the edges of G. However, $|A_3| < |B_1 \cup B_3|$ and so, $s(f_2) < s(f_0)$, contrary to the minimality of f_0.

This completes the proof of Theorem 17.

On the basis of the proof, the following procedure can be given to obtain a permutation P_0 and covering system of weights f_0 satisfying

$$\max_P c(P) = c(P_0) = s(f_0) = \min_f s(f)$$

for matrices with non-negative, integer elements (this restriction is adopted for the sake of convenience only). The procedure is formulated for the bipartite graph associated with the matrix as above. The matrix is deemed to have been augmented to make it a square matrix, i.e. $c_{ij} = 0$ if $i = m+1, m+2, \ldots, n$.

18. Algorithm *for finding a system of independent edges of maximal value in a bipartite graph, and a minimal system of weights covering the edges of the graph.*

Let, for a complete bipartite graph $G = G(A, B)$, the vertices in A be $a_1, a_2, \ldots, a_n$ and those in B be $b_1, b_2, \ldots, b_n$ and, further, let c_{ij} be the non-negative value of the edge $\{a_i, b_j\}$.

(1) Select the covering system of weights f as follows:
$f(a_i) = \max_j c_{ij}$ and $f(b_i) = 0$ for $i = 1, 2, \ldots, n$.

(2) Let $G_0 = G_0(A, B)$ be the subgraph of the graph G formed by those edges of G which are exactly covered by f.

Select a maximal number of independent edges in G_0 (for example with the aid of Algorithm 15, i.e. by the method of alternating paths).

Form the partition of the set of vertices of G_0 in accordance with figure 2.12 (now there are no more $A_2 B_1$-edges).

(3) If A_1 is not empty and $A_1 \cup A_2$ includes no vertex with zero weight,

then $\begin{cases} \text{decrease the weights of the vertices in } A_1 \cup A_2 \text{ by the value } \varepsilon_0 \\ \quad \text{defined above and, similarly,} \\ \text{increase the weights of the vertices in } B_2 \text{ (see Remark 1 below)} \\ \quad \text{and, further,} \\ \text{let the covering system of weights thus obtained take over the} \\ \quad \text{role of } f \text{ and} \\ \text{return to (2).} \end{cases}$

(4) If A_1 is not empty,

then $\begin{cases} \text{decrease the weights of the vertices in } B_1 \cup B_3 \text{ by } \varepsilon_0 \text{ and} \\ \text{increase the weights of the vertices in } A_3 \text{ and, further,} \\ \text{let the covering system of weights thus obtained take over the} \\ \quad \text{role of } f \text{ and} \\ \text{return to (2).} \end{cases}$

(5) Record the result: a maximal number of independent edges is constituted by the heavy lines. A minimal covering system of weights in G is formed by f.

Remarks. 1. In particular cases, ε_0 can be taken to be higher than defined: at the definition of ε_1 and ε_2, it is not always necessary to take into account all edges and all vertices with non-zero weight, but only those concerned in the reduction. The main point is that all edges must stay covered and no weight can be negative. If the number of steps is of no concern, the choice $\varepsilon_0 = 1$ is always suitable.

2. Owing to the choice $\varepsilon_0 = \min(\varepsilon_1, \varepsilon_2)$, the graph G_0 is augmented by at least one edge at each reduction of the weights, so Algorithm 15 need not be applied more than $e = n^2$ times. Observe that this estimate does not depend at all upon the values of the edges, and we did not use the fact that the values are integer or even rational. (It is, however, assumed that each operation on numbers is deemed to be a single step.) The above trick of reducing and increasing the weights has been called the *Hungarian method* by H W Kuhn [10]. Since this trick is based upon the method of alternating paths, it is customary to call the method of alternating paths itself, and even its generalisation, the method of augmentation bypasses used in Algorithm 6, as the Hungarian method.

Now, an application of Theorems 14 and 17 for solving economical problems will be pointed out. (Some of their other significant applications will be presented later.)

19. Optimal matching. *Assume that a firm has m persons at its disposal for carrying out n jobs. One job can be carried out by one person only and one person can carry out one job only and not all persons are qualified for all jobs. The problem is to design a plan ensuring the employment of as many persons as possible.*

Let us assign a bipartite graph $G(A, B)$ to the problem with its vertices in A corresponding to persons, its vertices in B to jobs and each AB-edge meaning that the person corresponding to its endpoint in A is qualified for carrying out the job corresponding to its endpoint in B. Now, the optimal plan is constituted by a set of independent edges of G with a maximal number of elements. This can be selected by the method of alternating paths.

As an example, consider the case shown in figure 2.13. The two heavy edges show that the persons corresponding to vertices a_1 and a_3 have already been given a job. Employing the method of alternating paths, let us investigate whether this matching is optimal. The steps of the investigation are shown in figure 2.14. On the basis of figure 2.13, the partition of the vertices corresponding to figure 2.12 is indicated in figure 2.14(a). Since an A_2B_1-edge has been found here, the set of heavy lines can be augmented. Let us interchange the heavy and thin edges along the alternating path a_4b_2. The partition after

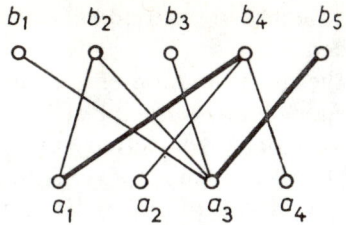

fig. 2.13

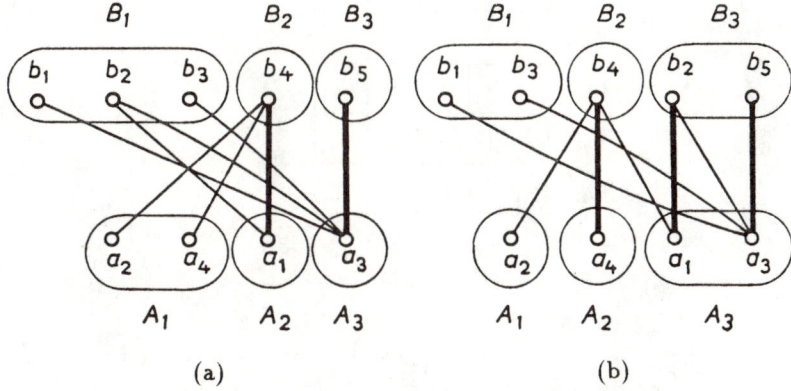

fig. 2.14

the interchange is illustrated in figure 2.14(b). Now, there is no A_2B_1-edge, so the three heavy lines shown here constitute an optimal employment plan.

20. Optimal assignment. *Let us retain the conditions of the previous problem with the modification that the efficiency of each person for each job is known and has been expressed by a non-negative integer (or rational) number. In particular, this can be the work done in unit time, the quantity of the goods produced, or $100 - s$ (where s is the percentage of faulty products), etc. If someone is not qualified for a job, his efficiency is defined as zero. Now, the problem is to design a plan of employment maximising the total efficiency.*

The problem is now associated with the bipartite graph $G(A, B)$ with all AB-edges existing and with the values of the AB-edges being the efficiencies. An optimal plan is yielded by a permutation P with $c(P)$ maximal. Such a permutation can be obtained with the aid of Algorithm 18. If $n < m$ then interchange the role of A and B. In an optimal plan, in the case $m \leq n$, all persons can be given a job and, in case $n < m$, all jobs can be carried out. It is, however, possible that some jobs are carried out with zero efficiency, which can, naturally, be disregarded.

The problem requiring the determination of a covering system of weights with a minimal sum of weights can be interpreted as follows. Let us stick to the previous problem with the modification that the efficiencies denote the hourly wages obtainable by particular persons at particular jobs. The question is, what is the minimal sum of money the firm has to reserve for a given period to be able to pay the wages for any possible assignment of jobs.

	b_1	b_2	b_3	b_4	b_5
a_1	5	3	7	4	2
a_2	8	5	8	6	4
a_3	4	0	4	3	0
a_4	6	3	6	5	0

fig. 2.15

The chart shown in figure 2.15 indicates the efficiencies in a particular case. Let us find an employment plan of optimal efficiency. Integer numbers have been employed for simplicity and to facilitate the application of Algorithm 18. Let us first augment the chart by the row a_5 consisting of zeros to make it a square matrix. The course of the solution is shown in figure 2.16. The vertices have been selected in the order of increasing subscripts. The non-zero weights have been written beside the vertices and beside the relevant bands of the over-covering charts. On the left-hand side of the relationships, the total efficiency of the heavy edges is shown, while the sum of weights is indicated on the right. In the first step, the assignment of weights has been done in accordance with step (1) of Algorithm 18. Then, according to (2), the graph G_0 formed by the exactly covered edges has been constructed and a maximal number of independent edges has been marked by heavy lines. Thereupon, according to (3), the weights have been modified with the choice $\varepsilon_0 = 1$. This yielded the second over-covering chart, and following (2) resulted in the second graph again. Finally, according to (3) again, the weights have been modified with the choice $\varepsilon_0 = 2$ leading to the third chart. Then, (2) gives the solution: see the third graph.

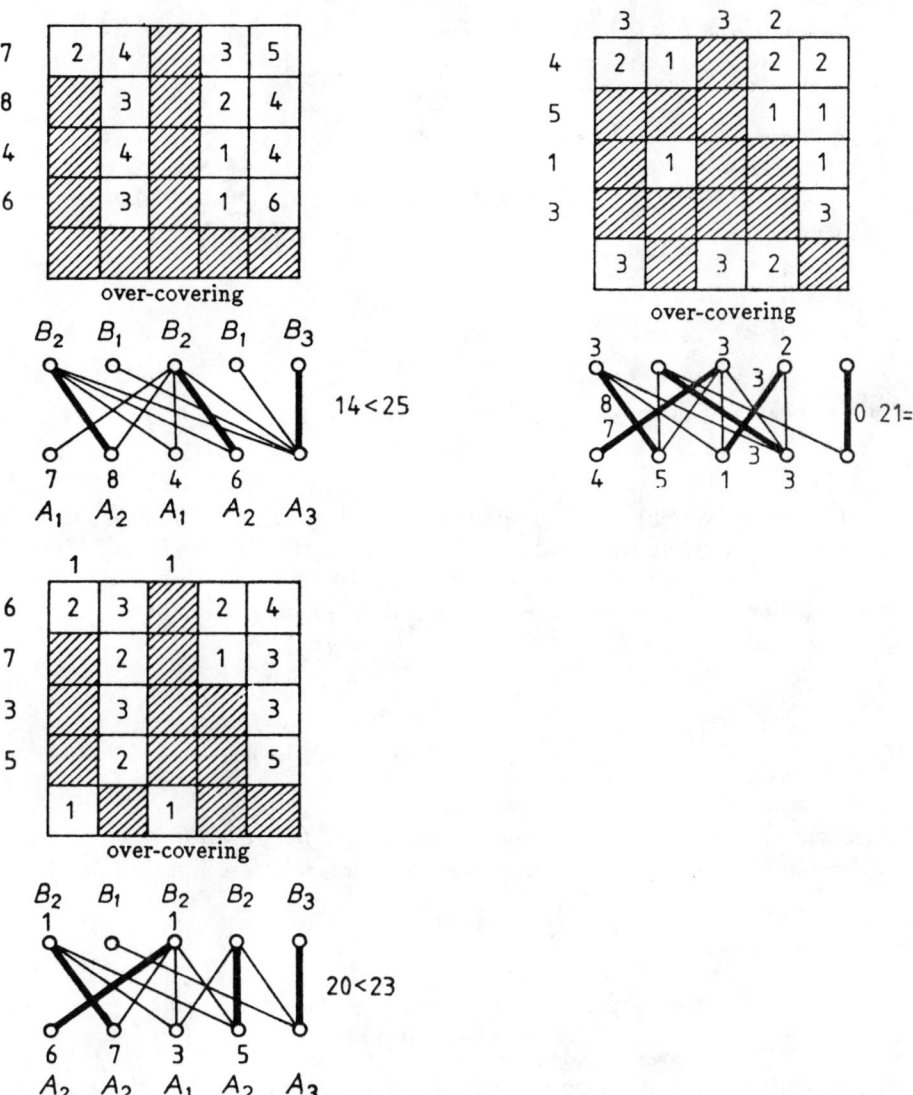

fig. 2.16

2.4 Max flow — min cut

In this section we solve the second problem presented in §2.1, a method will be shown for searching an xy-flow of maximal value in a graph $\overrightarrow{G} = (P, E, \overrightarrow{\mathcal{G}})$. A few notations are first introduced. If a real function g has been defined over the set E then the *value* of g *on the edge* $e = (p, q)$ will be denoted by $g(e)$

or $g(p,q)$. Let, in the case of $Q, R \subseteq P$,

$$g(Q,R) = \sum_{\substack{(q,r) \in E \\ q \in Q, r \in R}} g(q,r),$$

$$g(p,Q) = \sum_{\substack{(p,q) \in E \\ p \text{ fixed}, q \in Q}} g(p,q), \quad g(Q,p) = \sum_{\substack{(q,p) \in E \\ p \text{ fixed}, q \in Q}} g(q,p),$$

and if $E' \subseteq E$, then

$$g(E') = \sum_{e \in E'} g(e),$$

and further, if E' is empty, then for non-negative g:

$$g(E') = 0.$$

Two non-negative real functions have appeared in the present problem: the *edge-capacity function* κ defined over the set E and the *flow function* f 'limited by κ'. The number $\kappa(E')$ is called the *capacity* of E'. Recall the relations defining an xy-flow. Among these, (2°) and (3°) can be formulated with the aid of the value $\sigma = \sigma(f)$ of the flow as

$$\sum_{\substack{(p,q) \in E \\ p \text{ fixed}}} f(p,q) - \sum_{\substack{(q,p) \in E \\ p \text{ fixed}}} f(q,p) = \begin{cases} \sigma & \text{if } p = x, \\ 0 & \text{if } p \in P \text{ and } p \neq x, y, \\ -\sigma & \text{if } p = y. \end{cases}$$

With the previously introduced notations, an *xy-flow limited by κ in $\vec{G}$* (an xy-flow in short) can be defined by the following relationships:

$$0 \leq f(p,q) \leq \kappa(p,q), \quad \text{if} \quad (p,q) \in E, \tag{1}$$

$$f(p,P) - f(P,p) = \begin{cases} \sigma & \text{if } p = x, \\ 0 & \text{if } p \in P \text{ and } p \neq x, y, \\ -\sigma & \text{if } p = y. \end{cases} \tag{2}$$

According to this definition, the function f with $\sigma(f) = 0$ is also regarded as an xy-flow (there may still be edges in $\vec{G}$ with the non-zero value of the flow, e.g. along the edges of a directed circuit of $\vec{G}$) and, similarly so is the zero-valued xy-flow with zero value on each edge. An xy-flow with positive value on an $\overrightarrow{xy}$-path $\vec{L}$ in $\vec{G}$ and zero on all other edges is called an *xy-path flow* in $\vec{G}$ and $\vec{L}$ is the *support of the path flow*. Obviously, the value of an xy-path flow f on the edges of its support is $\sigma(f)$.

Our aim is to select an element f from the set $\mathcal{F}$ of xy-flows in $\vec{G}$ limited by κ, which maximises the value $\sigma(f)$ of the flow.

Let the edge $\vec{xy}$-cut $\vec{W}$ of a graph $\vec{G}$ induce the partition $\{X,Y\}$ of P with $x \in X$ and $y \in Y$ and let $\overleftarrow{W}$ denote the set of $\overrightarrow{YX}$-edges. Let f be an xy-flow of value σ in $\vec{G}$. Then, according to (2) and then to (1):

$$\sigma(f) = \sigma = f(x,P) - f(P,x) = \sum_{p \in X}(f(p,P) - f(P,p))$$
$$= f(X,P) - f(P,X) = f(X,X) + f(X,Y) - f(X,X) - f(Y,X)$$
$$= f(X,Y) - f(Y,X) = f(\vec{W}) - f(\overleftarrow{W}) \leq f(\vec{W}) \leq \kappa(\vec{W}).$$

Hence we obtain the following two statements:

21. *For any xy-flow f and edge $\vec{xy}$-cut $\vec{W}$ in a graph $\vec{G}$:*

$$\sigma(f) = f(\vec{W}) - f(\overleftarrow{W}) \leq \kappa(\vec{W})$$

and equality replaces the inequality only if

$$f(\overleftarrow{W}) = 0.$$

22. *If the set of xy-flows in a graph $\vec{G}$ is $\mathcal{F}$ and the set of edge $\vec{xy}$-cuts is $\vec{\mathcal{W}}$, then*

$$\max_{f \in \mathcal{F}} \sigma(f) \leq \min_{\vec{W} \in \vec{\mathcal{W}}} \kappa(\vec{W}).$$

Let us first restrict the problem to the special case when the edge-capacity function κ is integer valued. Let us construct the graph $\vec{G}'$ from the graph $\vec{G}$ as follows: if $\kappa(p,q) = k \neq 0$, then replace the edge $e = (p,q)$ by k edges (p,q) and if $k=0$, delete the edge e. Assume that f of value σ is an integer xy-flow in $\vec{G}$. If $f(p,q) = i$ in $\vec{G}$ then mark i of the edges (p,q) in $\vec{G}'$. This is possible in view of (1). Let $\vec{G}'_0$ denote the subgraph of $\vec{G}'$ consisting of the marked edges and their tails and heads. The formula (2) has the following meaning for $\vec{G}'_0$:

$$\varphi_{\text{out}}(p) - \varphi_{\text{in}}(p) = \begin{cases} \sigma & \text{if } p = x, \\ 0 & \text{if } p \in P \text{ and } p \neq x,y, \\ -\sigma & \text{if } p = y. \end{cases} \quad (2')$$

Let us start from x in $\vec{G}'_0$ and proceed along edges, following their directions. According to (2'), y must be reached, so an $\vec{xy}$-path has also been traversed. Let us delete the edges of the traversed $\vec{xy}$-path. For the remaining graph, (2') is still valid, but with $(\sigma - 1)$ written instead of σ. Repeating the traversal yields the existence of σ edge-disjoint $\vec{xy}$-paths in $\vec{G}'_0$ and hence in $\vec{G}'$, too. Deleting the edges of these edge-disjoint $\overrightarrow{xy}$-paths as well, a graph is obtained with

$$\varphi_{\text{out}}(p) - \varphi_{\text{in}}(p) = 0$$

for all of its vertices p. Such a graph can be decomposed into edge-disjoint circuits (see [2], 3.15) and simultaneously yields a zero-valued flow called *circulation*.

Clearly, the reverse is also true: σ edge-disjoint $\overrightarrow{xy}$-paths and certain directed circuits of $\vec{G}'$ together determine σ xy-path flows of unit value and a circulation of $\vec{G}'$ so that their support is pairwise edge-disjoint and these together form an integer xy-flow f of value σ in $\vec{G}$. In other words, f can be *decomposed into σ xy-path flows* of unit value and a circulation. This completes the proof of the following theorem:

23. *An xy-flow of integer value k in a graph $\vec{G}$ can be decomposed into k xy-path flows and a circulation.*

Evidently, an edge $\overrightarrow{xy}$-cut formed by κ edges in $\vec{G}'$ corresponds to any edge $\overrightarrow{xy}$-cut of integer capacity κ in $\vec{G}$. Consequently, on the one hand, the previous considerations, Theorem 22 and Theorem 11 of Menger yield the following:

24. *If the edge-capacity function κ is integer-valued in a graph $\vec{G}$, then among the maximal xy-flows limited by κ there exists at least one of integer value.*

On the other hand, in the case of integer-valued edge-capacity functions, the following theorem of L R Ford and D R Fulkerson ([11], page 11, Theorem 5.1) has been proved for arbitrary xy-flows:

25. *If the edge-capacity function in a graph $\vec{G}$ is κ the set of xy-flows limited by κ is $\mathcal{F}$ and the set of edge $\overrightarrow{xy}$-cuts is $\vec{\mathcal{W}}$, then*

$$\max_{f \in \mathcal{F}} \sigma(f) = \min_{\vec{W} \in \vec{\mathcal{W}}} \kappa(\vec{W}).$$

Remark. The Ford–Fulkerson theorem is a consequence of its special case proved above. Indeed, if Theorem 25 is valid in case of integer-valued edge-capacity functions, then it is also valid for rational edge-capacity functions since, for a finite number of rational numbers, there exists a unit rational number so that these rational numbers are its multiples. If, however, κ assumes real values on the edges of $\vec{G}$ then a sequence $\kappa_1, \kappa_2, \ldots$ can be considered with $\kappa_i(e)$ rational for each edge e of $\vec{G}$, and for all i, and

$$\lim_{i \to \infty} \kappa_i(e) = \kappa(e).$$

Now, applying Theorem 25 to the case of the edge-capacity functions κ_i, its validity follows for κ.

In spite of this remark, the theorem of Ford and Fulkerson will be proved independently to facilitate the presentation of the most suggestive variation of the method of augmenting paths employed for proving Theorems 5 and 11. This variation adapted to flows will be first illustrated by an example and then a few preparatory steps for the proof will be taken. The values of the edge-capacity function have been written on the edges of the directed graph shown in figure 2.17(a). In figure 2.17(b), an xy-flow f of value 3 has been indicated: in the pairs written on the edges, the second number denotes the value of f on the edges and the first one indicates the 'residual capacity'. The edges with a non-zero value of f have been drawn by heavy lines. f can be augmented to obtain the xy-flows f_1 or f_2, both of value 5. On the one hand, f is increased by 2 along the edges (x, q), (q, r) and (r, y), this yields f_1 (figure 2.18(a)). On the other hand, 'the transported quantity can be increased' also by carrying two units from the source to q along the edge (x, q), this can be carried to p from q along the edge (p, q) without carrying two units from p to q from among the three units previously brought there (i.e. two units are 'carried back' from q to p); and thereafter this is carried from p to the sink along the edge (p, y) (figure 2.18(b)).

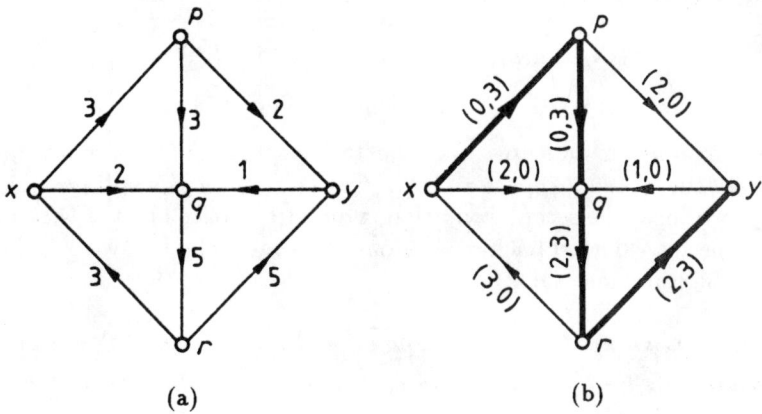

fig. 2.17

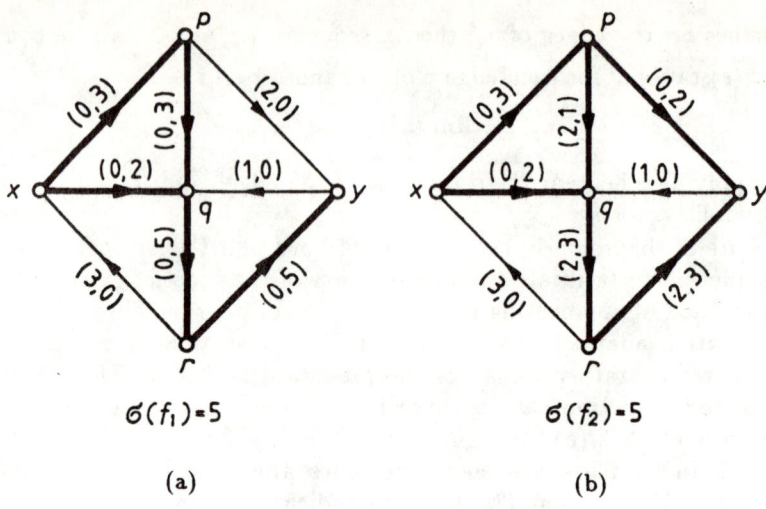

fig. 2.18

Let κ be an edge-capacity function in $\vec{G}$, f be an xy-flow of value σ in $\vec{G}$ and let $\vec{L}$ be a subgraph of the graph $\vec{G}$ where L is an xy-path in G. Let us traverse the edges of L starting from x and let us direct the edges of L according to the traversal. This yields the graph $\vec{L}'$. Obviously, the edge (p, q) in $\vec{L}$ is included in $\vec{L}'$ if and only if p is not a neighbour of q in $\vec{L}'$. The graph $\vec{L}$ is called a *flow augmenting path relative to* f if the following holds for any edge e of $\vec{L}$:

$$f(e) < \kappa(e) \qquad \text{if } e \text{ is included in } \vec{L}',$$
$$f(e) > 0 \qquad \text{if } e \text{ is not included in } \vec{L}'.$$

The first condition indicates that the transportation of a further quantity limited by the *residual capacity* $\kappa(e) - f(e)$ *relative to* f is still possible along the edge while, at the second condition, a quantity limited by the *back-capacity* $f(e)$ can be transported backwards along the edge. The following will now be proved (cf. Statement 4):

26. *If f is an xy-flow in a graph $\vec{G}$ and there exists a flow augmenting path relative to f in $\vec{G}$ then there is a flow in $\vec{G}$ with a value higher than $\sigma(f)$.*

Proof. Let the edge-capacity function in $\vec{G}$ be κ, let the flow augmenting path relative to f be $\vec{L}$ and let $\vec{L}'$ be the same as above. Let ε_1 be the least among the residual capacities of the edges in $\vec{L}$, i.e.

$$\varepsilon_1 = \min_e(\kappa(e) - f(e))$$

where e is included both in $\vec{L}$ and $\vec{L}'$. Let ε_2 be the least back-capacity, i.e.

$$\varepsilon_2 = \min_e f(e)$$

where e is included in $\vec{L}$ but not in $\vec{L}'$. Finally let

$$\varepsilon_0 = \min(\varepsilon_1, \varepsilon_2).$$

It can be stated on the basis of the above considerations that $\varepsilon_0 > 0$. Let us define the function f_1 over the edges of $\vec{G}$ as follows:

$$f_1(e) = \begin{cases} f(e) + \varepsilon_0 & \text{if } e \text{ is included both in } \vec{L} \text{ and } \vec{L}', \\ f(e) - \varepsilon_0 & \text{if } e \text{ is included in } \vec{L} \text{ but not in } \vec{L}', \\ f(e) & \text{if } e \text{ is not included in } \vec{L}. \end{cases}$$

In order to prove our assertion, we show that f_1 is an xy-flow in $\vec{G}$ and

$$\sigma(f_1) = \sigma(f) + \varepsilon_0 > \sigma(f).$$

Let the vertices of $\vec{L}$ be denoted in the order of the traversal of $\vec{L}'$ as $x = p_0, p_1, \ldots, p_k = y$. Owing to the appropriate choice of ε_0, the function f_1 satisfies the relation (1) in the definition of a flow. Let us examine the difference in (2) if f is replaced by f_1.

If $p = p_0 = x$ then the first member is increased or the second one is decreased by ε_0 depending whether the edge (p_0, p_1) in $\vec{L}'$ is in $\vec{L}$ or not. Therefore,

$$f_1(x, P) - f_1(P, x) = \sigma(f) + \varepsilon_0.$$

If $p = p_k = y$ then the second member is increased or the first one is decreased by ε_0 depending whether the edge (p_{k-1}, p_k) in $\vec{L}'$ is in $\vec{L}$ or not. Therefore,

$$f_1(y, P) - f_1(P, y) = -\sigma(f) - \varepsilon_0.$$

If $p = p_i$, $(1 \le i \le k-1)$ then four cases are obtained depending upon which of the edges (p_{i-1}, p_i) and (p_i, p_{i+1}) in $\vec{L}'$ are included in $\vec{L}$. It is easy to check that the difference in (2) does not change in any of the four cases although both members may change. If p is not a vertex of $\vec{L}'$, then, obviously, neither the first nor the second member changes in (2). This results in

$$f_1(p, P) - f_1(P, p) = 0 \quad \text{if } p \ne x, y.$$

The above considerations prove Statement 26.

Let κ be the edge-capacity function and f be an xy-flow in a graph $\vec{G}$. The subset X of the set P of vertices of the graph $\vec{G}$ is defined in the following recursive way:

$x \in X$;
if $p \in X$ and $f(p,q) < \kappa(p,q)$, then $q \in X$,
if $p \in X$ and $f(q,p) > 0$, then $q \in X$.

The vertices in X are called *accessible along a flow augmenting path relative to f* (cf. the concept of 'accessible along a bypass path relative to H'). The following statement will now be proved:

27. *If f is an xy-flow in a graph $\vec{G}$ and y is accessible along a flow augmenting path relative to f then there exists a flow augmenting path relative to f in $\vec{G}$.*

Proof. Let κ denote the edge-capacity function in $\vec{G}$. The definition of the set X implies that, starting from the source and proceeding along the edges of $\vec{G}$ with a one-fold traversal (not necessarily following the directions), y can be reached with $f(p,q) < \kappa(p,q)$ fulfilled if the direction of the traversal coincides with the orientation along the edge (p,q); and with $f(p,q) > 0$ if the traversal on the edge (p,q) is opposite to its orientation. Let us start from the source x and proceed as long as y has not been reached. If, moreover, the number of traversed edges is minimal, then no circuit in G can have been traversed. Hence, the edges of a flow augmenting path relative to f in $\vec{G}$ have been traversed. This completes the proof of Statement 27.

Proof of Theorem 25. Let f be an xy-flow of maximal value in the graph $\vec{G}$. Let $\{X, Y\}$ be the partition of the set of vertices of $\vec{G}$ where X is formed by the vertices accessible along a flow augmenting path relative to f. Then, $x \in X$. In any case, $y \in Y$ since otherwise, according to Theorems 26 and 27, the maximality of f would be impossible. Consequently, the set of $\overrightarrow{XY}$-edges constitutes an edge $\overrightarrow{xy}$-cut in $\vec{G}$, let this be $\vec{W}$. The definition of the set X implies that if $p_1 \in X$ and $q_1 \in Y$, for an edge (p_1, q_1) then

$$f(p_1, q_1) = \kappa(p_1, q_1)$$

and so

$$f(\vec{W}) = \kappa(\vec{W}).$$

If, however, $p_2 \in Y$ and $q_2 \in X$ for an edge (p_2, q_2) then

$$f(p_2, q_2) = 0$$

and so

$$f(\overleftarrow{W}) = 0.$$

So, by Theorem 21:

$$\sigma(f) = f(\overrightarrow{W}) = \kappa(\overrightarrow{W}).$$

Taking Theorem 22 into account, the assertion of Theorem 25 is obtained.

On the basis of Theorems 21 and 26, the following theorem has also been derived from Theorem 25 ([11], pages 12 and 13, corollaries 5.2 and 5.3):

28. *In a graph $\overrightarrow{G}$ with edge-capacity function κ, an xy-flow f is of maximal value if and only if there is no flow augmenting path relative to f in $\overrightarrow{G}$. Further, an edge $\overline{xy}$-cut W is of minimal capacity if and only if*

$$\sigma(f) = f(\overrightarrow{W}) = \kappa(\overrightarrow{W}) \quad \text{and} \quad f(\overleftarrow{W}) = 0$$

for any xy-flow f of maximal value in $\overrightarrow{G}$.

The following procedure is the result of the investigations carried out in connection with the proof of Theorem 25. It is applicable to graphs $\overrightarrow{G}$ with integer (or rational) edge-capacity function κ (cf. Algorithm 6.)

29. Algorithm *for finding an xy-flow of maximal value and an edge $\overline{xy}$-cut of minimal capacity:*

(1) If there is no $\overline{xy}$-path in $\overrightarrow{G}$

then $\begin{cases} \text{let the value of } f \text{ be zero on each edge of } \overrightarrow{G} \text{ and} \\ \text{jump to (5).} \end{cases}$

(2) Let the value of f be zero on each edge of $\overrightarrow{G}$. (The vertices of $\overrightarrow{G}$ accessible along a flow augmenting path relative to f will now be marked.)

Mark the vertex x.

(3) If y is marked, then jump to (4).

If there is a yet unselected edge $e = (p, q)$ in $\overrightarrow{G}$ with exactly one of its heads and tails marked,

then $\begin{cases} \text{select the edge } e \text{ and, further,} \\ \text{if } p \text{ is marked and } f(e) < \kappa(e), \text{ then mark the vertex } q, \text{ and} \\ \text{if } q \text{ is marked and } f(e) > 0, \text{ then mark the vertex } p \text{ and, finally,} \\ \text{return to (3).} \end{cases}$

Jump to (5).

(4) Mark the edge (r, y) or (y, r) in $\vec{G}$ whose selection resulted in marking y.

If $r \neq x$, then $\begin{cases} \text{let } r \text{ take over the role of } y \text{ and} \\ \text{return to (4). (This results in marking the edges of} \\ \text{a flow augmenting path relative to } f.) \end{cases}$

Let $\vec{L}$ denote that flow augmenting path relative to f whose edges have been marked.

Let $\vec{L}'$ and ε_0 be defined as above (the choice $\varepsilon_0 = 1$ is always suitable).

Let f_1 be the following function defined over the edges of the graph $\vec{G}$:

$$f_1(e) = \begin{cases} f(e) + \varepsilon_0 & \text{if } e \text{ is included both in } \vec{L} \text{ and } \vec{L}', \\ f(e) - \varepsilon_0 & \text{if } e \text{ is included in } \vec{L}, \text{ but not in } \vec{L}', \\ f(e) & \text{if } e \text{ is not included in } \vec{L}. \end{cases}$$

Let f_1 take over the role of f.

Consider the vertices marked and edges selected in (3) and (4) as unmarked and unselected, respectively.

Return to (3).

(5) Record the result: f is an xy-flow of maximal value and an edge $\overrightarrow{xy}$-cut of minimal capacity is formed by the edges whose tails are marked (accessible along a flow augmenting path relative to f) and their heads not.

The application of Algorithm 29 is illustrated by an example. We start from the graph shown in figure 2.19(a), the numbers are the capacities of the edges. Figures 2.19(b), (c), (d) and (e) illustrate the successive augmentations of the flow. The first numbers in the pairs written on the edges are the residual capacities, the second numbers are the back-capacities, i.e. the values of the flow on the edges. The value of the flow is zero on the edges drawn by dotted lines. The maximum value of the flow is 8. In figure 2.19(f), the vertices accessible by the flow are denoted by solid circles and the edges drawn by heavy lines form an edge cut of minimal capacity.

Remarks. 1. If the edge-capacity values are integers or rational numbers then Algorithm 29 terminates in a finite number of steps. The following upper bound can be given for the number of steps:

$$n^2 \sum_e \kappa(e).$$

However, two related problems emerge. First, an estimate not containing the edge-capacities would be more advantageous like that not containing the values of the edges in the case of Algorithm 18 (consider a case with very high edge-capacity values). Second, if the edge-capacities are irrational, the algorithm above may fail to converge in a finite number of steps and it may even

Optimal Flows

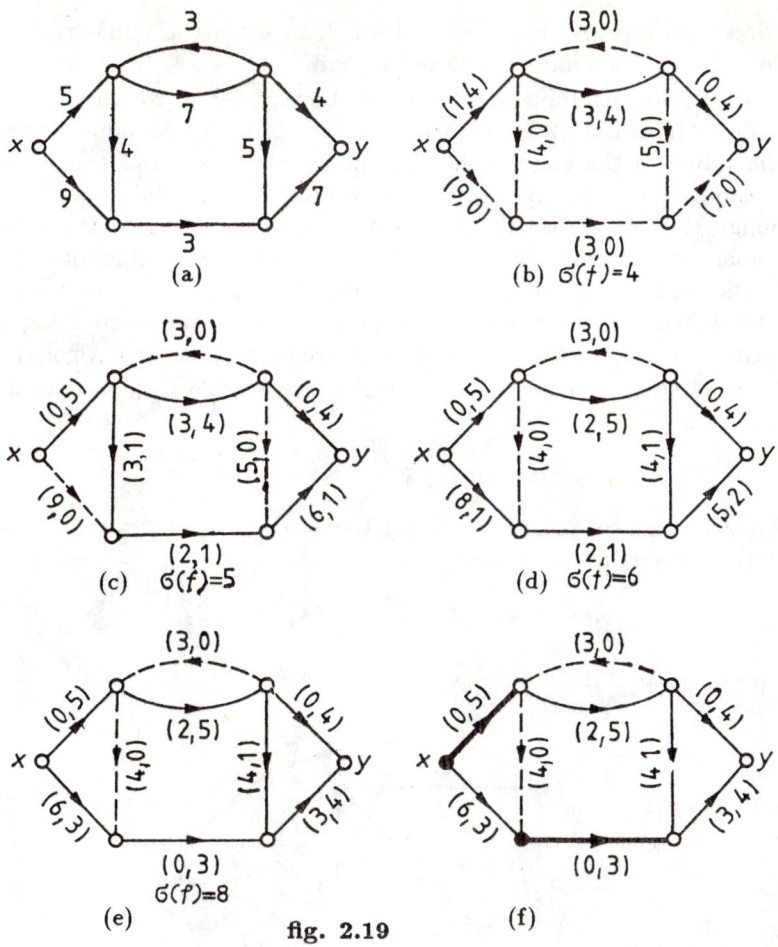

fig. 2.19

converge to a wrong value instead of the maximal flow in an infinite number of steps. An example for this awkward case has been first presented by L R Ford Jr and D R Fulkerson (see [11], page 21), but the example discussed below is somewhat simpler. Both misgivings were at once resolved by J Edmonds and R M Karp [12] when they proved that, selecting the shortest flow augmenting path at each augmentation, the procedure necessarily ends after about n^3 iterations in a graph of n vertices, irrespective of the magnitudes of the edge-capacities and of whether the edge-capacities are irrational or not. We also mention the algorithm of E A Dinic [13] and A V Karzanov [14] which requires n^2 iterations and performs each iteration in about n steps.

2. If a computer program is written on the basis of this algorithm, the restriction for integers or rational numbers means seemingly no limitation since only rational numbers can be represented in a computer. If, however,

an algorithm diverges for irrational numbers, then the number of steps will be high for rational numbers close to them.

The following example, due to László Lovász, serves to illustrate first that, provided Algorithm 29 is started by a flow of non-zero value, the sequence of the values of the augmented flows does not exceed a certain value (16 in our example) and the algorithm fails to terminate in a finite number of steps, although the edge-capacity values can be selected to make the value of the maximal flow arbitrarily high. Thereafter, selecting the values of certain edge-capacities to be irrational, the sequence of the values of the augmented flows will be shown not to converge to the maximal flow value even if Algorithm 29 is started by a flow of zero value (cf. the remarks following Algorithm 18).

Consider the equation $x^3 + x - 1 = 0$. It is easily seen to have a positive root a with

$$\frac{1}{2} < a < 1$$

and (according to Rolle's theorem) the number a is irrational. Keep the following easy relation in mind:

$$a^k - a^{k+1} = a^{k+3} \qquad (k = 0, 1, 2, \ldots).$$

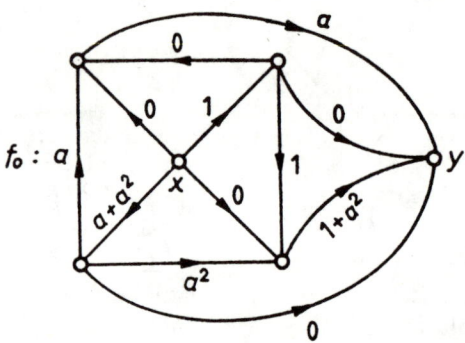

fig. 2.20

The directed graph in our example is shown in figure 2.20. The capacity of each edge is arbitrarily high, say infinite, but it suffices to say, for example, 100. The values of the initial flow f_0 have been written over the edges. Obviously:

$$\sigma(f_0) = 1 + a + a^2.$$

The path of each augmenting flow f_k will be selected according to the remainder of the division of k by 4. In the course of the kth augmentation, the flow

Optimal Flows

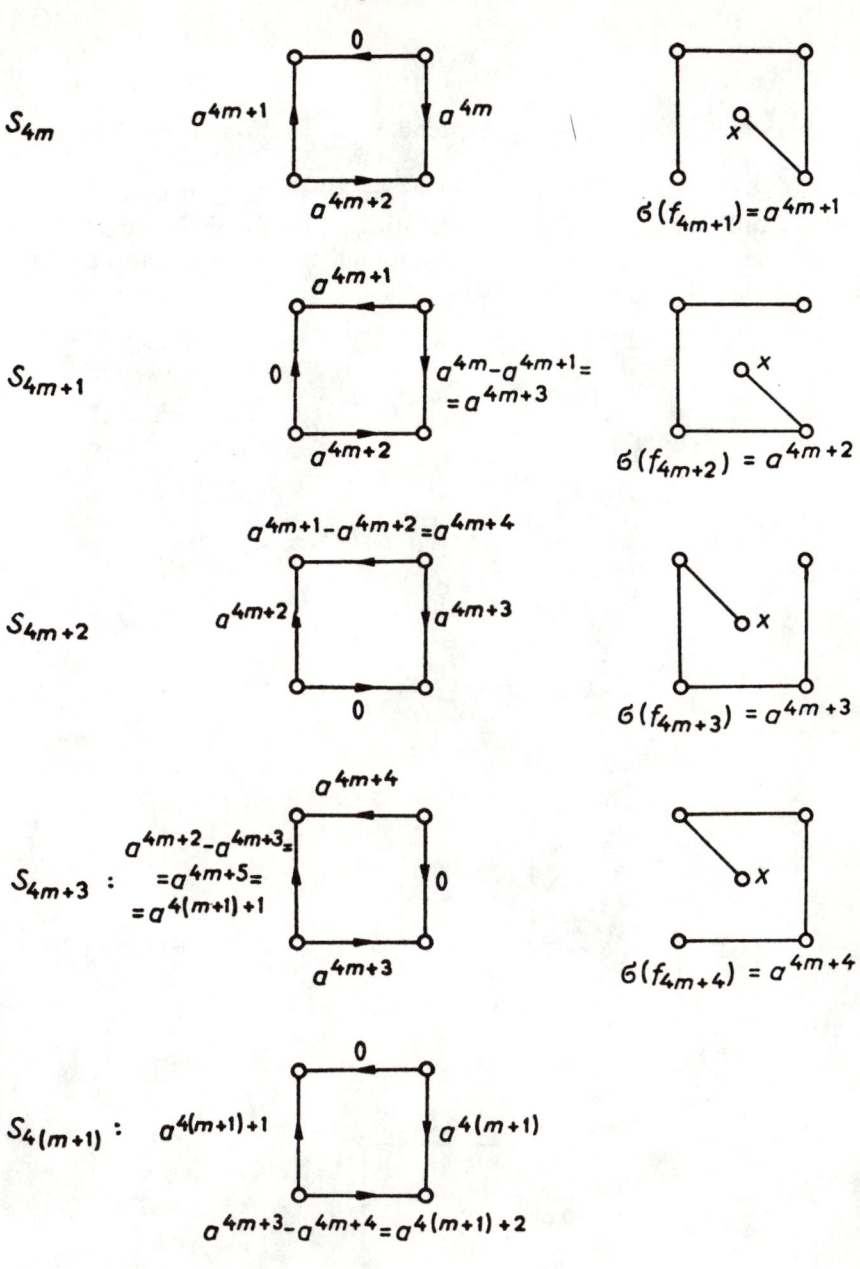

fig. 2.21

will be augmented by a^k ($k = 1, 2, \ldots$). A four-step phase of the augmentation is shown in figure 2.21. The notation

$$\sum_{k=0}^{r} \sigma(f_0) = S_r$$

is used. In each row the first diagram indicates the flow derived on the edges not incident with x or y and the second diagrams represent the subsequent flow augmenting paths without the edges incident with y. It is clear from the diagrams that, on the given flow augmenting paths, no higher augmentations are possible. Using the above relationship and that $a^2 < a$, the value of the flow after the rth augmentation is:

$$S_r = (1 + a + a^2) + a + a^2 + \ldots + a^r$$
$$= \frac{1}{a}(a + a^2 + a^3 + a^2 + a^3 + \ldots + a^{r+1})$$
$$= \frac{1}{a}(a + a^2 + (1 - a) + a^2 + a^3 + \ldots + a^{r+1})$$
$$= \frac{1}{a}(1 + 2a^2 + a^3 + \ldots + a^{r+1})$$
$$< \frac{1}{a}(1 + a + a^2 + a^3 + \ldots + a^{r+1} + \ldots)$$
$$= \frac{1}{a - a^2} = \frac{1}{a^4} < 16.$$

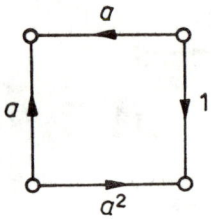

fig. 2.22

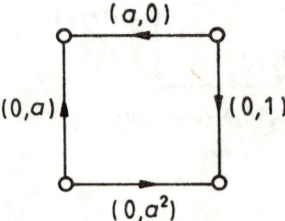

fig. 2.23

Let us now modify the capacities over the 'quadrangle' not incident to the vertices x or y of the original graph as shown in figure 2.22 and let the capacity values remain large on the edges incident with the source and the sink. Let us start with a flow of zero value. Thereafter, flows of value 1, a and a^2 are also started, they are together denoted by f_0 again. This is shown in figure 2.23 where the notations of figure 2.19 have been used. Thereafter, the augmentations by the flows $f_1, f_2, \ldots$ are as above. The fact that the augmentations are permitted by the reduced capacities can be verified in figure 2.24. The above calculation remains valid for the values of the augmenting flow and this fact proves that, in case of irrational edge-capacities, the sequence of the values of the flow augmented in accordance with Algorithm 29 does not converge to the value of the maximal flow which can be selected to be as high as possible by the appropriate specification of the capacities of the edges incident with x and y.

In what follows we show how Theorem 25 can be applied to solve similar max–min problems. First, Theorem 11 will be shown to be a special case of Theorem 25. Indeed, let us define the edge-capacity function κ in the graph $\overrightarrow{G}$ of Theorem 11 to have $\kappa(e) = 1$ for each edge e in $\overrightarrow{G}$. Let $\overrightarrow{W}$ be an edge $\overrightarrow{xy}$-cut of minimal capacity and $\kappa(\overrightarrow{W}) = \kappa$. Since $\kappa(\overrightarrow{W}) = |\overrightarrow{W}|$, $\overrightarrow{W}$ is a minimal edge $\overrightarrow{xy}$-cut in $\overrightarrow{G}$. According to Theorem 25, $\sigma(f) = \kappa$ for all xy-flows f of maximal value. Let us select an xy-flow of maximal value as described in Theorem 23. This determines κ edge-disjoint $\overrightarrow{xy}$-paths in $\overrightarrow{G}$ by means of the supports in the relevant decomposition. Since it is not possible to select more than κ edge-disjoint $\overrightarrow{xy}$-paths in $\overrightarrow{G}$, Theorem 11 clearly holds.

We saw in §2.1, where the problem of a maximal flow was raised, that even if the flow in a non-directed graph is requested, a directed graph must be treated in any case. The reason is that any flow from the source towards the sink must be regarded as directed since not only the value of a flow but also its direction must be specified for every edge. The way of conversion to a directed graph has also been given: take the edges (p, q) and (q, p) instead of the edge $e = \{p, q\}$ in the graph G and assign to both the capacity of edge e. Let us find a flow f of maximal value in the graph obtained. Finally, delete one of the two edges associated with the edge e according to the following choice: if f has turned out to be zero on one of these edges then this should be deleted and if $0 < f(p,q) \leq f(q,p)$, then delete (p,q) (the one whose flow value is lesser) and modify the value of the flow on edge (q,p) to $f(q,p) - f(p,q)$. Thus the original undirected graph G has been regained and the orientations mark the flow f of maximal value.

The above considerations also show that Theorem 5 is a special case of Theorem 25. Indeed, let us define the edge-capacity function κ in the graph G of Theorem 5 to be $\kappa(e) = 1$ for each edge e in G. Replace G by the directed graph $\overrightarrow{G}$ as above. Let $\overrightarrow{W}$ be an edge $\overrightarrow{xy}$-cut of minimal capacity in $\overrightarrow{G}$ and

Graph Theory: Flows, Matrices

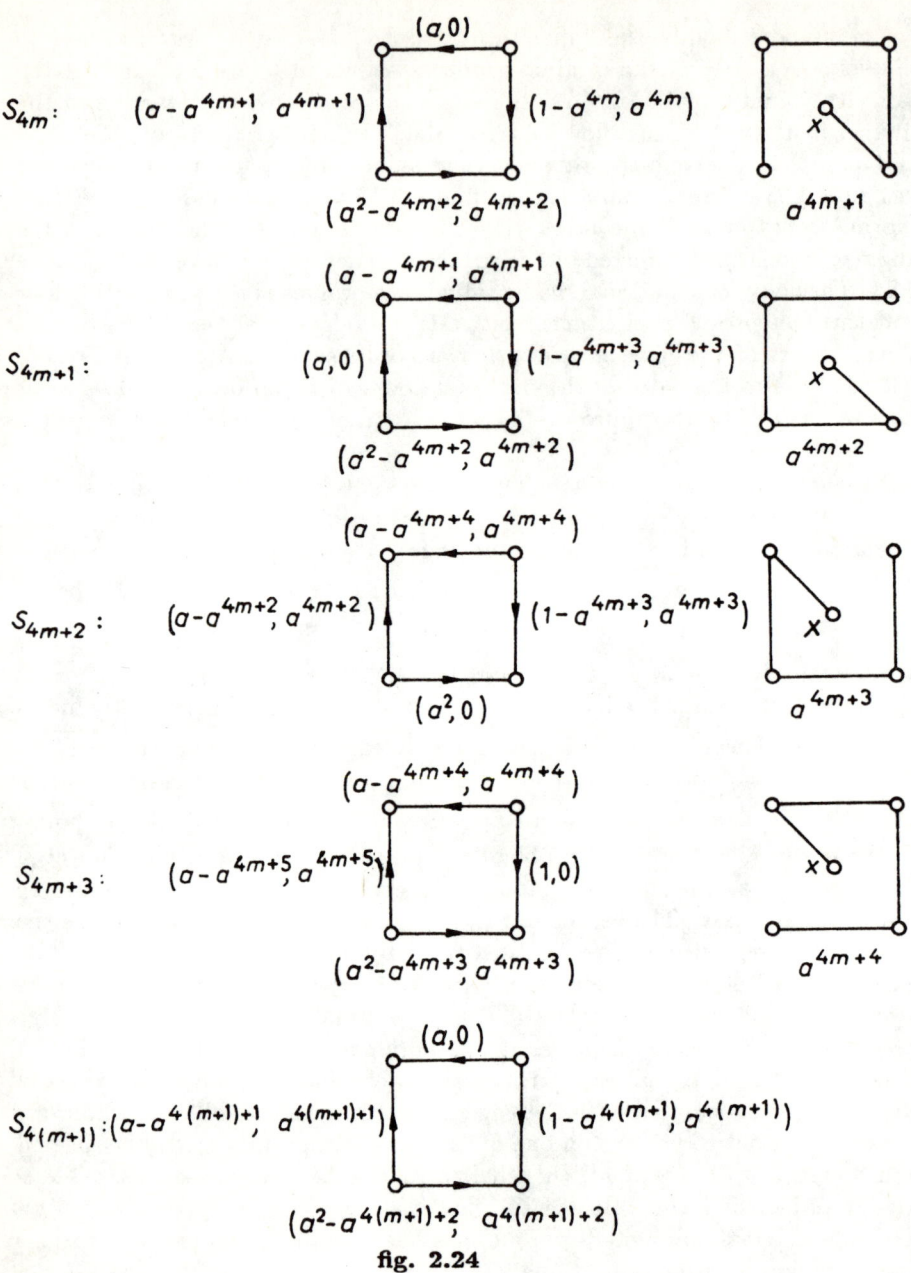

fig. 2.24

$\kappa(\overrightarrow{W}) = |\overrightarrow{W}| = \kappa$. According to Theorem 25, $\sigma(f) = \kappa$ for all xy-flows f of maximal value in $\overrightarrow{G}$. Let us select an xy-flow f of maximal value in $\overrightarrow{G}$ as described in Theorem 23 so that the total number of edges in these κ supports of the xy-path flows appearing in the decomposition of f should be minimal. Then the value of the flow cannot be 1 (i.e. non-zero) over both of the edges (p,q) and (q,p) corresponding to the edge $\{p,q\}$ in G, i.e. both edges cannot be included among the edges of the supports. Indeed, should the edge (p,q) belong to L_1 and (q,p) to L_2 (figure 2.25), then the supports L_1 and L_2 could

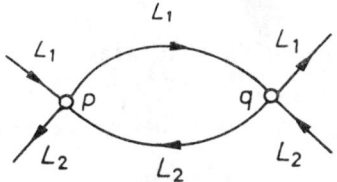

fig. 2.25

be replaced by supports consisting of fewer edges: one could be obtained from the $\overrightarrow{xp}$ part of L_1 and of the $\overrightarrow{py}$ part of L_2 and the other from the $\overrightarrow{xq}$ part of L_2 and of the $\overrightarrow{qy}$ part of L_1. This would imply that the edges (p,q) and (q,p) can be omitted from the supports, contrary to our choice. Consequently, the edges of these κ supports can be retained when returning to G and so, κ edge-disjoint xy-paths are obtained by disregarding the orientations. Now, according to Theorem 28, $f(\overrightarrow{W}) = \kappa$ and $f(\overleftarrow{W}) = 0$ and so the edges in $\overrightarrow{W}$ are retained while the edges in $\overleftarrow{W}$ are deleted when we return to G. If the directions of the edges in $\overrightarrow{W}$ are now disregarded, an edge xy-cut W is obtained in G with $|W| = \kappa$. Consequently, Theorem 5 is obtained, taking Theorem 3 also into account.

Let us return to the transportation problem raised in §2.1 whose solution is yielded by Theorem 25 and Algorithm 29, provided the transportation is limited by the given capacities of the road-sections only but not by the nodes. Let us now examine the case when the transportation is limited by the capacities of both the road-sections and the nodes. This case also includes the situation when the flow is limited by the nodes only (for example, in case of sea or air transport it may suffice to consider the capacities of the ports or airports only), since then a sufficiently high capacity can be prescribed for each road-section (e.g. the maximum of the node capacities). Such capacity values not really limiting the flow are also denoted by ∞.

Let us define the non-negative, real edge-capacity function κ over the edges of a graph $\overrightarrow{G} = (P, E, \overrightarrow{\mathcal{G}})$ and the non-negative, real *vertex-capacity function* ν over the set P. The *capacity of a vertex p* is the number $\nu(p)$. A real function f defined over the set E is called an xy-*flow limited by* κ *and* ν provided the following relations hold:

$$0 \leq f(p,q) \leq \kappa(p,q) \quad \text{if} \quad (p,q) \in E,$$
$$f(p,P) \leq \nu(p) \quad \text{if} \quad p \in P \quad \text{and} \quad p \neq x,y,$$
$$f(P,p) \leq \nu(p) \quad \text{if} \quad p \in P \quad \text{and} \quad p \neq x,y,$$

$$f(p,P) - f(P,p) = \begin{cases} \sigma & \text{if} \quad p = x \\ 0 & \text{if} \quad p \in P \quad \text{and} \quad p \neq x,y \\ -\sigma & \text{if} \quad p = y. \end{cases}$$

The number $\sigma = \sigma(f)$ is called the *value of the flow* f. Now, denoting the set of the xy-flows limited by κ and ν in $\vec{G}$ by $\mathcal{F}$, the solution of the relevant problem is yielded by the specification of a function f_0 in $\mathcal{F}$ with

$$\sigma(f_0) = \max_{f \in \mathcal{F}} \sigma(f).$$

Let us assume that $\vec{G}$ contains no loop and $(x,y), (y,x) \notin E$. This constitutes no substantial restriction on applications but renders the following discussion simpler.

The problem can be reduced to the case when the maximal flow must be found among the xy-flows limited by the edge-capacity function only and so the solution will again be provided by Theorem 25 and Algorithm 29. How can the role of the vertex-capacities be the same as that of the edge-capacities? Let us 'expand' the vertex of capacity ν into an edge of the same capacity. More precisely, if the vertex p ($\neq x,y$) in P is the head of the edges $(q_1,p), (q_2,p), \ldots, (q_m,p)$ and the tail of the edges $(p,r_1), (p,r_2), \ldots, (p,r_n)$, let the vertex p be deleted from $\vec{G}$ together with the edges incident to it and let us substitute them by the vertices p' and p'' and the edges $(p',p''), (q_1,p')$, $(q_2,p'), \ldots, (q_m,p'), (p'',r_1), (p'',r_2), \ldots, (p'',r_n)$. Let, further,

$$\kappa^*(p',p'') = \nu(p),$$

$$\kappa^*(q_i,p') = \kappa(q_i,p) \quad \text{and} \quad \kappa^*(p'',r_i) = \kappa(p,r_i)$$

for all i. (The modification is demonstrated in figures 2.26(a) and (b) in the case $m = 3$ and $n = 2$.) Let us carry out the modification for all vertices ($\neq x,y$) of $\vec{G}$. This yields the graph $\vec{G}^* = (P^*, E^*, \mathcal{G}^*)$ with the edge-capacity function κ^*. Such a modification is illustrated in figure 2.27. Let $\mathcal{F}$ and $\mathcal{F}^*$ be the sets of the xy-flows in $\vec{G}$ limited by κ and ν and of the xy-flows in $\vec{G}^*$ limited by κ^*, respectively. It only remains to be shown that a one-to-one correspondence between the elements f of $\mathcal{F}$ and the elements f^* of $\mathcal{F}^*$ exists with

$$\sigma(f) = \sigma(f^*).$$

This correspondence is
$$f^*(p', p'') = f(p, P)$$
for the edge expanded from the vertex p ($\neq x, y$) and is given in a natural way for the other edges. Thereafter the proof is nothing but a sequence of tedious calculations which is omitted.

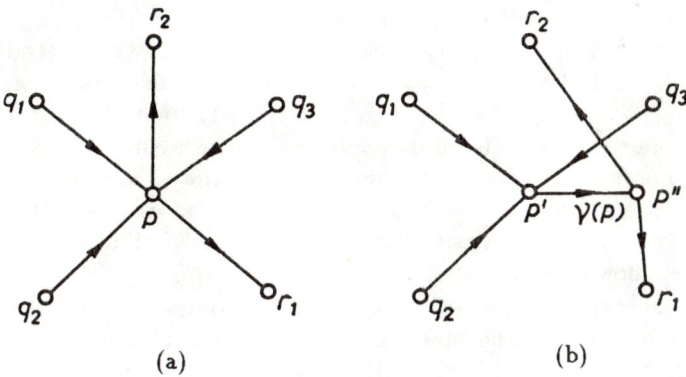

fig. 2.26

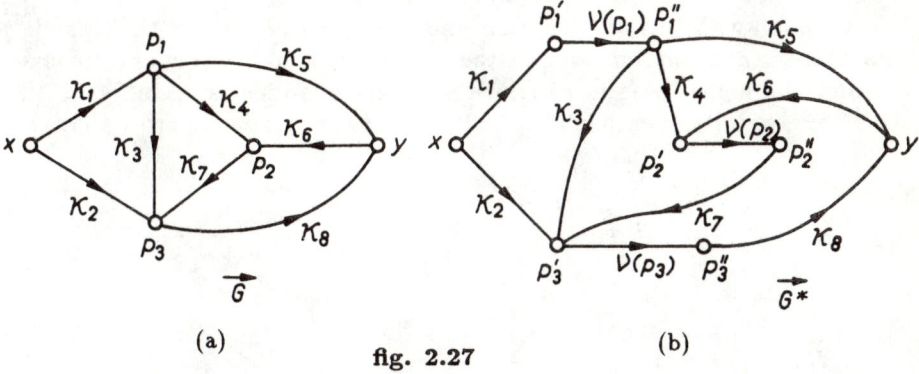

fig. 2.27

Without going into details we mention that the above idea of converting $\vec{G}$ into $\vec{G}^*$ can be used to deduce Theorem 12 from Theorem 11. To this end, the useful way of selecting the edge-capacity function κ and the vertex-capacity function ν in $\vec{G}$ of Theorem 12 is the following: $\kappa(p, q) = \infty$ for all edges (p, q) in $\vec{G}$, $\nu(x) = \nu(y) = \infty$ and $\nu(p) = 1$ for all vertices $p \neq x, y$. Simultaneously, using Theorem 12, we can deduce Theorem 8 in the same way as Theorem 5 has been deduced from Theorem 25 (or could have been from Theorem 11). Further, Theorem 11 can be deduced from Theorems 12 and

25 with the following idea. Let a new vertex be included along each edge of the graph $\vec{G}$, i.e. each edge (p_1, p_2) of $\vec{G}$ is substituted by a new vertex p' and by two edges (p_1, p') and (p', p_2). Let us introduce the vertex-capacity functions ν and edge-capacity functions κ with the following property: the value of ν is 1 in each new vertex, the value of ν in the old vertices and that of κ on the edges is ∞. Details of these are left to the reader.

Our transportation problem, leading to the task of finding a maximal flow, has been restricted in §2.1 to the case involving a single source and a single sink. If a maximal flow is requested in a graph with sources $x_1, x_2, \ldots, x_m$ and sinks $y_1, y_2, \ldots, y_n$, then, leaving the definition of the new flow to the reader, the following idea is helpful in reducing the problem to the case already discussed: augment the graph by the vertices x and y as well as the edges (x, x_i) and (y_j, y) $(i = 1, 2, \ldots, m;\ j = 1, 2, \ldots, n)$ and associate the capacity ∞ to the new edges. Obviously, the solution of the problem is provided by finding an xy-flow of maximal value in the new graph.

Let us now consider another transportation problem involving more than one source and one sink: the task is now to construct an $x_i y_i$-flow for each of several kinds of goods. Let the problem be to find a set of $x_i y_i$-flows f_i in a graph $\vec{G}$ equipped with an edge-capacity function κ so that the total flow value $\sum_i \sigma(f_i)$ be maximal. The concept of edge-cut needs to be modified, too; it must be a set of edges E' 'cutting' each $x_i y_i$-flow. We only wish to note in connection with this problem that now the maximal value of the total flow may be less than the minimum of the capacities of the edge-cuts (of course, it cannot be higher even now). Such a situation is illustrated in figure 2.28 where the capacity of each edge is 1. It is left to the reader to prove that here

$$\max \sum_{i=1}^{3} \sigma(f_i) = \frac{3}{2} \quad \text{and} \quad \min \kappa(E') = 2.$$

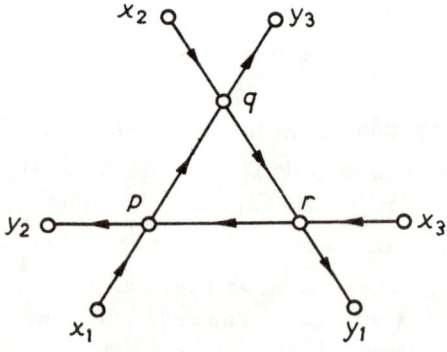

fig. 2.28

In any case, the total flow value $\frac{3}{2}$ is attainable with the choice $\sigma(f_i) \equiv \frac{1}{2}$, and the edges (p,q) and (q,r) together constitute an edge-cut of capacity 2 (cf. Problem 49).

As noted before, Theorems 5, 8, 10, 11, 12 and 13 due to K Menger were already known in 1927. Theorem 25 of Ford and Fulkerson has been seen to be derivable from Theorem 11 in the case of an integer edge-capacity function and even the idea of the proof of Theorem 25 and the fundamentals of Algorithm 29 have been constituted by the ideas in the proof of Theorem 11. It is interesting that, in spite of this, the theorems of Ford and Fulkerson, complicated by the capacity function, became known as late as during the nineteen-fifties, since it was only then that their appearance was forced by the practical problems whose solution became urgent.

2.5 Dynamic flow. The mobilisation problem

The dynamic transportation problem raised in §2.1 will now be solved by reducing it to the static case (the idea of Ford and Fulkerson [15]). Assume therefore that an army must solve the following mobilisation problem. The various corps camp at different locations and, on a mobilisation order issued at a given moment, all of them must move to a single place. The problem is to devise a transportation plan realising this concentration in minimum time with the limitation determined by the given network of roads. Obviously, the problem has many 'peaceful' applications, too.

The solution is presented on an extremely simple example. The discussion, however, is carried out to yield the general case, too. The corps located at the camps $z_1, z_2, \ldots$ connect at the nodes $x_1, x_2, \ldots$ to the network of roads given by a directed loopless graph $\overrightarrow{G}$. The connection between z_i and x_i can be assumed not to require any significant amount of time since x_i may be selected to be at the 'throat' of z_i. Measuring the size of the corps in some unit, let a_i denote the measure of the corps stationed at z_i. Let the point of concentration be y_1. Let us select some time unit, for example 1 hour. With respect to this, let the capacity of the node x_i be $\nu(x_i)$, the capacity of y_1 be ∞ and the capacity of the edge e_i be $\kappa(e_i)$. The speed of the transport vehicles and the length of each road-section (edge) are taken into account by specifying a period of transit on each edge: let this be the integer $\tau(e_i)$ on edge e_i. We shall be able to determine the minimal duration of the concentration and the itineraries realising this. Let us now consider the example shown in figure 2.29 to follow the discussion. Here $a_1 = \nu(x_1) = 10$, $a_2 = \nu(x_2) = 4$, each edge-capacity is 2 and the values of τ have been given in parentheses. Let zero be the time when the execution of the mobilisation order starts. The network of roads is 'expanded' in time: let the 'state' of x_i and y_1 at t o'clock be denoted by the vertices $x_i(t)$ and $y(t)$ $(t = 0, 1, 2, \ldots)$; $y(0)$ can be omitted. For each t, let us include the edges $(x_i(t), x_i(t+1))$ with capacity $\nu(x_i)$ and, for each edge $e_i = (p_i, q_i)$, the edge $(p_i(t), q_i(t + \tau(e_i)))$ with capacity $\kappa(p_i, q_i)$. Let us further include the source x, the sink y and, for all i, the edges $(x, x_i(0))$ with capacity a_i and the edges $(y(t), y)$ with capacity ∞. Thus we obtain from $\vec{G}$ the graph $\vec{G}_1$ at $t = 0, 1$, the graph $\vec{G}_2$ at $t = 0, 1, 2$, the graph $\vec{G}_3$ at $t = 0, 1, 2, 3$, etc. all of which are provided with edge-capacity functions. The graphs $\vec{G}_2$ and $\vec{G}_4$ obtained from figure 2.29 are shown in figure 2.30. Observe that an xy-flow limited by the edge-capacity function constructed in $\vec{G}_t$ indicates the extent of the progress of the concentration realisable in t

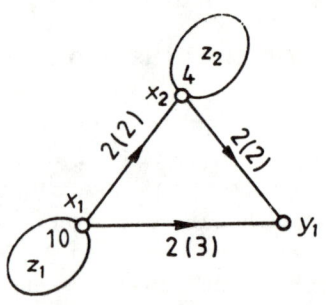

fig. 2.29

hours. If this is perceived, the solution of the problem is at hand. Let us seek xy-flows of maximal value in the graphs $\vec{G}_k$ at $k = 1, 2, \ldots$. Let m be the minimal number so that the value of the maximal xy-flow in the graph $\vec{G}_m$ is $\sum_i a_i$. Now, the minimal time required for the concentration is m hours, and a concentration plan executable in m hours is provided by any of the maximal xy-flows in $\vec{G}_m$.

Let us follow the procedure in the above example. Denoting a maximal xy-flow in $\vec{G}_k$ by f_k, it is easy to verify that $\sigma(f_1) = 0$, $\sigma(f_2) = 2$, $\sigma(f_3) = 6$, $\sigma(f_4) = 10$ and, for the flow marked by the heavy edges in figure 2.31, $\sigma(f_5) = 14$. Here, the first number in the pairs written over the edges is the residual capacity and the second one is the flow value. Therefore, in this example, $m = 5$.

This easily comprehensible but lengthy procedure can be simplified by an idea of L R Ford Jr and D R Fulkerson (see [11], page 142). According to this, it is not necessary to 'expand' the graph, but it suffices to find a minimal cost flow in the original graph: see §2.9 below.

Optimal Flows

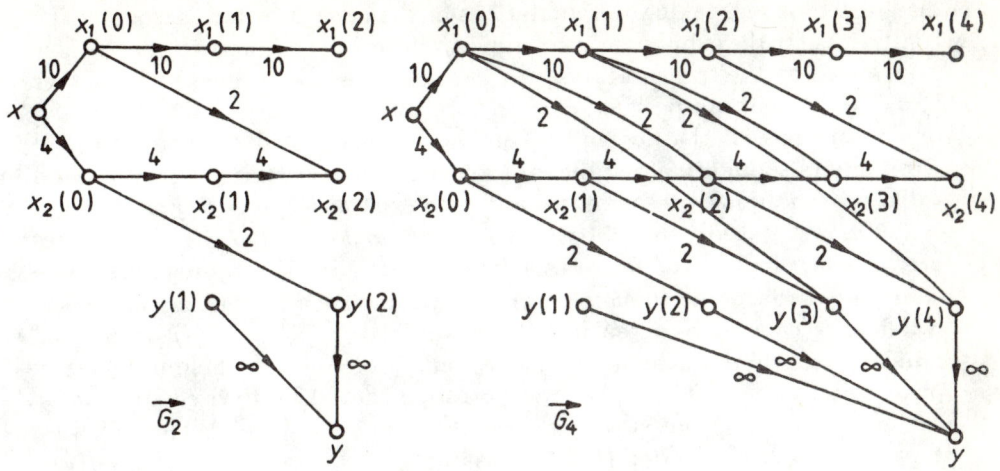

fig. 2.30

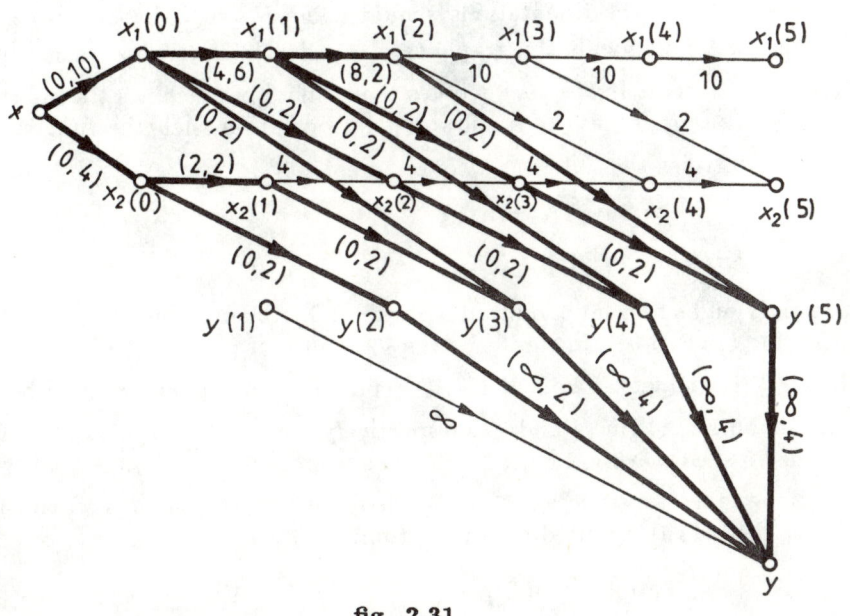

fig. 2.31

2.6 The synthesis of flow problems

In the transportation problems so far, the question of how much goods can or must be transported, i.e. how great is the *supply* or *demand*, has been in fact neglected. Our only question concerned how the network of roads can be maximally loaded, taking the limitations of the capacities into account. The constraints for the supply and demand introduce a new limitation for the flow determined by the transportation. In this case it is not even certain whether flows exist which can be 'fed' from or 'exhaust' the supply or which 'satisfy' or do not 'exceed' the demand. This section is devoted to such, so-called feasibility (also known as synthesis) problems. More precisely, necessary and sufficient conditions are sought for a flow to exist at these new limitations.

Let us first consider the following, so-called *warehousing problem*. Machines are manufactured in the factories $x_1, x_2, \ldots, x_m$ and the products are to be transported to the warehouses $y_1, y_2, \ldots, y_n$. For each factory x_i, we fix the warehouses capable of accepting transports from x_i. Let $\alpha(x_i)$ denote the number of machines manufactured at x_i and let $\beta(y_j)$ be the holding capacity of y_j in unit time. Then, the transportation should be devised to exhaust the 'supplies' $\alpha(x_i)$ but not to exceed the 'demands' $\beta(y_j)$. The following bipartite graph $G = G(A, B)$ is associated with the problem: the vertices x_i corresponding to the factories constitute the set A and the vertices y_j corresponding to the warehouses form the set B. An edge $\{x_i, y_j\}$ is included if and only if the transportation from x_i to y_j is possible. The graph G is directed so that the orientations should coincide with the directions of the transport. This yields the graph $\vec{G} = \vec{G}(A, B)$ all of whose edges are, therefore, $\vec{AB}$-edges. If $f(x_i, y_j)$ denotes the number of machines transported along edge (x_i, y_j) in case of a suitable transportation then the function f necessarily satisfies the following two 'requirements' in $\vec{G}$:

$$f(x, B) = \alpha(x) \quad \text{if} \quad x \in A, \tag{1}$$

$$f(A, y) \leq \beta(y) \quad \text{if} \quad y \in B. \tag{2}$$

A necessary and sufficient condition is required for the existence of a function f satisfying (1) and (2), in short, for the *feasibility* of (1) and (2).

Let $A' \subseteq A$ and let $G(A')$ and $\vec{G}(A')$ denote the sets of vertices in B being the neighbours of A' in G and $\vec{G}$, respectively. Naturally, $G(A') = \vec{G}(A')$. Since the finished machines must be removed from every vertex of A' and they can only be taken to vertices in $\vec{G}(A')$, the following relation is necessary for the feasibility of the required transportation:

$$\alpha(A') \leq \beta(B') \quad \text{if} \quad B' = \vec{G}(A'). \tag{3}$$

It will be proved that (1) and (2) are feasible if (3) is satisfied for all subsets A' of A, i.e. the following theorem holds:

30. *Let the non-negative, integer functions α and β be defined over the sets of vertices A and B, respectively of a graph $\vec{G} = \vec{G}(A, B)$. The requirements*

$$f(x, B) = \alpha(x) \quad \text{if} \quad x \in A \tag{1}$$

$$f(A,y) \leq \beta(y) \quad \text{if} \quad y \in B. \tag{2}$$

are feasible by a non-negative, integer function f defined over the edges of $\vec{G}$ if and only if for all subsets A' of A

$$\alpha(A') \leq \beta(B') \quad \text{where} \quad B' = \vec{G}(A'). \tag{3}$$

The proof of the theorem will be obtained in several steps, by reducing it to other problems. First, a different problem is associated with this one as follows. Each vertex x of G in A is replaced by $\alpha(x)$ vertices and each vertex y in B by $\beta(y)$ vertices. If, and only if, x is a neighbour of y in G then each of the vertices replacing x is made adjacent by an edge to each of the vertices replacing y. This yields the graph

$$G^* = G^*(A^*, B^*) \quad \text{with } |A^*| = \alpha(A) \text{ and } |B^*| = \beta(B).$$

The following will be proved: (1) and (2) are feasible in $\vec{G}$ if and only if there are $|A^*|$ independent A^*B^*-edges in $\vec{G}^*$, i.e. there is a set of independent edges covering A^* in G^*. In order to prove the sufficiency of this condition, let H be the set of independent edges covering A^* in the graph G^*. Now, considering the edge (x,y) in $\vec{G}$, let X^* replace x and Y^* replace y and let $f(x,y)$ be the number of X^*Y^*-edges in H. Then f is easily seen to satisfy the requirements (1) and (2). Similarly, vice versa, considering the edge (x,y), we include $f(x,y)$ independent X^*Y^*-edges in H. A similar procedure is followed for the rest of the edges, the only point to take care of is that the newly selected edges should form a set of independent edges with the ones previously included in H. It is easy to verify that this is possible in view of the way G^* has been constructed. Figure 2.32 demonstrates the above correspondence in a simple case. The values of α and β have been written beside the vertices, those of f are on the edges. The edges in H are heavy.

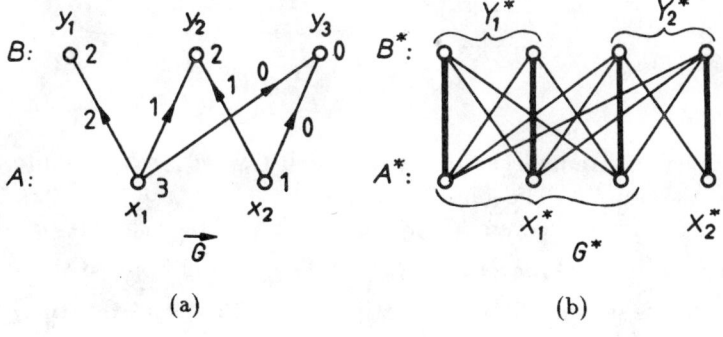

fig. 2.32

Therefore, the new problem refers to the graph G^* and involves finding a necessary and sufficient condition for the existence of a set of independent edges covering A^* in G^*. Such a condition is contained in the Theorem below which is due to P Hall ([16], Theorem 7.3.3, [17] and [2], 5.19) and which is proved using Theorem 14.

31. *In a bipartite graph $G(A, B)$, there exists a set of independent edges covering A if and only if*

$$|A'| \leq |B'| \quad \text{if } B' = G(A')$$

for each subset A' of A.

Proof. The necessity of the condition is evident. In order to prove its sufficiency, assume that the condition is satisfied in the bipartite graph $G = G(A, B)$. Let the minimal number of covering vertices of G be denoted by s. Since both A and B are sets of covering vertices of G, $|A| \geq s$. If $|A| = s$ then the theorem immediately follows from Theorem 14. Assume therefore that $|A| > s$. Let R denote a minimal set of covering vertices of G, i.e. a set of covering vertices with cardinality $|R|$. Let $R \cap A = A_1$, $R \cap B = B_1$, $A_2 = A - A_1$ and $B_2 = B - B_1$. Such a partition is shown in figure 2.33. In view of the properties of the set R, G cannot contain any $A_2 B_2$-edges. Therefore, the neighbours of all vertices in A_2 are in B_1, i.e. $G(A_2) \subseteq B_1$ and so $|G(A_2)| \leq |B_1|$. Since $|A_1 \cup B_1| = s$ and $|A_1 \cup A_2| = |A| > s$, we obtain $|B_1| < |A_2|$. Comparing with the above relation: $|G(A_2)| \leq |B_1| < |A_2|$.

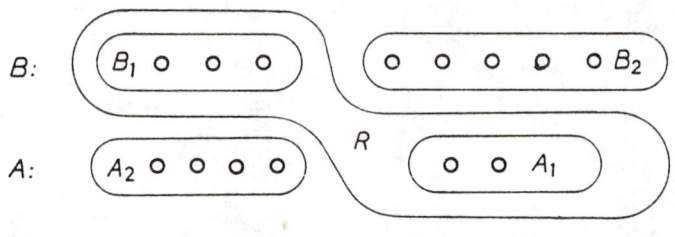

fig. 2.33

This, however, contradicts the assumption that G satisfies the condition. This completes the proof of Theorem 31.

In order to prove Theorem 30, it is only left to show that the condition of Theorem 30 is satisfied for a graph $\vec{G} = \vec{G}(A, B)$ if and only if the condition of Theorem 31 is satisfied for the bipartite graph G^* associated with $\vec{G}$. Let us first assume that (3) holds for $\vec{G}$. Let $A_1^* \subseteq A^*$ be arbitrary. Let us augment

A_1^* as follows: if $x^* \in A_1^*$ and x^* is among the vertices in X^* replacing the vertex x in A, then all vertices of X^* are appended to A_1^*. Let us carry out this augmentation for all vertices in A_1^*. This results in the set A_1^* having been augmented to obtain the set A_2^*. In any case, $A_1^* \subseteq A_2^* \subseteq A^*$ and, in view of the construction of G^*, $G^*(A_1^*) = G^*(A_2^*)$. Let $A_2 \subseteq A$ be the set whose vertices have been replaced by A_2^*. Then the vertices of $\vec{G}(A_2)$ have been replaced by the vertices of $\vec{G}(A_2^*)$. Now, the construction of G^* and (3) imply that

$$|A_1^*| \leq |A_2^*| = \alpha(A_2) \leq \beta(\vec{G}(A_2)) = |G^*(A_2^*)| = |G^*(A_1^*)|,$$

i.e. G^* satisfies the condition of Theorem 31. The reverse statement can be obtained similarly and then even the augmentation is unnecessary.

This completes the proof of Theorem 30.

In the above warehousing problem, the transportation was not limited by the capacities of the roads. As a further requirement, an edge-capacity function will now be also taken into account. Let, therefore, κ be the non-negative, integer edge-capacity function in the graph $\vec{G} = \vec{G}(A,B)$. $\vec{G}(A,B)$ can be assumed to be *complete*, i.e. each vertex in B is a neighbour of all vertices in A by means of the appropriate $\vec{AB}$-edges, since if no transportation from x to y is possible in case of $x \in A$ and $y \in B$ then this can also be expressed by $\kappa(x,y) = 0$. Our new requirement is the following:

$$0 \leq f(x,y) \leq \kappa(x,y) \quad \text{for all edges } (x,y) \text{ in } \vec{G}. \tag{4}$$

Assume that the requirements (1), (2) and (4) are feasible. Let $A_1 \subseteq A$, $B_1 \subseteq B$ and $B_2 = B - B_1$. Obviously, the supply in A_1 cannot exceed the quantity transportable from A_1. But the quantity transported from A_1 to B_1 cannot be higher than the demand of B_1 and the quantity transported from A_1 to B_2 cannot exceed $\kappa(A_1, B_2)$ in view of (4). Therefore, a necessary condition of the feasibility of these three requirements is

$$\alpha(A_1) \leq \beta(B_1) + \kappa(A_1, B_2) \quad \text{if } B_2 = B - B_1 \tag{5}$$

for all $A_1 \subseteq A$ and $B_1 \subseteq B$. This condition will be proved to be also sufficient for the feasibility of the requirements (1), (2) and (4), i.e. the following theorem is valid:

32. *Let the non-negative integer functions α, β and κ be defined over the sets of vertices A and B of a graph $\vec{G} = \vec{G}(A,B)$ and over the set of edges of $\vec{G}$, respectively. The requirements*

$$0 \leq f(x,y) \leq \kappa(x,y) \quad \text{for all edges } (x,y) \text{ in } \vec{G}. \tag{4}$$
$$f(x,B) = \alpha(x) \quad \text{if } x \in A \tag{1}$$
$$f(A,y) \leq \beta(y) \quad \text{if } y \in B. \tag{2}$$

are feasible by an integer function f defined over the edges of G if and only if

$$\alpha(A_1) \leq \beta(B_1) + \kappa(A_1, B_2) \qquad \text{if } B_2 = B - B_1. \tag{5}$$

for all sets $A_1 \subseteq A$ and $B_1 \subseteq B$.

This theorem will be proved using Theorem 30. (Although it would be simpler to deduce it from Theorem 25 of Ford and Fulkerson, but the former fits better with the construction of this book.) The connection between the two theorems can be established by modifying the problem as follows. (The idea is due to G B Danzig, L Kantorovich and M K Gavurin, see [18] and [19].) Let the vertices in A and B of the graph $\overrightarrow{G} = \overrightarrow{G}(A, B)$ be denoted as $x_1, x_2, \ldots, x_m$ and $y_1, y_2, \ldots, y_n$, respectively. Two vertices are added on each edge of the graph $\overrightarrow{G}$ and each edge is replaced by three edges as shown in figure 2.34. This yields the graph $\overrightarrow{G'} = \overrightarrow{G'}(A', B')$ from $\overrightarrow{G}$. The case with $m = 2$ and $n = 3$ is illustrated in figure 2.35. The sets A^* and B^* are

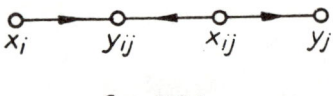

fig. 2.34

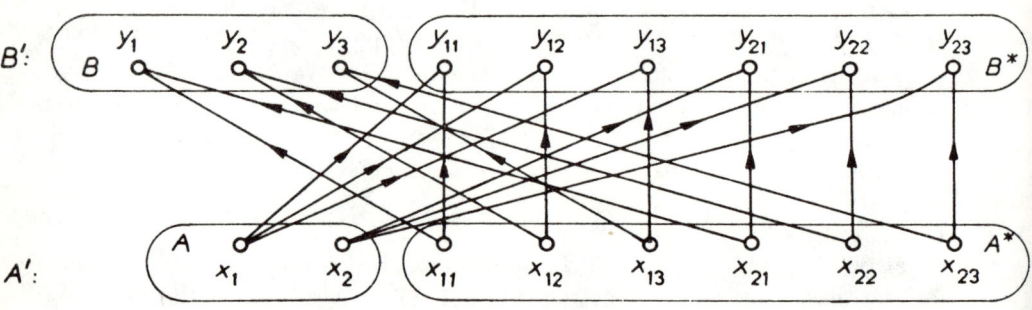

fig. 2.35

formed by the vertices x_{ij} and y_{ij}, respectively. $A' = A \cup A^*$, $B' = B \cup B^*$ and, obviously, $|A^*| = |B^*| = m \cdot n$. Let us define the functions α and β in $\overrightarrow{G'}$ as follows: the values $\alpha(x_i)$ and $\beta(y_j)$ are the same as in $\overrightarrow{G}$ and, for the new vertices,

$$\alpha(x_{ij}) = \beta(y_{ij}) = \kappa(x_i, y_j) \quad \text{for all pairs } i, j.$$

Let the function f' defined over the edges of the graph $\overrightarrow{G'}$ be associated with the function f defined over the edges of the graph $\overrightarrow{G}$ as follows (see figure 2.36, too):

$$f'(x_i, y_{ij}) = f'(x_{ij}, y_j) = f(x_i, y_j)$$

and

$$f'(x_{ij}, y_{ij}) = \kappa(x_i, y_j) - f(x_i, y_j).$$

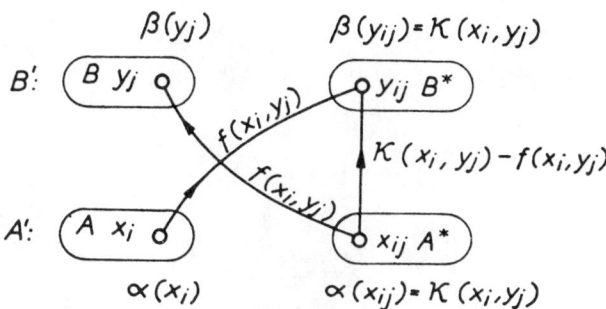

fig. 2.36

It is hence evident that if (1), (2) and (4) are feasible in $\overrightarrow{G}$ by f, then (1) and (2) are feasible in $\overrightarrow{G'}$ by f' (naturally with respect to A' and B'). Conversely, assume that (1) and (2) are feasible in $\overrightarrow{G'}$ by f^*. Then

$$f^*(x_{ij}, y_j) + f^*(x_{ij}, y_{ij}) = \kappa(x_i, y_j),$$
$$f^*(x_i, y_{ij}) + f^*(x_{ij}, y_{ij}) \leq \kappa(x_i, y_j).$$

These two relations imply that

$$f^*(x_i, y_{ij}) \leq f^*(x_{ij}, y_j).$$

Let in $\overrightarrow{G}$

$$f(x_i, y_j) = f^*(x_i, y_{ij}).$$

(1), (2) and (4) will be shown to be feasible in $\overrightarrow{G}$ by f. Indeed:

$$f(x_i, B) = f^*(x_i, B^*) = \alpha(x_i),$$
$$\beta(y_j) \geq f^*(A^*, y_j) \geq \sum_i f^*(x_i, y_{ij}) \geq f(A, y_j),$$
$$f(x_i, y_j) = f^*(x_i, y_{ij}) \leq f^*(x_i, y_{ij}) + f^*(x_{ij}, y_{ij}) \leq \kappa(x_i, y_j).$$

Thus, (1), (2) and (4) have been shown to be feasible in $\overrightarrow{G}$ if and only if (1) and (2) are feasible in $\overrightarrow{G'}$.

In order to prove Theorem 32, it is to be verified that if (1), (2) and (4) are not feasible in $\vec{G}$ then there are $A_1 \subseteq A$ and $B_1 \subseteq B$ not satisfying (5). Assume, therefore, that (1), (2) and (4) are not feasible in $\vec{G}$. Then (1) and (2) are not feasible in $\vec{G}'$. Consequently, according to Theorem 30, there is an $A'' \subseteq A'$ in $\vec{G}'$ with

$$\alpha(A'') > \beta(B'') \quad \text{if } B'' = \vec{G}'(A'').$$

Let us introduce the following notations:

$$A'' \cap A = A_1, \quad B'' \cap B = B_1 \quad \text{and} \quad B_2 = B - B_1.$$

It will be shown that in $\vec{G}$

$$\alpha(A_1) > \beta(B_1) + \kappa(A_1, B_2).$$

So

$$\alpha(A_1) + \alpha(A'' \cap A^*) = \alpha(A'') > \beta(B'') = \beta(B_1) + \beta(B'' \cap B^*).$$

Hence

$$\alpha(A_1) - \beta(B_1) > \beta(B'' \cap B^*) - \alpha(A'' \cap A^*).$$

The quantity on the right-hand side denoted by δ is to be shown to equal the value $\kappa(A_1, B_2)$ relating to $\vec{G}$.

Consider figure 2.36. Since, at each edge (x_{ij}, y_{ij}) in $\vec{G}'$

$$\beta(y_{ij}) - \alpha(x_{ij}) = \kappa(x_i, y_j) - \kappa(x_i, y_j) = 0,$$

introducing the notation

$$B_3 = B'' \cap B^* - \vec{G}'(A'' \cap A^*),$$

the following can be written:

$$\delta = \beta(B_3) = \sum_{y_{ij} \in B_3} \beta(y_{ij}) = \sum_{y_{ij} \in B_3} \kappa(x_i, y_j).$$

If $y_{ij} \in B_3$, then $x_{ij} \notin A''$ so, as shown in figure 2.36, on the one hand, $y_{ij} \in B''$ is possible only due to x_i and thus $x_i \in A_1$ and, on the other hand, $y_j \notin B''$ and thus $y_j \in B_2$. In short, if $y_{ij} \in B_3$, then the edge (x_i, y_j) in $\vec{G}$ is an $\overrightarrow{A_1 B_2}$-edge.

If, however, (x_i, y_j) is an $\overrightarrow{A_1 B_2}$-edge in $\vec{G}$ then $x_i \notin A_1$ and $y_{ij} \in B'' \cap B^*$, but $x_{ij} \notin A''$ and so $y_{ij} \in B_3$. It follows from these considerations that

$$\delta = \kappa(A_1, B_2)$$

and this completes the proof of Theorem 32.

In order to decide whether the requirements in Theorem 30 are feasible, it suffices to consider a single subset of B for each subset of A and to verify condition (3). However, the condition in Theorem 32, relates to all subsets of B at each subset of A. It will be now shown that for the employment of condition (5), too, it suffices to consider a 'well selected' subset of B at each subset of A. Obviously, $B_1' \subseteq B$ is well selected to $A_1 \subseteq A$ if the right-hand side of (5) is minimal at B_1'. Let, for $A_1 \subseteq A$, B_1' consist of exactly those vertices y in B which meet

$$\beta(y) \leq \kappa(A_1, y).$$

The right-hand side of (5) will be shown to be in fact minimal at B_1'. Indeed, let $A_2 = A - A_1$, $B_2' = B - B_1'$ and let $\{B_1, B_2\}$ be an arbitrary partition of B. Let further $B_{11} = B_1 \cap B_1'$, $B_{12} = B_1 \cap B_2'$, $B_{21} = B_1' - B_{11}$ and $B_{22} = B_2' - B_{12}$ (see figure 2.37). Taking

$$\beta(y) > \kappa(A_1, y)$$

into account for all vertices y in B_2', we obtain

$$\beta(B_{21}) \leq \kappa(A_1, B_{21})$$

$$\kappa(A_1, B_{21}) < \beta(B_{12}).$$

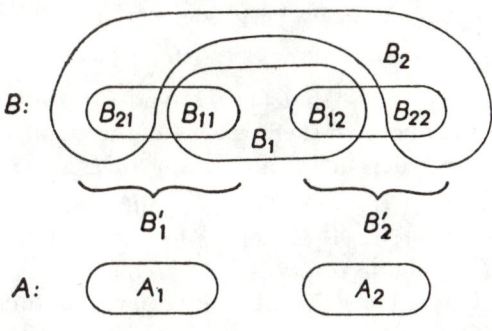

fig. 2.37

Using these relations,

$$\beta(B_1') + \kappa(A_1, B_2') = \beta(B_{21}) + \beta(B_{11}) + \kappa(A_1, B_{12}) + \kappa(A_1, B_{22})$$

$$< \kappa(A_1, B_{21}) + \beta(B_{11}) + \beta(B_{12}) + \kappa(A_1, B_{22}) = \beta(B_1) + \kappa(A_1, B_2).$$

Therefore, if A_1 is fixed, the minimum of the right-hand side of (5) is indeed attained at the partition $\{B_1', B_2'\}$ of B. Consequently, the following theorem can also be stated:

33. *Let the non-negative integer functions α, β and κ be defined over the sets of vertices A and B of a complete graph $\overrightarrow{G} = \overrightarrow{G}(A,B)$ and over the set of edges of $\overrightarrow{G}$, respectively. The requirements*

$$0 \leq f(x,y) \leq \kappa(x,y) \quad \text{for all edges } (x,y) \text{ in } \overrightarrow{G}, \tag{4}$$

$$f(x,B) = \alpha(x) \quad \text{if } x \in A, \tag{1}$$

$$f(A,y) \leq \beta(y) \quad \text{if } y \in B \tag{2}$$

are feasible by an integer function f defined over the edges of $\overrightarrow{G}$ if and only if the following holds for all subsets A_1 of A: if B_1' denotes the set of those vertices $y \in B$ which satisfy

$$\beta(y) \leq \kappa(A_1, y)$$

then

$$\alpha(A_1) \leq \beta(B_1') + \kappa(A_1, B_2') \quad \text{where} \quad B_2' = B - B_1'. \tag{6}$$

This theorem implies Theorem 31 in the following way: consider the bipartite graph $G = G(A,B)$ in Theorem 31. Hence, the complete graph $\overrightarrow{G} = \overrightarrow{G}(A,B)$ is obtained by assigning directions and including new edges. Let $\kappa(x,y) = \infty$ for all 'old' edges (x,y) and $\kappa(x,y) = 0$ for all 'new' edges in $\overrightarrow{G}$ and let further $\alpha(x) = 1$ and $\beta(y) = 1$ for all vertices $x \in A$ and $y \in B$, respectively. It is easy to see that (1), (2) and (4) are feasible in $\overrightarrow{G}$ by an integer function f if and only if there is a set of independent edges covering A in G (a set of edges is independent if the value of f is 1 on the corresponding edges in $\overrightarrow{G}$). In view of the necessary and sufficient condition of feasibility stated in Theorem 33, the set B_2' selected to A_1 is formed by the vertices not adjacent in G to any vertex of A_1. Consequently, $\kappa(A_1, B_2') = 0$. Since $\alpha(A_1) = |A_1|$ and $\beta(B_1') = |B_1'|$, the necessary and sufficient condition in Theorem 33 is equivalent to the one stated in Theorem 31 with respect to G.

Among the transportation problems raised at the beginning of this section we have so far only studied the feasibility conditions of the so-called warehousing problem. If, however, the demand of consumers must be satisfied by transportations starting from factories or warehouses, a flow is to be found which can be fed from the supply and which satisfies the demands. More precisely, keeping the edge-capacity limits, the feasibility of the following requirements is to be investigated in the complete graph $\overrightarrow{G} = \overrightarrow{G}(A,B)$ besides (4):

$$f(x,B) \leq \alpha(x) \quad \text{if } x \in A, \tag{7}$$

$$f(A,y) = \beta(y) \quad \text{if } y \in B. \tag{8}$$

This problem is easily reduced to those already discussed. Indeed, think of the transportation satisfying the demand as a film and imagine the 'film' being played back in reverse. To this end, the direction must be reversed on all edges. For the graph $\overrightarrow{G}(B, A)$ thus obtained, (4) is yielded by the interchange of the symbols x and y, and (8) and (7) are modified as follows:

$$f(y, A) = \beta(y) \quad \text{if } y \in B,$$

$$f(B, x) \leq \alpha(x) \quad \text{if } x \in A.$$

Consequently, Theorems 32 or 33 yield the necessary and sufficient condition of the feasibility of this problem, too. Indeed, there is something to play back in reverse if and only if there was something at all on the film.

Let us now investigate the transportation problem which exhausts the supply and satisfies the demand, i.e. the one constituting the following requirements in the complete graph $\overrightarrow{G} = \overrightarrow{G}(A, B)$:

$$0 \leq f(x, y) \leq \kappa(x, y) \quad \text{for all edges } (x, y) \text{ in } \overrightarrow{G}, \tag{4}$$

$$f(x, B) = \alpha(x) \quad \text{if } x \in A, \tag{1}$$

$$f(A, y) = \beta(y) \quad \text{if } y \in B. \tag{8}$$

The satisfaction of (5) for all sets $A_1 \subseteq A$ and $B_1 \subseteq B$ is obviously necessary for the feasibility of these requirements, too. But the following relation is also evidently necessary:

$$\alpha(A) = \beta(B). \tag{9}$$

If, however, (5) is satisfied for all sets $A_1 \subseteq A$ and $B_1 \subseteq B$ then, according to Theorem 32, there is a function f satisfying (4), (1) and (2) and, in view of (2),

$$\alpha(A) = \sum_{y \in B} f(A, y) \leq \sum_{y \in B} \beta(y) = \beta(B).$$

Therefore, if (9) is satisfied, then (8) is necessarily also valid. So, the following theorem has been obtained:

34. *Let the non-negative integer functions α, β and κ be defined over the sets of vertices A and B of a complete graph $\overrightarrow{G} = \overrightarrow{G}(A, B)$ and over the set of edges of $\overrightarrow{G}$, respectively. The requirements*

$$0 \leq f(x, y) \leq \kappa(x, y) \quad \text{for all edges } (x, y) \text{ in } \overrightarrow{G}, \tag{4}$$

$$f(x, B) = \alpha(x) \quad \text{if } x \in A, \tag{1}$$

$$f(A, y) = \beta(y) \quad \text{if } y \in B \tag{8}$$

are feasible by an integer function f defined over the edges of $\vec{G}$ if and only if
$$\alpha(A) = \beta(B) \tag{9}$$
and for all sets $A_1 \subseteq A$ and $B_1 \subseteq B$
$$\alpha(A_1) \leq \beta(B_1) + \kappa(A_1, B_2) \qquad \text{if } B_2 = B - B_1. \tag{5}$$

Naturally, (6) may replace (5) in this theorem if the appropriate part of the text is also replaced.

In the above feasibility problems, we have only studied whether a flow f exists in the bipartite graph $\vec{G} = \vec{G}(A, B)$ from the vertices in A to those in B, i.e. one realising the transportation with the given requirements. However, in the case of feasibility, no reference was made to the way of finding a suitable function f. If the graph $\vec{G}$ is augmented as below, a suitable function f can be found by the flow-constructing procedure presented in §2.4. Let us add the source vertex x and sink vertex y to $\vec{G}$ as well as the edges (x, x_i) and (y_i, y) at each $x_i \in A$ and $y_i \in B$. Let the edge-capacity function κ be defined for the new edges as follows:
$$\kappa(x, x_i) = \alpha(x_i) \qquad \text{if } x_i \in A,$$
$$\kappa(y_i, y) = \beta(y_i) \qquad \text{if } y_i \in B.$$
It is easy to see that any xy-flow of maximal value in the augmented graph yields a suitable function f in $\vec{G}$.

In order to further generalise the feasibility problem, the 'map' of the network of roads serving for the transportation will not be restricted to a bipartite graph. Let, therefore, $\vec{G} = (P, E, \vec{\mathcal{G}})$ be an arbitrary directed graph. The disallowance of loops is no substantial restriction. The assumption that there are no two edges with both their tails and heads in common, i.e. there are no *multiple edges* in $\vec{G}$, results in no loss of generality. It can also be assumed that, for each pair of vertices p, q, there is exactly one (p, q)-edge and one (q, p)-edge in $\vec{G}$, since, if this results in the introduction of new edges, then the capacity along them will be chosen to be zero. The vertices of the graph represent the places with supply (the factories), the places with demand (the warehouses) and the nodes touched by the transportation. Let us partition the set of vertices P of the graph in the above order as $\{F, W, R\}$. Let γ denote the integer function defined on the elements of P which gives both the values of the supplies and the demands (the negative supplies), i.e. let
$$\gamma(p) > 0 \qquad \text{if } p \in F,$$
$$\gamma(p) < 0 \qquad \text{if } p \in W,$$

$$\gamma(p) = 0 \quad \text{if } p \in R,$$

Let, further, the non-negative edge-capacity function in $\vec{G}$ be κ. Now, the necessary and sufficient condition for the existence of an integer function f defined over the edges of $\vec{G}$ is sought with the following requirements:

$$0 \leq f(p,q) \leq \kappa(p,q) \quad \text{if } (p,q) \in E \tag{10}$$

$$f(p,P) - f(P,p) = \gamma(p) \quad \text{if } p \in P. \tag{11}$$

This problem is only seemingly more difficult than the previous one, since it can be reduced to that. How should one construct the above bipartite graph from $\vec{G}$? Let us 'bisect' each vertex p to make 'one half of it' a tail only and its 'other half' a head only. To be more precise, if the edges (q_i, p) and (p, r_j)

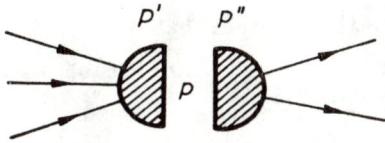

fig. 2.38

are incident with p then let the vertices p' and p'' replace p and let the edges incident with these two vertices be (q_i, p') for all i and (p'', r_j) for all j (see figure 2.38). If, further, an edge (p'', p') is introduced for each vertex p, the complete graph $\vec{G}^* = \vec{G}^*(A, B)$ is obtained from $\vec{G}$ with $p'' \in A$ and $p' \in B$ for each $p \in P$. The question is, what functions α, β, κ^* and f^* should be selected in $\vec{G}^*$ so that (10) and (11) are feasible in $\vec{G}$ by f if and only if (4), (1) and (8) are feasible in $\vec{G}^*$ by f^* (and κ^*). It will be shown that if h is selected large enough, e.g. as

$$h = \kappa(P,P) + \sum_{p \in P} |\gamma(p)|,$$

then the following functions are suitable (see figure 2.39):

$$\alpha(p'') = h + \gamma(p);$$
$$\beta(p') = h;$$
$$\kappa^*(p'', q') = \begin{cases} \kappa(p,q) & \text{if } p \neq q, \\ \infty & \text{if } p = q; \end{cases}$$
$$f^*(p'', q') = \begin{cases} f(p,q) & \text{if } p \neq q, \\ h - f(P,p) & \text{if } p = q. \end{cases}$$

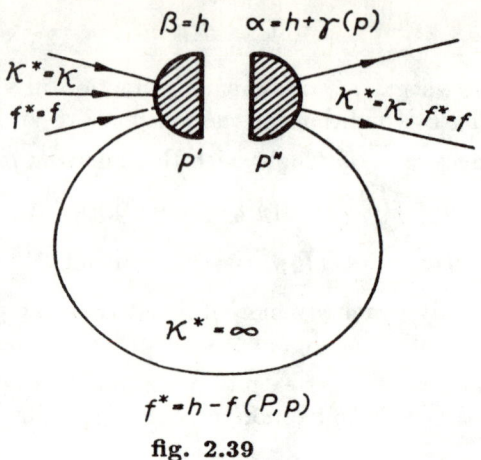

fig. 2.39

It is already evident from the diagram that if (4), (1) and (8) are feasible in $\vec{G}^*$ by f^* (and κ^*) then (10) and (11) are feasible in $\vec{G}$ by f. Let us now assume the latter. Then (4) is obviously satisfied in $\vec{G}^*$ by f^* and κ^*. But (1) is also satisfied because, for an arbitrary vertex p,

$$f^*(p'', B) = f(p, P) + h - f(P, p) = h + \gamma(p) = \alpha(p''),$$

and (8) is also satisfied since

$$f^*(A, p') = f(P, p) + h - f(P, p) = h = \beta(p').$$

Now we have to decide what is the effect of (9) and (6) with respect to the graph $\vec{G}$. Since

$$\alpha(A) = |P| \cdot h + \gamma(P)$$

and

$$\beta(B) = |P| \cdot h,$$

it follows from (9) that

$$\gamma(P) = 0 \tag{12}$$

holds in $\vec{G}$. Let us now consider the condition (6) in $\vec{G}^*$. Let $A_1 \subseteq A$ and let $\{B_1', B_2'\}$ be the corresponding partition of the set B as given in Theorem 33. Accordingly, $p' \in B_1'$ if and only if

$$\beta(p') \leq \kappa^*(A_1, p'),$$

i.e.

$$h \leq \kappa^*(A_1, p').$$

Taking figure 2.39 and the definition of h into account, one can see that this latter is valid if and only if $p'' \in A_1$. Therefore, $A_1 \cup B_1'$ corresponds to a subset X of P as follows: $p \in X$ if and only if $p'' \in A_1$ and $p' \in B_1'$. Evidently

$$|A_1| = |B_1'| = |X|.$$

Observe that, using the notation $Y = P - X$,

$$\kappa^*(A_1, B_2') = \kappa(X, Y).$$

Consequently, (6) can be rewritten as

$$|A_1| \cdot h + \gamma(X) \leq |B_1'| \cdot h + \kappa(X, Y),$$

i.e.

$$\gamma(X) \leq \kappa(X, Y).$$

This completes the proof of the following theorem of D Gale [20]:

35. *Let γ be an integer function defined over the set of vertices P of a graph $\overrightarrow{G} = (P, E, \overrightarrow{\mathcal{G}})$ without loops and multiple edges and let κ be a non-negative integer function defined over the set of edges E of $\overrightarrow{G}$. Then the requirements*

$$0 \leq f(p, q) \leq \kappa(p, q) \qquad \text{if } (p, q) \in E, \tag{10}$$

$$f(p, P) - f(P, p) = \gamma(p) \qquad \text{if } p \in P \tag{11}$$

are feasible by an integer function f defined over the set E if and only if

$$\gamma(P) = 0 \tag{12}$$

and, for any partition $\{X, Y\}$ of P,

$$\gamma(X) \leq \kappa(X, Y). \tag{13}$$

If (11) is modified in the previous problem as

$$f(p, P) - f(P, p) \geq \gamma(p) \qquad \text{if } p \in P,$$

then the only difference between this necessary and sufficient condition of the feasibility and that of Theorem 35 is that the relation (12) should be omitted. The verification of this is left to the reader.

As further generalisation of the last problem let us also introduce a not necessarily zero lower bound on the flow representing the transportation over the edges of the graph. So, let the integer functions κ_1 and κ_2 defined over the set E satisfy

$$0 \leq \kappa_1(p, q) \leq \kappa_2(p, q) \qquad \text{if } (p, q) \in E.$$

Using these functions and γ, our requirements for f are

$$0 \leq \kappa_1(p, q) \leq f(p, q) \leq \kappa_2(p, q) \qquad \text{if } (p, q) \in E, \tag{14}$$

$$f(p,P) - f(P,p) = \gamma(p) \qquad \text{if } p \in P \tag{11}$$

Let us find a necessary and sufficient condition for the feasibility of the new requirements. We wish to obtain the answer from Theorem 35. Therefore, the two capacity functions are modified to make the lower one always zero. Let, therefore,

$$\kappa^*(p,q) = \kappa_2(p,q) - \kappa_1(p,q) \qquad \text{if } (p,q) \in E.$$

The function f should be replaced by a function f^* satisfying the inequality (10), i.e.

$$0 \le f^*(p,q) \le \kappa^*(p,q) \qquad \text{if } (p,q) \in E.$$

To this end, f is to be reduced by κ_1, therefore

$$f^*(p,q) = f(p,q) - \kappa_1(p,q) \qquad \text{if } (p,q) \in E.$$

Let us find the new function γ^* satisfying (11) with f^*. This can be obtained from the following:

$$\begin{aligned} f^*(p,P) - f^*(P,p) &= f(p,P) - \kappa_1(p,P) - f(P,p) + \kappa_1(P,p) \\ &= \gamma(p) - (\kappa_1(p,P) - \kappa_1(P,p)). \end{aligned}$$

Let, therefore

$$\gamma^*(p) = \gamma(p) - (\kappa_1(p,P) - \kappa_1(P,p)).$$

Let us now apply Theorem 35 to the functions with an asterisk. Then, according to (12),

$$0 = \gamma^*(P) = \gamma(P) - \sum_{p \in P} (\kappa_1(p,P) - \kappa_1(P,p)).$$

However, the value κ_1 of each edge appears twice in the summation but with different signs, so the sum is zero. Therefore, (12) is still valid. If (13) is applied to the functions with an asterisk, we obtain

$$\gamma^*(X) \le \kappa^*(X,Y) = \kappa_2(X,Y) - \kappa_1(X,Y).$$

However

$$\gamma^*(X) = \gamma(X) - \kappa_1(X,P) + \kappa_1(P,X).$$

Comparing the two relations

$$\begin{aligned} \gamma(X) &\le \kappa_2(X,Y) - \kappa_1(X,Y) + \kappa_1(X,P) - \kappa_1(P,X) \\ &= \kappa_2(X,Y) - \kappa_1(X,Y) + \kappa_1(X,X) + \kappa_1(X,Y) \\ &\quad - (\kappa_1(Y,X) + \kappa_1(X,X)) \\ &= \kappa_2(X,Y) - \kappa_1(Y,X). \end{aligned}$$

So, the following theorem has been obtained as a consequence of Theorem 35:

36. Let γ be an integer function defined over the set of vertices P of a graph $\vec{G} = (P, E, \vec{\mathcal{G}})$ without loops and multiple edges and $\kappa_1 \leq \kappa_2$ be non-negative integer functions defined over the set of edges E of $\vec{G}$. Then the requirements

$$0 \leq \kappa_1(p,q) \leq f(p,q) \leq \kappa_2(p,q) \qquad \text{if } (p,q) \in E, \tag{14}$$

$$f(p,P) - f(P,p) = \gamma(p) \qquad \text{if } p \in P \tag{11}$$

are feasible by an integer function f defined over the set E if and only if

$$\gamma(P) = 0 \tag{12}$$

and, at any partition $\{X, Y\}$ of P

$$\gamma(X) \leq \kappa_2(X, Y) - \kappa_1(Y, X). \tag{15}$$

As a further modification of our problem, let us assume that there are no sources feeding the flow and no sinks consuming it, the flow is considered 'closed', i.e.

$$\gamma(p) = 0 \qquad \text{if } p \in P.$$

The solution of this so-called *circulation problem* immediately follows from the previous theorem:

37. Let $\kappa_1 \leq \kappa_2$ be non-negative integer functions defined over the set of edges E of a graph $\vec{G}(P, E, \vec{\mathcal{G}})$ without loops and multiple edges. Then the requirements

$$0 \leq \kappa_1(p,q) \leq f(p,q) \leq \kappa_2(p,q) \qquad \text{if } (p,q) \in E, \tag{14}$$

$$f(p,P) - f(P,p) = 0 \qquad \text{if } p \in P$$

are feasible by an integer function f defined over the set E if and only if at any partition $\{X, Y\}$ of P

$$\kappa_1(Y, X) \leq \kappa_2(X, Y). \tag{15}$$

As a further generalisation, let us retain (14) in the previous problem but let us modify (11) by introducing two functions as follows:

$$\gamma_1(p) \leq f(p, P) - f(P, p) \leq \gamma_2(p) \qquad \text{if } p \in P. \tag{16}$$

Two necessary conditions of the feasibility are immediate. Namely, for a partition $\{X, Y\}$ of P, using (14) we obtain

$$\gamma_1(X) \leq f(X, P) - f(P, X) = f(X, X) + f(X, Y) - f(Y, X) - f(X, X)$$
$$= f(X, Y) - f(Y, X) \leq \kappa_2(X, Y) - \kappa_1(Y, X),$$

and similarly

$$\gamma_2(X) \geq \kappa_1(X, Y) - \kappa_2(Y, X).$$

Reducing the problem to a circulation problem with a little trick, the two above necessary conditions can be proved to be also sufficient for the feasibility. This is not detailed here. So, the following theorem due to A J Hoffman [21] is valid:

38. *Let $\gamma_1 \leq \gamma_2$ be integer functions defined over the set P of vertices of a graph $\overrightarrow{G} = (P, E, \overrightarrow{\mathcal{G}})$ without loops and multiple edges and let $\kappa_1 \leq \kappa_2$ be non-negative integer functions defined over the set of edges E of $\overrightarrow{G}$. Then the requirements*

$$0 \leq \kappa_1(p,q) \leq f(p,q) \leq \kappa_2(p,q) \qquad \text{if } (p,q) \in E, \tag{14}$$

$$\gamma_1(p) \leq f(p, P) - f(P, p) \leq \gamma_2(p) \qquad \text{if } p \in P \tag{16}$$

are feasible by an integer function f defined over the set E if and only if for any partition $\{X, Y\}$ of P,

$$\gamma_1(X) \leq \kappa_2(X, Y) - \kappa_1(Y, X)$$

and

$$\kappa_1(X, Y) - \kappa_2(Y, X) \leq \gamma_2(X).$$

Now, Theorem 35 will be used to yield the special case of Theorem 25 where the edge-capacity function and the xy-flow are integers. Consider the graph $\overrightarrow{G} = (P, E, \overrightarrow{\mathcal{G}})$ in Theorem 25 and the set of xy-flows limited by κ. Assume that $\overrightarrow{G}$ contains no loops or multiple edges. Let the function γ in $\overrightarrow{G}$, as in (11), be the following with some integer k:

$$\gamma(p) = \begin{cases} k & \text{if } p = x, \\ 0 & \text{if } p \in P \text{ and } p \neq x, y, \\ -k & \text{if } p = y. \end{cases}$$

Let us apply Theorem 35 to $\overrightarrow{G}$ and consider the necessary and sufficient condition of feasibility. (12) is obviously satisfied now. For (13), it suffices to regard the partitions $\{X, Y\}$ of P with $\gamma(X) > 0$ since, otherwise, (13) is obviously satisfied. However, $\gamma(X) > 0$ (namely, $\gamma(X) = k$) if and only if $x \in X$ and $y \in Y$.

Therefore, the set of $\overrightarrow{XY}$-edges is always an edge $\overrightarrow{xy}$-cut in $\overrightarrow{G}$. So, (13) becomes

$$k \leq \kappa(X, Y).$$

Let $\{X_0, Y_0\}$ be the partition of the set P (with, of course, $\gamma(X_0) > 0$) and with $\kappa(X_0, Y_0)$ minimal and let

$$k_0 = \kappa(X_0, Y_0).$$

Now, if $k_0 = \gamma(x)$ is selected then by Theorem 35, the requirements (10) and (11) are feasible by the function f which satisfies

$$\sigma(f) = k_0$$

according to (11). The above considerations complete the proof of this special case of Theorem 25. Observe that the feasibility theorems of this section remain valid if we omit the restriction of the relevant functions to be 'integer'. In the present case, these restrictions came from the fact that each feasibility theorem, although indirectly, has been deduced from Theorem 31. At the end of this sequence we reached Theorem 25 as well. The reader may already suspect that all the presented feasibility theorems can be deduced from Theorem 25. This way of discussion is found in the first three paragraphs of Chapter II in [11]. Further variations of the feasibility theorems can also be discovered there.

Let us return to the problem described in Theorem 34. Let us select the value of κ to be 1 in each edge. Assume the constraints (4), (1) and (8) to be feasible by f. Then the value of f is either 0 or 1 in each edge. Let us consider the subgraph $\vec{G}'$ of $\vec{G}$ consisting of those edges of $\vec{G}$ where the value of f is 1. Now, for any vertex $x \in A$ and $y \in B$ in $\vec{G}'$:

$$\varphi_{\text{out}}(x) = \alpha(x) \quad \text{and} \quad \varphi_{\text{in}}(y) = \beta(y).$$

If the orientations of the edges are now disregarded, Theorem 34 yields the solution of the following so-called *valency problem*: if two non-negative integer functions $\alpha(x)$ and $\beta(y)$ are given, find a necessary and sufficient condition of the existence of a subgraph of the complete bipartite graph $G = G(A, B)$ with

$$\varphi(p) = \begin{cases} \alpha(p) & \text{if } p \in A, \\ \beta(p) & \text{if } p \in B. \end{cases}$$

The condition is either (9) and (5) or (9) and (6) of Theorem 33, formulated for $\vec{G}(A, B)$ provided with a function κ equal to 1 on all edges. Let us reformulate these conditions for the complete bipartite graph $G(A, B)$. Let the vertices in A and B be denoted as $x_1, x_2, \ldots, x_m$ and $y_1, y_2, \ldots, y_n$, respectively. Let

$$\alpha(x_i) = \alpha_i, \quad i = 1, 2, \ldots, m$$

and

$$\beta(y_j) = \beta_j, \quad j = 1, 2, \ldots, n.$$

Let the subscripts be selected so that

$$\alpha_1 \geq \alpha_2 \geq \ldots \geq \alpha_m \quad \text{and} \quad \beta_1 \geq \beta_2 \geq \ldots \geq \beta_n.$$

What is the meaning of conditions (9) and (6) in the present case? (9) requires that

$$\sum_{i=1}^{m} \alpha_i = \sum_{j=1}^{n} \beta_j.$$

This is, however, obvious. Let us consider condition (6) in the case when the value of κ is 1 on all edges in $\overrightarrow{G}$. This condition requires that, for any set $A_1 \subseteq A$ in $\overrightarrow{G}(A,B)$, if we denote the set of the vertices $y \in B$ satisfying $\beta(y) \leq \kappa(A_1, y) = |A_1|$ by B_1' and the set $B - B_1'$ by B_2', the following is satisfied:

$$\alpha(A_1) \leq \beta(B_1') + \kappa(A_1, B_2'),$$

i.e.

$$\alpha(A_1) \leq \beta(B_1') + |A_1| \cdot |B_2'|. \tag{9'}$$

Hence, for any fixed A_1, the partition $\{B_1', B_2'\}$ of B depends clearly upon $k = |A_1|$ only. Therefore, the right-hand side of (9') depends upon k only. Consequently, in order to satisfy (9') at any fixed $k\ (= 1, 2, \ldots, m)$, it suffices to select an A_1 with $|A_1| = k$ and $\alpha(A_1)$ maximal. Since the numbers α_i are monotonically decreasing for any fixed k, it suffices to examine (9') for the set A_1 with elements $x_1, x_2, \ldots, x_k$. Then B_1' consists of the vertices y_j satisfying

$$\beta_j \leq k$$

holding. Since the numbers β_j also decrease monotonously, there is a subscript j' determining the partition $\{B_1', B_2'\}$, the one which satisfies

$$\beta_{j'} > k \geq \beta_{j'+1}.$$

Consequently, in case of k, (9') becomes

$$\alpha_1 + \alpha_2 + \ldots + \alpha_k \leq \beta_n + \beta_{n-1} + \ldots + \beta_{j'+1} + k \cdot j',$$

i.e.

$$\sum_{i=1}^{k} \alpha_i \leq \sum_{j=1}^{n} \min(\beta_j, k), \qquad k = 1, 2, \ldots, m. \tag{9'}$$

This condition can be simplified as follows. Let us place vertices in n rows along a rectangular grid as follows: place one point to each of the first β_j positions of the jth row $(j = 1, 2, \ldots, n)$. Let β_i^* denote the number of points placed in the ith column. A possible case is illustrated in figure 2.40. One can easily see from the diagram that the right-hand side of (9') for a fixed k equals the number of points placed in the first k columns. Consequently, (9') is just

$$\sum_{i=1}^{k} \alpha_i \leq \sum_{i=1}^{k} \beta_i^*, \qquad k = 1, 2, \ldots, m.$$

This completes the proof of the following theorem:

39. *Let α and β be non-negative, integer functions. Then the complete bipartite graph $G(A, B)$ has a subgraph with*

$$\varphi(p) = \begin{cases} \alpha(p) & \text{if } p \in A, \\ \beta(p) & \text{if } p \in B \end{cases}$$

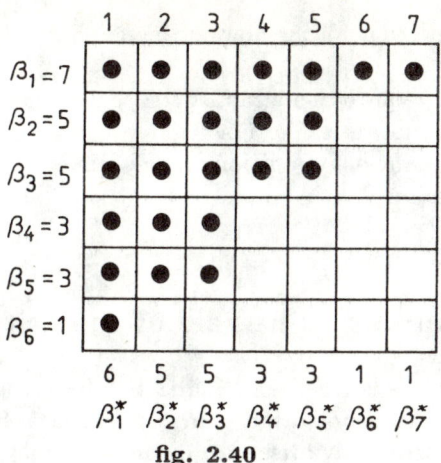

fig. 2.40

if and only if, using the above order notations,

$$\sum_{i=1}^{k} \alpha_i \leq \sum_{i=1}^{k} \beta_i^*, \qquad k = 1, 2, \ldots, m.$$

This theorem is due to D Gale and H J Ryser ([20] and [22]) in the following equivalent matrix formulation. A matrix $M_{m,n}$ is desired with each entry equal 0 or 1. The requirement is that its ith row contains α_i ones and its jth column contains β_j ones ($i = 1, 2, \ldots, m$, $j = 1, 2, \ldots, n$). We wish to find a necessary and sufficient condition to meet these requirements. We relate a subgraph of the complete bipartite graph $G(A, B)$ to $M_{m,n}$ by the correspondence illustrated in figure 2.11.

Flow problems are suitable for solving further valency problems. Another one is mentioned as an example. If two non-negative, integer functions α and β are given, find a necessary and sufficient condition for the existence of a subgraph of a graph $\vec{G} = (P, E, \vec{\mathcal{G}})$ with

$$\varphi_{\text{out}}(p) = \alpha(p) \quad \text{and} \quad \varphi_{\text{in}}(p) = \beta(p) \qquad \text{if } p \in P.$$

Let us associate the following graph $\vec{G}^*$ to $\vec{G}$. Two vertices p' and p'' are associated to each vertex $p \in P$, and an edge (p', q'') to each edge (p, q). Furthermore, we include the vertices x and y, and the edges (x, p'), (p'', y) for each vertex $p \in P$. The edge-capacity function κ is defined in this graph $\vec{G}^*$ as follows:

$$\kappa(p', q'') = 1 \qquad \text{if } (p, q) \in E,$$
$$\kappa(x, p') = \alpha(p) \qquad \text{if } p \in P,$$
$$\kappa(p'', y) = \beta(p) \qquad \text{if } p \in P.$$

It can be easily verified that all xy-flows of integer value $\sum_{p \in P} \alpha(p)$ limited by κ in $\vec{G}^*$ determine a required subgraph of $\vec{G}$, corresponding to the edges where the value of the flow is 1 (i.e. not 0).

The necessary and sufficient condition for the existence of this flow is found in the previous discussion.

2.7 Optimal planning. The role of the critical path

So far the required flows have been limited by the capacities of the given network of roads. If a transportation plan is required according to the flow, several efficiency problems may arise. For example, it may be desirable to execute the transportation along an itinerary as short as possible or to minimise the cost of the transportation. In what follows, optimisation problems of this kind will be treated. In this section, a procedure will be presented for finding a path of minimal or maximal length connecting two points of a network of roads; how it is applied to planning problems will also be shown.

Let us define a real function ϱ over the set of edges E of a graph $\vec{G} = (P, E, \vec{\mathcal{G}})$. The number $\varrho(e) = \varrho(p, q)$ is called the value of the edge $e = (p, q) \in E$ (in accordance with the terminology used in §2.4). The *value of a subgraph* $\vec{G}' = (P', E', \vec{\mathcal{G}})$ of $\vec{G}$ is defined as the number

$$\varrho(\vec{G}') = \sum_{e \in E'} \varrho(e).$$

With the vertices x and y of the graph $\vec{G}$ fixed, we require an xy-path of minimal value in $\vec{G}$. If $\varrho(e)$ means the length of the section of the network of roads represented by e, we arrive at the above problem. If

$$\varrho(e) = 1 \quad \text{for all } e \in E$$

then any $\vec{xy}$-path of minimal value is simultaneously an $\vec{xy}$-path of minimal length, i.e. one containing a minimal number of edges.

The essence of our procedure for finding an $\vec{xy}$-path of minimal value is the following: zero is associated to the vertex x and a number λ associated to any vertex $p \neq x$ so that it is large enough to ensure the existence of an $\vec{xp}$-path of value less than λ in $\vec{G}$ provided there is an $\vec{xp}$-path in $\vec{G}$ at all. (The existence of an $\vec{xy}$-path in $\vec{G}$ is naturally presumed.) This requirement can be fulfilled by the following choice:

(1) $$\lambda(p) = 1 + \sum_{e \in E} |\varrho(e)| \quad \text{if } x \neq p \in P.$$

Thereupon, the numbers associated to the edges are decreased on the following principle. If the number associated with the vertex $q \in P$ is λ_1, if q can certainly be reached from x along a path whose value is not higher than λ_2, and if $\lambda_2 < \lambda_1$, then the number associated to q is decreased to λ_2. The last number associated to y in the course of these decreases will yield the minimum of the values of the $\overrightarrow{xy}$-paths and an $\overrightarrow{xy}$-path of minimal value will also be found.

The instruction for decreasing is the following:

(2) *If there is an edge* $(p, q) \in E$ *with*
$$\lambda(q) - \lambda(p) > \varrho(p, q),$$
then let the number associated with q be
$$\lambda(p) + \varrho(p, q)$$
instead of $\lambda(q)$.

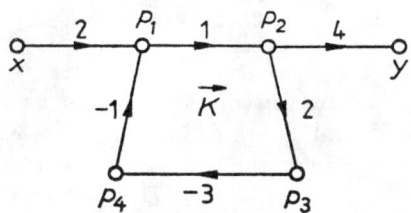

fig. 2.41

The question is whether the sequence of these instructions will always terminate. Unfortunately, the answer is negative. Consider figure 2.41. The values have been written on the edges. Initially $\lambda(x) = 0$ and, at any i, $\lambda(y) = \lambda(p_i) = 6$. The decreasing can be executed at the vertices p_i of the circuit $\overrightarrow{K}$ in the diagram any number of times; observe also that
$$\varrho(\overrightarrow{K}) < 0.$$

The question is whether circuits of negative value can be the only reason if instruction (2) can be repeated infinitely. The answer is affirmative since the following statement is true:

40. *Instruction (2) can be repeated for a graph $\overrightarrow{G}$ any number of times if and only if there is a circuit $\overrightarrow{K}$ in $\overrightarrow{G}$ with*
$$\varrho(\overrightarrow{K}) < 0.$$

Proof. Assume first that $\varrho(\vec{K}) < 0$ for a circuit $\vec{K}$ in $\vec{G}$.

$$p_1, p_2, \ldots, p_k, p_{k+1} = p_1$$

denote the vertices of $\vec{K}$ (in this order, in accordance with its orientation). Indirectly assume that the process of (2) terminates. Let λ_i denote the number last associated with p_i ($i = 1, 2, \ldots, k$). Then

$$\lambda_2 - \lambda_1 \leq \varrho(p_1, p_2),$$
$$\lambda_3 - \lambda_2 \leq \varrho(p_2, p_3),$$
$$\ldots$$
$$\lambda_k - \lambda_{k-1} \leq \varrho(p_{k-1}, p_k),$$
$$\lambda_1 - \lambda_k \leq \varrho(p_k, p_1),$$

Summing these relations we obtain

$$0 \leq \sum_{i=1}^{k} \varrho(p_i, p_{i+1}) = \varrho(\vec{K}),$$

contrary to $\varrho(\vec{K}) < 0$.

Let us now assume that the instruction (2) can be repeated for $\vec{G}$ any number of times. Both P and E are finite sets, therefore there is a $q_1 \in P$ so that infinitely many numbers can be associated to q_1 by means of (2). Consequently, there is an edge $(q_2, q_1) \in E$ so that infinitely many of the numbers associated with q_1 are obtained as

$$\lambda(q_2) + \varrho(q_2, q_1).$$

Therefore, there are infinitely many pairs of numbers, so that any such pair λ_1, λ_2 has the following property: at a certain stage of the application of (2), λ_2 is associated with the vertex q_2, at a later stage, λ_1 is associated with the vertex q_1 and

$$\lambda_2 + \varrho(q_2, q_1) = \lambda_1.$$

Similarly, there is an edge $(q_3, q_2) \in E$ and infinitely many triplets of numbers so that any such triplet $\lambda_1, \lambda_2, \lambda_3$ satisfies the following: $\lambda_3, \lambda_2, \lambda_1$ are associated with the vertices q_3, q_2, q_1, respectively, in this order of the stages, and

$$\lambda_1 - \lambda_2 = \varrho(q_2, q_1),$$
$$\lambda_2 - \lambda_3 = \varrho(q_3, q_2).$$

By repeating the above reasoning, sooner or later we reach a vertex already encountered. The subscripts of the vertices can be rearranged to have $q_{n+1} = q_1$ and $q_1 = q_{n+1}, q_n, \ldots, q_2, q_1$ can be assumed to constitute the vertices of a circuit $\vec{K}'$ in $\vec{G}$ in this order, according to its orientation. Moreover, after (2) has been employed sufficiently many times, the numbers $\lambda_1, \lambda_2, \ldots, \lambda_n, \lambda_{n+1}$ are obtained with

$$\lambda_1 - \lambda_2 = \varrho(q_2, q_1),$$
$$\lambda_2 - \lambda_3 = \varrho(q_3, q_2),$$
$$\ldots$$
$$\lambda_{n-1} - \lambda_n = \varrho(q_n, q_{n-1}),$$
$$\lambda_n - \lambda_{n+1} = \varrho(q_1, q_n),$$

and since both λ_{n+1} and λ_1 are associated with the vertex q_1, but λ_1 is at a later stage, the following inequality holds:

$$\lambda_{n+1} - \lambda_1 > 0.$$

The above relationships lead to

$$0 > \sum_{i=1}^{n} \varrho(q_{i+1}, q_i) = \varrho(\overrightarrow{K}').$$

This completes the proof of Statement 40.

Hence, the following condition is required for the applicability of our procedure:

(*) *The graph $\overrightarrow{G}$ contains no circuit $\overrightarrow{K}$ with*

$$\varrho(\overrightarrow{K}) < 0.$$

The following statement will now be proved:

41. *If the course of the successive applications of instruction (2) terminates and λ is the number last associated with y then*

$$\varrho(\overrightarrow{L}) \geq \lambda$$

for any $\overrightarrow{xy}$-path $\overrightarrow{L}$ in $\overrightarrow{G}$.

Proof. Let $x = p_0, p_1, p_2, \ldots, p_m = y$ denote the vertices of $\overrightarrow{L}$ in this order, following the orientation, and let the numbers last associated with them be $\lambda_0, \lambda_1, \lambda_2, \ldots, \lambda_m = \lambda$, respectively. Then

$$\lambda_1 - \lambda_0 \leq \varrho(x, p_1),$$
$$\lambda_2 - \lambda_1 \leq \varrho(p_1, p_2),$$
$$\ldots$$
$$\lambda - \lambda_{m-1} \leq \varrho(p_{m-1}, y).$$

The sum of the inequalities is

$$\lambda - \lambda_0 \leq \sum_{i=0}^{m-1} \varrho(p_i, p_{i+1}) = \varrho(\overrightarrow{L}).$$

This proves the assertion since $\lambda_0 \leq 0$.

The way of finding an $\overrightarrow{xy}$-path of value λ in $\overrightarrow{G}$ will be discovered while proving the following statement:

42. *If the course of the successive applications of instruction (2) terminates and λ is the number last associated with y, and there is an $\overrightarrow{xy}$-path in $\overrightarrow{G}$, then there is an $\overrightarrow{xy}$-path $\overrightarrow{L}_0$ in $\overrightarrow{G}$ satisfying*

$$\varrho(\overrightarrow{L}_0) = \lambda.$$

Proof. At first we show that if the course of the successive applications of (2) terminates and if the number finally associated with a vertex p different from x is $\lambda(p)$ and there is an $\overrightarrow{xp}$-path in $\overrightarrow{G}$, then

$$\lambda(p) < 1 + \sum_{e \in E} |\varrho(e)|.$$

In other words, for any vertex p different from x at least one decreasing occurs if p is accessible from x along a directed path. This assertion is obvious since, even if the $\overrightarrow{xp}$-path includes all edges of $\overrightarrow{G}$, $\lambda(p)$ cannot be greater than

$$\sum_{e \in E} |\varrho(e)|$$

in view of the termination.

Let us now consider the number λ finally associated with y. Since λ has been obtained by means of decreases, there is a vertex $q_1 \in P$ and an edge $(q_1, y) \in E$ with

$$\lambda - \lambda_1 = \varrho(q_1, y)$$

where λ_1 is the number associated with q_1 at some stage of (2). Had λ_1 changed later, it could only have decreased. But then λ associated with y could also have been decreased. Therefore λ_1 is the number finally associated with q_1. Let us now repeat this reasoning with respect to λ_1. This yields the vertices $q_2, q_3, \ldots$ which are all different since, otherwise, the reasoning in the second part of the proof of Statement 40 would yield a directed circuit of negative value, contrary to assumption (*).

Our investigations imply that the sequence $y, q_1, q_2, \ldots$ can terminate at x only. The sequence does in fact terminate at x, otherwise there would be a repetition. This, however, implies that the number zero, initially associated with x does not change in the course of the application of (2). So, let our sequence be:

$$y = q_0, q_1, \ldots, q_n = x.$$

This yields the vertices of an $\vec{xy}$-path $\vec{L}_0$ in $\vec{G}$ in the order opposite to its orientation. If the numbers associated with the elements of this sequence in this order, are denoted by

$$\lambda = \lambda_0, \lambda_1, \ldots, \lambda_n = 0,$$

then the following equalities hold:

$$\lambda_{n-1} - 0 = \varrho(x, q_{n-1}),$$
$$\lambda_{n-2} - \lambda_{n-1} = \varrho(q_{n-1}, q_{n-2}),$$
$$\ldots$$
$$\lambda - \lambda_1 = \varrho(q_1, y).$$

The sum of these results is

$$\lambda = \varrho(\vec{L}_0),$$

i.e. that $\vec{L}_0$ is an $\vec{xy}$-path of minimal value in $\vec{G}$. This completes the proof of Statement 42.

Our considerations have also led to the following procedure (the 'maximal' one will later be explained):

43. Algorithm *for finding an $\vec{xy}$-path of minimal (maximal) value* in a graph $\vec{G} = (P, E, \vec{\mathcal{G}})$ containing no directed circle of negative value. The real function defined over the set E is ϱ.

(1) If there is no $\vec{xy}$-path in $\vec{G}$, then the procedure has terminated.

(2) Let us define the real function λ over the set P as follows:

$$\lambda(p) = \begin{cases} 0 & \text{if } p = x, \\ 1 + \sum_{e \in E} |\varrho(e)| & \text{if } p \in P \text{ and } p \neq x. \end{cases}$$

(3) Let us form the number $\lambda(q) - \lambda(p)$ for each edge $(p, q) \in E$. If there is an edge $(p, q) \in E$ with $\lambda(q) - \lambda(p) > \varrho(p, q)$,

then $\begin{cases} \text{select such an edge,} \\ \text{let } \lambda_1(r) = \begin{cases} \lambda(p) + \varrho(p, q) & \text{if } r = q, \\ \lambda(r) & \text{if } r \in P \text{ and } r \neq q; \end{cases} \\ \text{let } \lambda_1 \text{ take the role of the function } \lambda \text{ and} \\ \text{return to (3).} \end{cases}$

(4) Let us mark an edge (p, y) of $\vec{G}$ with $\lambda(y) - \lambda(p) = \varrho(p, y)$.

If $p \neq x$ then $\begin{cases} \text{let } p \text{ take the role of } y \text{ and} \\ \text{return to (4).} \end{cases}$

(5) Record the result: the marked edges constitute the edges of an $\vec{xy}$-path of minimal value in $\vec{G}$ and this minimal value is $\lambda(y)$.

An $\vec{xy}$-path of maximal value in $\vec{G}$ can be obtained by means of Algorithm 43 by using the function ϱ_1 instead of ϱ so that

$$\varrho_1(e) = -\varrho(e) \qquad \text{if } e \in E.$$

Then, in order to preserve the applicability of the algorithm, an assumption similar to (*) must be made:

($\overset{*}{*}$) *The graph $\vec{G}$ contains no circuit $\vec{K}$ with*

$$\varrho(\vec{K}) > 0.$$

As an example, an $\vec{xy}$-path of minimal value has been marked in figure 2.42 by heavy lines. The values of the function ϱ are on the edges, the numbers last associated with the vertices are in parentheses.

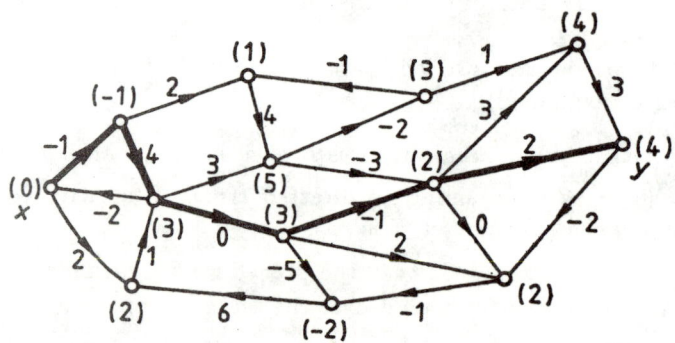

fig. 2.42

It has been previously noted that if $\varrho(e)$ is chosen as 1 over all edges of the graph then our procedure yields an $\vec{xy}$-path of minimal length in $\vec{G}$. It is true that no circuit of negative value may exist at this choice but it is possible that there exists not even a single $\vec{xy}$-path in $\vec{G}$ (so far, the existence of such a path has been assumed). It is easy to verify that an $\vec{xy}$-path exists in $\vec{G}$ if and only if our decreasing procedure leads to a decreasing of $\lambda(y)$.

Algorithm 43 is due to L R Ford [23]. Other methods suitable for constructing extremal paths or even trees in graphs are also known (see, for example, [24], [25] and [26]).

It will be shown next how Algorithm 43 can be applied to the optimisation of planning problems. Imagine that the tasks arising in the course of building a house are to be planned. A lot of *jobs* have to be taken into account in our plan, such as the drawing of drafts, obtaining the building permit, the conclusion of the contract, masonry, carpentry, plumbing, etc. We wish to determine from our plan, how much time is required for the total programme. To this end, the time required for the particular jobs as well as the necessary sequence of the jobs (for example that plumbing must be preceded by masonry, by the conclusion of the contract, etc.) must be known. Let us assume that all these data are at our disposal. Then, our plan can be modelled by a graph as follows.

Let us introduce a directed edge corresponding to each job so that neither the tail nor the head of any edge can be identical with the tail or head of any other edge. If the job (p_1, q_1), i.e. the job represented by the edge (p_1, q_1), must precede the job (p_2, q_2), then let us introduce a *virtual job* (q_1, p_2). Thus, the necessary sequence of jobs can be indicated by directed paths. If a $\overrightarrow{pq}$-path already exists then the possible virtual job (p, q) need not be introduced. It is possible that a job can start upon the execution of a certain part of a job (q, r) but not earlier. Then, instead of the job (q, r), its appropriate parts are treated as jobs. Finally, let us introduce the vertices x and y denoting the start and end, respectively of the total program and, further, introduce the virtual job (x, p) if $\varphi_{\text{in}}(p) = 0$ and the virtual job (q, y) if $\varphi_{\text{out}}(q) = 0$. This yields the so-called *program graph* $\overrightarrow{G} = (P, E, \overrightarrow{\mathcal{G}})$ corresponding to our problem. Let us introduce the non-negative real function ϱ defined over the set E so that $\varrho(p, q)$ denotes the duration of executing the job (p, q) in some time units. For any virtual job (p, q), naturally, $\varrho(p, q) = 0$.

What is the earliest time for an *event* $p \in P$ to occur? Evidently, if the maximum of the values ϱ of the $\overrightarrow{xp}$-paths in $\overrightarrow{G}$ is $\varrho_{\max}$ then the event p can occur no earlier than $\varrho_{\max}$ time units after the start of the entire program. Therefore, the duration of executing the entire program is given by the maximum of the values ϱ of the $\overrightarrow{xy}$-paths in $\overrightarrow{G}$. The $\overrightarrow{xy}$-paths of maximal value in the program graph are customarily called *critical paths*. If the execution of the total program is desired not to be too protracted, the time limits of the jobs in the critical paths, of the so-called *critical jobs*, must strictly be adhered to. The program graph $\overrightarrow{G}$ contains no directed circle since, otherwise, there would be a job which had to be finished before its start. Therefore, a critical path $\overrightarrow{L}$ can be found in $\overrightarrow{G}$ by means of Algorithm 43. The algorithm should be applied to the function ϱ_1 satisfying

$$\varrho_1(e) = -\varrho(e) \quad \text{if } e \in E.$$

If the number finally associated to the vertex $p \in P$ is $\lambda_1(p)$ then let

$$\lambda(p) = -\lambda_1(p).$$

Obviously, $\lambda(x) = 0$, $\lambda(y)$ is the time required for the execution of the entire program and for any edge (p, q) in $\overrightarrow{L}$:

$$\lambda(q) - \lambda(p) = \varrho(p, q).$$

In order to discover the critical jobs, all critical paths have to be found. In the manner described below even more information can be gained: the 'loose spots' of the program as well as their acceptable delay will be discovered. Let us associate the numbers $\lambda_a(p)$ and $\lambda_b(p)$ to each vertex $p \in P$ as follows:

$$\lambda_a(p) = \begin{cases} 0 & \text{if } p = x; \\ \text{the maximum of the values } \varrho \text{ of the } \overrightarrow{xp}\text{-paths} & \text{if } p \neq x; \end{cases}$$

$$\lambda_b(p) = \begin{cases} \lambda(y) & \text{if } p = y, \\ \lambda(y) - (\text{the maximum of the values } \varrho \text{ of the } \overrightarrow{py}\text{-paths}) & \text{if } p \neq y. \end{cases}$$

The numbers $\lambda_a(p)$ and $\lambda_b(p)$ have been determined in the program graph shown in figure 2.43. The values ϱ have been written on the edges. The dotted edges are virtual jobs, the value ϱ on these is zero. The numbers in

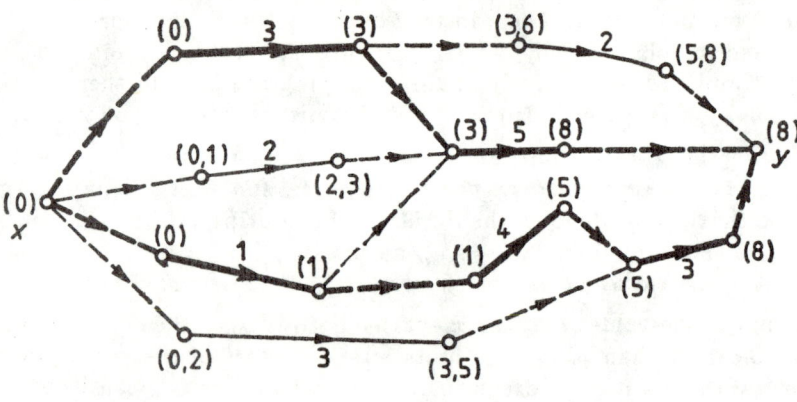

fig. 2.43

parentheses beside the vertices are the values $\lambda(p)$ or the pairs $\lambda_a(p), \lambda_b(p)$. There are two critical paths in this diagram, their edges have been drawn by heavy lines.

The meaning of the numbers $\lambda_a(p)$ and $\lambda_b(p)$ is the following: the job (p, q) cannot be started and ended sooner than $\lambda_a(p)$ and $\lambda_a(q)$ time units, respectively after the start of the program and, if a delay in the entire program should be avoided, the job (p, q) must not be started and ended later than $\lambda_b(p)$ and $\lambda_b(q)$ time units, respectively, after the start of the program. The number

$$\lambda_a(q) - \lambda_b(p) - \varrho(p, q)$$

indicates the maximum time delay in beginning job (p, q) that does not offend the spare time of other jobs. The beginning of job (p, q) can by no means be delayed by more than

$$\lambda_b(q) - \lambda_b(p) - \varrho(p, q)$$

time units. The quantity
$$\lambda_a(q) - \lambda_a(p) - \varrho(p,q)$$
also yields information about certain spare time. The number
$$\lambda_b(p) - \lambda_a(p)$$
can be regarded as the spare time of event p. These numbers are all non-negative and they may serve as measures of 'looseness' from different aspects.

Obviously, for critical jobs or for events contained in critical paths these numbers are all zero and if p is such an event then
$$\lambda_a(p) = \lambda_b(p) = \lambda(p).$$
It will be now shown that the reverse is also true: if the numbers above are all zero for an $\overrightarrow{xy}$-path $\overrightarrow{L}_1$ then $\overrightarrow{L}_1$ is a critical path and then, of course,
$$\lambda_a(p) = \lambda_b(p) = \lambda(p)$$
for any vertex p in $\overrightarrow{L}_1$; hence all critical paths are easily discovered. For the proof, let us denote the vertices of $\overrightarrow{L}_1$ in this order by
$$x = p_0, p_1, p_2, \ldots, p_n = y.$$
Then
$$\lambda_a(p_i) + \varrho(p_i, p_{i+1}) = \lambda_a(p_{i+1}) \qquad (i = 0, 1, \ldots, n-1).$$
Summing these relations we obtain
$$\sum_{i=0}^{n-1} \lambda_a(p_i) + \varrho(\overrightarrow{L}_1) = \sum_{i=1}^{n} \lambda_a(p_i),$$
i.e.
$$\varrho(\overrightarrow{L}_1) = \lambda_a(y) - \lambda_a(x).$$
Since
$$\lambda_a(y) - \lambda_a(x) = \lambda_b(y) - \lambda_b(x) = \lambda(y),$$
$\overrightarrow{L}_1$ is really a critical path and our assertion is immediate.

The so-called PERT-method (the abbreviation for Program Evaluation Research Task or Program Evaluation and Review Technique; see [27]) is based upon the critical path method. Its main point is that it applies probabilistic tools after substituting the unknown execution times of the jobs by statistical averages. The method has been extended to cover changes due to unforeseen circumstances, since it permits adjustments 'improving' the program during the execution (see [28]).

2.8 Minimal cost transportation

Among the transportation problems raised in §2.6 we shall now consider those where the requirements for the supply and the demands can be satisfied. Let $G(A, B)$ denote the complete bipartite graph representing the transportation.

We can also consider the directed graph $\overrightarrow{G}(A, B)$ indicating the direction of transportation. Let us denote the vertices of A and B by $a_1, a_2, \ldots, a_m$ and $b_1, b_2, \ldots, b_n$, respectively. Let α and β be positive integer functions representing the supply and demand, respectively. They are defined over the sets A and B, respectively. The non-negative integer function f is defined over the set of edges E of the graph and $f(a_i, b_j) = f_{ij}$ expresses the quantity transported along the edge (a_i, b_j). The function f is also called *transportation* in short. Let ϱ denote the non-negative, integer *cost function* defined over the set E: $\varrho(a_i, b_j) = \varrho_{ij}$ is the transporting cost of a unit quantity transported along edge (a_i, b_j). Let M be the *cost matrix*

$$M = M_{m,n} = [\varrho_{ij}].$$

The *transportation cost* of f is

$$\varrho(f) = \sum_{i=1}^{m}\sum_{j=1}^{n} f(a_i, b_j) \cdot \varrho(a_i, b_j) = \sum_{e \in E} f(e) \cdot \varrho(e).$$

Our aim is to determine a minimum cost transportation f which can be fed from the supply and which satisfies the demand. So the function f is bounded by the following requirements:

$$f(a, B) \leq \alpha(a) \quad \text{if } a \in A,$$
$$f(A, b) \leq \beta(b) \quad \text{if } b \in B.$$

Hence

$$\alpha(A) \geq \beta(B).$$

The problem can be reduced as follows to the case when this inequality is replaced by an equality: let us augment the graph into another complete bipartite graph including a new vertex b_{n+1} in B with a demand

$$\beta(b_{n+1}) = \alpha(A) - \beta(B),$$

and let the cost on the new edges be zero, i.e.

$$\varrho(a, b_{n+1}) = 0 \quad \text{if } a \in A.$$

The definition of f is extended by

$$f(A, b_{n+1}) = \beta(b_{n+1}).$$

Then, in the augmented graph $G' = G'(A, B')$,

$$\alpha(A) = \beta(B').$$

and the transportation cost of f is not changed by the 'surplus transportation' to b_{n+1} (of course, this surplus transportation need not be executed). For the required transportation over the edges of G':

$$\beta(B') \leq f(A,B) \leq \alpha(A),$$

and so $\alpha(A) = \beta(B')$ implies that

$$f(a, B') = \alpha(a) \quad \text{if } a \in A,$$
$$f(A, b) = \beta(b) \quad \text{if } b \in B'.$$

Hence we can assume also for the initial graph $G(A, B)$ that

$$\alpha(A) = \beta(B). \tag{1}$$

Then the required minimal cost transportation must be 'maximal', i.e. one satisfying the following two equalities:

$$f(a, B) = \alpha(a) \quad \text{if } a \in A, \tag{2}$$
$$f(A, b) = \beta(b) \quad \text{if } b \in B. \tag{3}$$

Since the graph $G(A, B)$ is complete the transportation is possible from any edge in A to any edge in B. If transportation is, in fact, impossible from a to b then it suffices to choose the cost $\varrho(a, b)$ 'very large', for example larger than the sum of the costs on the possible edges. Indeed, if the required transportation is feasible on the possible edges and f is of minimal cost, then obviously $f(a, b) = 0$. The symbol ∞ is also used to represent such very high costs.

Before the discussion of the problem we show that it can be treated as the generalisation of the problem of §2.3 in which finding an independent system of edges of maximal value in a bipartite graph with integer values on the edges (see Theorem 17 and Algorithm 18). Let us return to that problem: at $m \leq n$, a permutation of m distinct numbers chosen from the numbers $1, 2, \ldots, n$ was denoted by P and the ith number herein by $P(i)$. The value c_{ij} of an edge $\{a_i, b_j\}$ in the complete graph $G(A, B)$ was at least zero. A permutation P maximising the sum

$$c(P) = \sum_{i=1}^{m} c_{iP(i)}$$

was required. Let K be a sufficiently large number to have

$$K > c_{ij}$$

for any pair i, j and let

$$d_{ij} = K - c_{ij}.$$

Then

$$d(P) = \sum_{i=1}^{m} d_{iP(i)} = mK - \sum_{i=1}^{m} c_{iP(i)} = mK - c(P).$$

mK being constant, $c(P)$ is maximal if and only if $d(P)$ is minimal. Now, in our transportation problem let $m \leq n$ and

$$\alpha(a) = \beta(b) = 1 \quad \text{if } a \in A \text{ and } b \in B.$$

Then, f being integer,

$$f(e) = 0 \text{ or } 1$$

for any edge e. So, in the case of a maximal transportation, the edges with $f = 1$ are the elements of a set of independent edges in the graph $G(A, B)$ and therefore

$$\varrho(f) = \sum_{i=1}^{m} \varrho_{iP(i)}.$$

Hence, with the substitution $\varrho_{ij} = d_{ij}$, we really obtain the previous problem.

Having established the connection between the two problems, it is no wonder that the Hungarian method employed by Kuhn plays a role in finding a transportation of minimal cost. Although the solution below makes use of Theorem 30 based on Theorem 31 of Hall, of Theorem 28 implied by Theorem 25 of Ford and Fulkerson, and of Algorithm 29, the solution of F L Hitchcock (his name is frequently associated with the problem) originates earlier [29].

Consider a maximal transportation f, i.e. a function f satisfying (2) and (3). Let us subtract 1 from each element in the ith row of the cost matrix $M = [\varrho_{ij}]$, in short, let us subtract 1 from the ith row of M. Let M_1 denote the matrix thus obtained from M. So

$$M_1 = \begin{bmatrix} \varrho_{11} & \varrho_{12} & \cdots & \varrho_{1n} \\ \cdots & & & \\ \varrho_{i1} - 1 & \varrho_{i2} - 1 & \cdots & \varrho_{in} - 1 \\ \cdots & & & \\ \varrho_{m1} & \varrho_{m2} & \cdots & \varrho_{mn} \end{bmatrix}.$$

How is the cost of f modified if the cost function ϱ_1 in M_1 is used? Simple calculations show that the original cost is decreased by $f(a_i, B)$ i.e., using (2),

$$\varrho_1(f) = \varrho(f) - \alpha(a_i).$$

If 1 is added to the ith row of M, then the new cost of f is

$$\varrho_2(f) = \varrho(f) + \alpha(a_i).$$

Similarly, if 1 is added to or subtracted from the jth column of M, then the modified cost of f is

$$\varrho_3(f) = \varrho(f) \pm \beta(b_j).$$

Observe that the numbers added to or subtracted from $\varrho(f)$ are independent of f. Hence, for a function f, $\varrho(f)$ is minimal if and only if all values $\varrho_k(f)$ are minimal, too. This fact suggests the idea of the solution: let us try to obtain 'many' zero costs by the above modification of the cost matrix M taking care to avoid negative costs. Let $G_0(A, B)$ be the subgraph of G formed by the edges of G with zero cost after the modification. Now, if (2) and (3) are feasible by f in G_0, or in $\vec{G}_0$ formed in accordance with the direction of the transportation, then $\varrho(f)$ is obviously minimal since the cost of f in G_0 is zero.

The appearance of at least one zero cost in each row and column of the cost matrix M can always be obtained by subtractions applied to its rows and columns without introducing negative costs. Let this modified cost matrix be M_0 and the cost function in M_0 be ϱ_0. Let $G_0 = G_0(A, B)$ denote the subgraph of G formed by the edges of G with zero cost according to M_0. Let $A' \subseteq A$ and let $B' = G_0(A')$ denote the set of those vertices of B which are adjacent in G_0 to the vertices in A'. Let further $A'' = A - A'$ and $B'' = B - B'$. Then there is no $A'B''$-edge with zero cost. Let us subtract 1 from each row of M_0 corresponding to a vertex in A' and add 1 to each column of M_0 corresponding to a vertex in B'. How does this modify the costs of the edges in G? The following phenomena can easily be verified:

the costs of the $A'B'$-edges and $A''B''$-edges do not change;
the cost of each $A'B''$-edge is decreased by 1;
the cost of each $A''B'$-edge is increased by 1.

No edge with negative cost can result since there is no $A'B''$-edge with zero cost. For the cost of the transportation f we have

$$\varrho_0(f) = \varrho(f) - \alpha(A') + \beta(B'). \qquad (4)$$

Now, if

$$\alpha(A') > \beta(B'),$$

then the cost of f can be decreased by modifying M_0 as above. If, however,

$$\alpha(A') \leq \beta(B') \qquad \text{where } B' = \vec{G}_0(A'), \qquad (5)$$

then we show that (2) and (3) are feasible in $\vec{G}_0$ by f, i.e. f is a maximal transportation.

Assume, therefore, that (5) holds in $\vec{G}_0$ for any $A' \subseteq A$. Then, according to Theorem 30, the following requirements are feasible in $\vec{G}_0$ by f:

$$f(a, B) = \alpha(a) \quad \text{if } a \in A, \qquad (2)$$
$$f(A, b) \leq \beta(b) \quad \text{if } b \in B. \qquad (6)$$

Hence, in $\vec{G}_0$ we have

$$\alpha(A) = f(A, B) \leq \beta(B).$$

Now, taking (1) and (6) into account, the feasibility of (3) in G_0 by f is obtained.

So our problem can be solved by constructing a graph $\vec{G}_0$ which satisfies (5) and by finding a maximal transportation f in $\vec{G}_0$. The problem will be reduced to a maximum flow problem and this will result in a procedure yielding f.

Assume that there is at least one zero cost in each row and column of the matrix M_0 obtained by modifying the original cost matrix M. Let $G_0 = G_0(A, B)$ be the subgraph of the original graph G formed by the edges of G with zero cost according to M_0. The graph $\vec{G}_0$ is derived by the usual orientations. Let us introduce the following notation:

$$\delta(A') = \alpha(A') - \beta(B') \quad \text{where } B' = \vec{G}_0(A').$$

Since $\delta(A') = 0$ at $A' = A$,

$$\max_{A' \subseteq A} \delta(A') \geq 0.$$

Let us augment the graph $\vec{G}_0$ as follows: include the source x, the sink y and connect an edge (x, a) to each vertex $a \in A$ and an edge (b, y) to each vertex $b \in B$. This yields the graph $\vec{G}_0^+$ from $\vec{G}_0$. Let us define the capacity-function κ over the edges of $\vec{G}_0^+$ as follows:

$$\begin{aligned}
\kappa(a, b) &= \infty && \text{for the } \overrightarrow{AB}\text{-edges,} \\
\kappa(x, a) &= \alpha(a) && \text{if } a \in A, \\
\kappa(b, y) &= \beta(b) && \text{if } b \in B.
\end{aligned}$$

Obviously, for any xy-flow f^+ in $\vec{G}_0^+$ limited by κ:

$$\begin{aligned}
f^+(a, B) &\leq \alpha(a) && \text{if } a \in A, \\
f^+(A, b) &\leq \beta(b) && \text{if } b \in B,
\end{aligned}$$

and so the value of f^+ satisfies

$$\sigma(f^+) \leq \alpha(A).$$

In view of the previous considerations we can state that if

$$\max_{A' \subseteq A} \delta(A') = 0.$$

Then (5) is necessarily satisfied in $\vec{G}_0$ and so the existence of f realising the feasibility of (2) and (3) in $\vec{G}_0$ is ensured. The xy-flow f^* in $\vec{G}_0^+$ limited by κ is defined by the function f as follows:

$$\begin{aligned}
f^*(a, b) &= f(a, b) && \text{for the } \overrightarrow{AB}\text{-edges,} \\
f^*(x, a) &= \alpha(a) && \text{if } a \in A, \\
f^*(b, y) &= \beta(b) && \text{if } b \in B.
\end{aligned}$$

f^* is an xy-flow of maximal value. The reverse statement is obvious: (2) and (3) are feasible in $\vec{G}_0$ by any xy-flow of maximal value in $\vec{G}_0^+$ and this yields a transportation of minimal cost.

But how can we establish whether $\max \delta(A')$ is zero and what should we do if it is not? Let us find an xy-flow of maximal value in $\vec{G}_0^+$, let this be f^+. Let $\vec{W}$ denote an edge $\overline{xy}$-cut of minimal capacity in $\vec{G}_0^+$. Since the capacity of each $\overrightarrow{AB}$-edge is ∞, $\vec{W}$ cannot include any $\overrightarrow{AB}$-edge. Let A'' be the set formed by the heads of the edges in $\vec{W}$ excluding y and B' the set of the tails of the edges in $\vec{W}$ excluding x. Obviously, $A'' \subseteq A$ and $B' \subseteq B$. Let, further, $A' = A - A''$ and $B'' = B - B'$. In the partition $\{X, Y\}$ of the set of vertices of the graph $\vec{G}_0^+$ induced by $\vec{W}$ we have $x \in X$ and $y \in Y$. Clearly, $A', B' \subseteq X$ and $A'', B'' \subseteq Y$. (A particular case is illustrated in figure 2.44 where the edges of $\vec{W}$ are heavy and the sets X and Y are encircled by dotted lines.) Now, according to Theorem 28 and to the definition of flow augmenting paths:

$$\sigma(f^+) = \kappa(\vec{W}) = \kappa(x, A'') + \kappa(B', y)$$

$$= \alpha(A'') + \beta(B'), \quad f^+(x, a) = \kappa(x, a) \quad \text{if} \quad a \in A'' \tag{7}$$

$$f^+(b, y) = \kappa(b, y) \quad \text{if} \quad b \in B', \tag{8}$$

$$f^+(a, b) = 0 \quad \text{if} \quad a \in A'' \quad \text{and} \quad b \in B'. \tag{9}$$

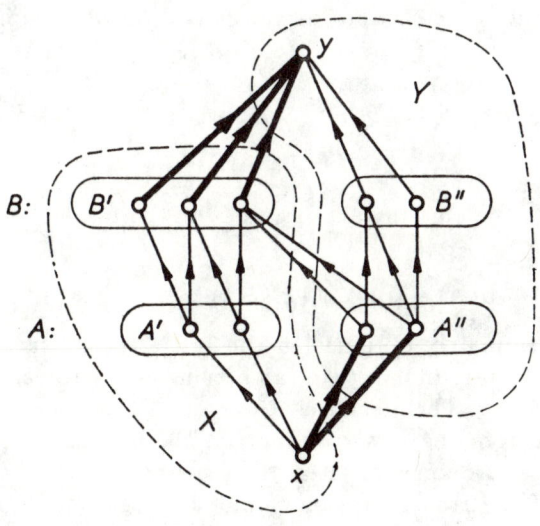

fig. 2.44

Since $\overrightarrow{W}$ contains no $\overrightarrow{AB}$-edge, the endpoint in B of each $\overrightarrow{A'B}$-edge is in B'. Therefore, $\overrightarrow{G}_0(A') \subseteq B'$. We show now that the two sets coincide. Let us assume the opposite, i.e. that $b_1 \in B'$ is the heads of $\overrightarrow{AB}$-edges only with their tails in A''. Recall that $\beta(b) > 0$ for all vertices $b \in B$. According to equation (8), we have $f^+(b_1, y) = \kappa(b_1, y) = \beta(b_1) > 0$. Therefore, there exists an edge (a_2, b_1) with $f^+(a_2, b_1) > 0$. But $a_2 \in A''$, contradicts (9).

Let us decompose $\delta(A')$ as follows:
$$\delta(A') = \alpha(A') - \beta(B') = \alpha(A) - \alpha(A'') - \beta(B'),$$
i.e.
$$\delta(A') = \alpha(A) - (\alpha(A'') + \beta(B')).$$
According to (7),
$$\delta(A') = \alpha(A) - \kappa(\overrightarrow{W}).$$

If, however, $A' \subseteq A$ is arbitrary and $B' = \overrightarrow{G}_0(A')$, then there is no $\overrightarrow{A'B''}$-edge in $\overrightarrow{G}_0^+$ and so $\alpha(A'') + \beta(B')$ yields the capacity of an edge $\overrightarrow{xy}$-cut in $\overrightarrow{G}_0^+$. Consequently, $\delta(A')$ is maximal if and only if $\alpha(A'') + \beta(B')$ is the capacity of an edge $\overrightarrow{xy}$-cut of minimal capacity. So, according to Theorem 28,
$$\sigma(f^+) = \alpha(A'') + \beta(B')$$
holds if and only if $\alpha(A'') + \beta(B')$ is the capacity of an edge $\overrightarrow{xy}$-cut of minimal capacity. Now, if
$$\sigma(f^+) = \alpha(A),$$
then f^+ immediately determines a transportation of minimal cost. Then, naturally,
$$\max \delta(A') = 0.$$
If, conversely,
$$\sigma(f^+) < \alpha(A),$$
then, establishing the partitions of A and B on the basis of an edge $\overrightarrow{xy}$-cut of minimal capacity,
$$\delta(A') = \alpha(A) - (\alpha(A'') + \beta(B')) = \alpha(A) - \sigma(f^+) > 0.$$

Consider the cost matrix obtained in the last step; let us subtract 1 from each of its rows corresponding to A' and let us add 1 to each of its columns corresponding to B'. This decreases the cost of the transportation in view of (4). Starting from the new cost matrix, let us construct the new graph $G_0 = G_0(A, B)$, etc. Sooner or later, the sequence of repetitions terminates, since the cost of the transportation cannot be negative after the decreases. However, on termination, the last xy-flow of maximal value simultaneously provides the required transportation.

Optimal Flows

So the procedure leading to a maximal transportation of minimal cost can be formulated as follows:

44. Algorithm *for searching a transportation of minimal cost in the complete bipartite graph* $G = G(A, B)$.

For the supply: $\alpha(a) > 0$ if $a \in A$; for the demands: $\beta(b) > 0$ if $b \in B$; and, further, $\alpha(A) = \beta(B)$.

(1) If all costs are positive in the ith row or column of the cost matrix $M_{m,n}$,

then $\begin{cases} \text{subtract 1 (or even the minimum, at once) from the } i\text{th row or} \\ \text{column of } M_{m,n}, \text{ and} \\ \text{return to (1).} \end{cases}$

Let the modified cost matrix be M_0.

(2) Let us define the subgraph $G_0 = G_0(A, B)$ of G as follows: an edge of G is included in G_0 if and only if its cost according to M_0 is zero.

(3) Let us construct the graph $\overrightarrow{G}_0^+$ by augmenting $\overrightarrow{G}_0(A, B)$ as follows: let us introduce the source x and sink y; connect an edge (x, a) to each vertex $a \in A$ and an edge (b, y) to each vertex $b \in B$.

Let the capacity function κ defined over the edges of $\overrightarrow{G}_0^+$ be the following:

$\kappa(a, b) = \infty$ for the AB-edges,
$\kappa(x, a) = \alpha(a)$ if $a \in A$ and
$\kappa(b, y) = \beta(b)$ if $b \in B$.

Let us find an xy-flow of maximal value in $\overrightarrow{G}_0^+$ (see Algorithm 29); let this be f^+.

If $\sigma(f^+) = \alpha(A)$, then jump to (5).

(4) Let us find an edge $\overrightarrow{xy}$-cut of minimal capacity in $\overrightarrow{G}_0^+$ (see Algorithm 29); let this be $\overrightarrow{W}$.

Let A'' be the set of the heads of the edges in $\overrightarrow{W}$ excluding y and let B' be the set of the tails of the edges in $\overrightarrow{W}$ excluding x. Let $A' = A - A''$.
Let us subtract 1 from each row of M_0 corresponding to a vertex in A'.
Let us add 1 to each column of M_0 corresponding to a vertex in B'.
Let the new matrix obtained from M_0 take over the role of M_0.
Return to (2).

(5) Record the result: a transportation of minimal cost is given by the following function f (let $a \in A$ and $b \in B$):

$$f(a, b) = \begin{cases} 0 & \text{if } \overrightarrow{G}_0 \text{ has no edge } (a, b) \\ f^+(a, b) & \text{if } (a, b) \text{ is an edge of } \overrightarrow{G}_0. \end{cases}$$

116   Graph Theory: Flows, Matrices

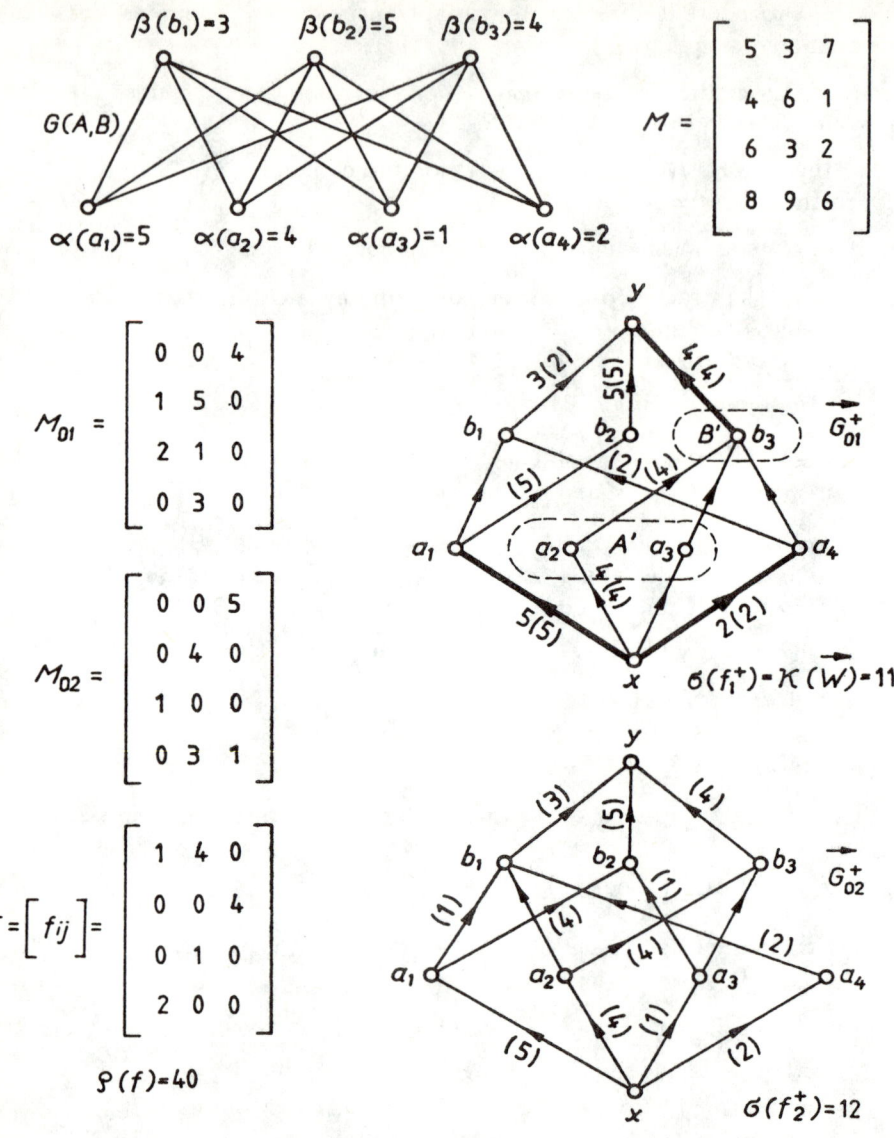

fig. 2.45

Let us follow the algorithm starting from the graph $G(A,B)$ and the cost matrix M shown in figure 2.45: let us subtract 3, 1, 2, 6 and 2 from rows 1, 2, 3, 4 and column 1 of M, respectively. This yields the modified cost matrix M_{01}. In the associated graph $\vec{G}_{01}^{+}$, f_1^+ is a flow of maximal value. Details of how to find it are omitted. The capacity values if different from ∞

and, in parentheses, the non-zero flow values have been written on the edges. $\sigma(f_1^+) = 11 < \alpha(A) = 12$. The set A' consists of a_2 and a_3, and the set B' b_3. 1 has been subtracted from rows 2 and 3 of the matrix M_{01} and added to its column 3. This led to the matrix M_{02}. In the associated graph $\vec{G}_{02}^+$, the flow f_2^+ yields a requested transportation f. The values of the function f are given by the matrix F as $f(a_i, b_j) = f_{ij}$.

Once the matrices $M = [\varrho_{ij}]$ and F are known, the cost of the minimal cost transportation f can be calculated

$$\varrho(f) = \sum_{i=1}^{4} \sum_{j=1}^{3} f_{ij} \cdot \varrho_{ij} = 40.$$

2.9 Minimal cost flows

A method has been found in §2.4 for establishing an xy-flow of maximal value limited by an edge-capacity function κ in a graph $\vec{G} = (P, E, \vec{\mathcal{G}})$. The xy-flow of maximal value is, in general, not unique. If a non-negative, real cost function ϱ is defined over the set E, the question is obvious: how to find, among the xy-flows of maximal value, one of minimal cost. So, an xy-flow f limited by κ is required with $\sigma(f)$ maximal and the *cost*

$$\varrho(f) = \sum_{e \in E} f(e) \cdot \varrho(e)$$

of f minimal. In the case of a non-negative, integer function κ, a procedure leading to f will be established by combining Algorithms 29 and 43. The essence of the procedure is the following: let us find an $\overline{xy}$-path of minimal cost in $\vec{G}$. This yields a minimal cost xy-flow of value 1. If, at some integer k, a minimal cost xy-flow f of value k could already be established then a path of minimal cost is selected from the flow augmenting paths relative to f. Augmenting f by this results in a minimal cost xy-flow of value $k+1$. This augmentation process terminates on arriving at a minimal cost xy-flow of maximal value. The interesting feature of the algorithm will be that, although formally we look for a minimal cost flow among the ones of maximal value, actually a sequence of flows of increasingly higher value will always be found among the minimal cost flows.

The augmentation procedure requires careful inspection. The xy-flow of the required maximal value σ is obtained by successively establishing minimal cost flows of value k with $k = 1, 2, \ldots, \sigma$. The reasoning becomes clearer by replacing $\vec{G}$ by the graph $\vec{G}' = (P, E', \mathcal{G}')$ by substituting each edge $e = (p, q)$ by $\kappa(p, q)$ edges (p, q) (this has also been used in proving Theorem

25 in the case of integer edge-capacity functions) and taking the value of the cost-function in $\vec{G}'$ to be $\varrho(e)$ on all of these edges. In other words, the value of the capacity-function in $\vec{G}'$ is 1 on any edge. So, the subgraph $\vec{H}$ in $\vec{G}'$ corresponding to the minimal cost xy-flow f of value k in $\vec{G}$ is formed by k edge-disjoint $\vec{xy}$-paths. Although in §2.4 $\vec{H}$ has only been shown to contain at least k edge-disjoint $\vec{xy}$-paths, recall that f is of minimal cost. This implies that the k edge-disjoint $\vec{xy}$-paths contain all edges of $\vec{H}$. Obviously, there is a flow augmenting path relative to f in $\vec{G}$ if and only if y is accessible in $\vec{G}'$ along a bypass path relative to $\vec{H}$. The cost of the flow f is

$$\varrho(f) = \varrho(\vec{H}).$$

How can we carry out this augmentation? Regarding the graph $\vec{G}'$, a bypass $\vec{xy}$-path $\vec{L}$ relative to $\vec{H}$ is required. This may include any edge E' not belonging to $\vec{H}$, but if the edge (p,q) is in $\vec{H}$ then it can only appear in $\vec{L}$ with opposite direction, as an edge (q,p). Now, if it is at all possible, let us augment so that the former edges in $\vec{L}$ are included in $\vec{H}$ and the edges in $\vec{H}$ corresponding to the latter ones are deleted. So, the cost of $\vec{H}$ is increased by the cost of the former ones and it is decreased by the cost of the latter ones. This augmentation and change of cost can also be established in the following way.

Considering $\vec{G}'$ and $\vec{H}$, let us introduce the graph $\vec{G}'(\vec{H})$ with its set of vertices being P and with the edges of $\vec{G}'(\vec{H})$ and the cost function ϱ' on them defined as follows: if an edge (p,q) in $\vec{G}'$ is not contained in $\vec{H}$, then let it be included in $\vec{G}'(\vec{H})$ with

$$\varrho'(p,q) = \varrho(p,q).$$

The edges of $\vec{H}$ should not be included in $G'(\vec{H})$, but each of these edges (p,q) should be substituted by exactly one edge (q,p) in $\vec{G}'(\vec{H})$ with

$$\varrho'(q,p) = -\varrho(p,q).$$

Thereupon, determine a minimal cost $\vec{xy}$-path $\vec{L}$ in $\vec{G}'(\vec{H})$. This yields a bypass path relative to $\vec{H}$ which is used as usual to augment $\vec{H}$ into $\vec{H}'$. By means of the graph $\vec{H}'$, an xy-flow of value $k+1$ is obtained in $\vec{G}$ with its cost being

$$\varrho(\vec{H}') = \varrho(\vec{H}) + \varrho'(\vec{L}).$$

The following points of the above process need clarification: Algorithm 43 is intended to be used for finding $\vec{L}$, but this is only possible if $\vec{G}'(\vec{H})$ contains no directed circuit of negative cost. Since the graph is not necessarily without edges of negative cost, there is a question whether $\vec{G}'(\vec{H})$ contains no directed circuit of negative cost? If there is no such circuit then Algorithm 43 does yield a minimal cost $\vec{xy}$-path in the graph and this can be used to augment $\vec{H}$ into $\vec{H}'$, but it is questionable whether the flow of value $k+1$ obtained by means of $\vec{H}'$ is of minimal cost. Since the answer to both questions will be favourable the procedure will successively provide minimal cost $\vec{xy}$-flows of values $1, 2, \ldots$ and, on termination, the flow obtained will be of maximal value.

The following is proved first:

45. *If $\vec{H}$ is a minimal cost subgraph of $\vec{G}'$ containing k edge-disjoint $\vec{xy}$-paths then $\vec{G}'(\vec{H})$ contains no directed circuit of negative value.*

Proof. Let us assume, indirectly, that

$$\varrho'(\vec{K}) < 0$$

for a circuit $\vec{K}$ in $\vec{G}'(\vec{H})$. (A particular case with $k = 3$ is shown in figure 2.46. Only the edges of $\vec{H}$ and $\vec{K}$ have been indicated here; those of $\vec{K}$ are heavy. If the non-directed edges are oriented in accordance with the direction from x towards y we obtain edges of $\vec{H}$, while with the opposite direction the edges are in $G'(\vec{H})$.) Let E_1 denote the set of those edges in $\vec{H}$ whose correspondents appear in $\vec{K}$ with opposite direction and let E_2 be formed by the edges in $\vec{K}$ not originating from E_1. Then

$$\varrho'(\vec{K}) = \varrho(E_2) - \varrho(E_1) < 0.$$

Let $\vec{H}^*$ be the subgraph of $\vec{G}'$ with its edges obtained from the edges of $\vec{H}$ by augmenting them with those in E_2 and deleting those in E_1. The vertices of $\vec{H}^*$ are the tails and heads of its edges (such as $\vec{H}^*$ obtained from figure 2.46 is shown in figure 2.47). It is easy to check for $\vec{H}^*$ (by means of the argument preceding Theorem 25) that

$$\varphi_{\text{out}}(x) - \varphi_{\text{in}}(x) = \varphi_{\text{in}}(y) - \varphi_{\text{out}}(y) = k, \tag{1}$$

and

$$\varphi_{\text{out}}(p) - \varphi_{\text{in}}(p) = 0 \tag{2}$$

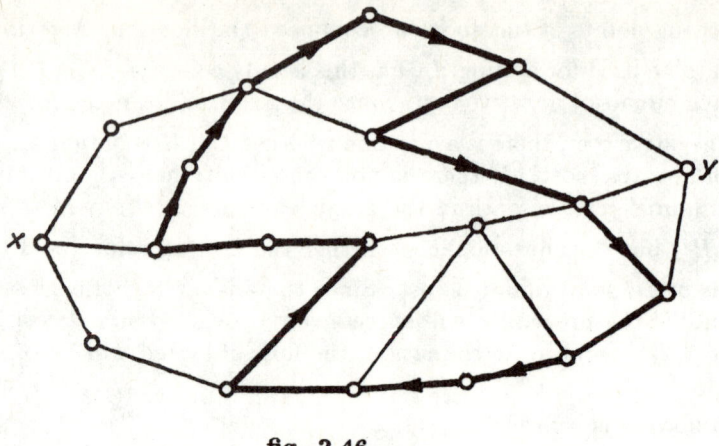

fig. 2.46

for all other vertices p. Hence $\vec{H}^*$ contains k edge-disjoint $\overline{xy}$-paths. Consider the costs of $\vec{H}^*$:

$$\varrho(\vec{H}^*) = \varrho(\vec{H}) + \varrho(E_2) - \varrho(E_1) < \varrho(\vec{H}).$$

But this contradicts that $\vec{H}$ is of minimal cost and hence completes the proof of Statement 45.

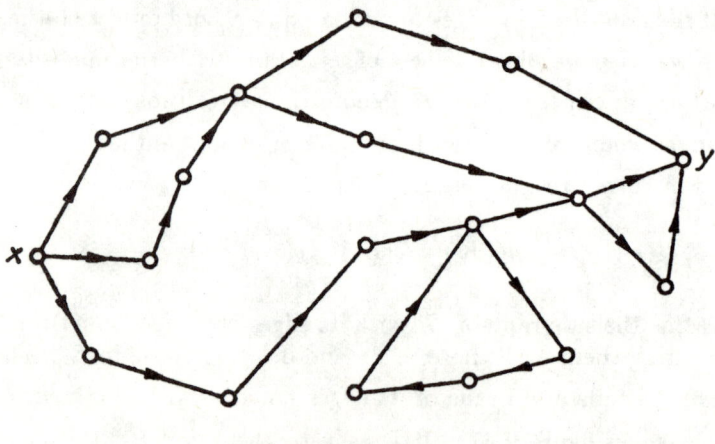

fig. 2.47

The proof of the following theorem requires much more consideration:

46. *If $\vec{H}$ is a minimal cost subgraph of $\vec{G}'$ containing k edge-disjoint $\overline{xy}$-paths and $\vec{H}$ is augmented into $\vec{H}'$ by means of the minimal cost $\overline{xy}$-flow $\vec{L}$*

in $\vec{G}'(\vec{H})$ then $\vec{H}'$ is of minimal cost among the subgraphs in $\vec{G}'$ containing $k+1$ edge-disjoint $\overline{xy}$-paths.

Proof. Let us assume, indirectly, that the subgraph $\vec{H}^*$ of $\vec{G}'$ is a minimal cost subgraph of $\vec{G}'$ containing $k+1$ edge-disjoint $\overline{xy}$-paths and

$$\varrho(\vec{H}^*) < \varrho(\vec{H}').$$

Since $\vec{H}^*$ is of minimal cost, it consists of $k+1$ edge-disjoint $\overline{xy}$-paths. Obviously, there is an edge in $\vec{H}'$ not included in $\vec{H}^*$. Let us delete the edges in $\vec{H}'$ included in $\vec{H}^*$ and let us reverse the direction of the rest of the edges in $\vec{H}'$. With 'flows' in mind, this results in establishing a circulation. More precisely, let $\vec{G}^*$ be the graph formed by the edges of $\vec{H}^*$ not in $\vec{H}'$ and the edges of $\vec{H}'$ not in $\vec{H}^*$ (with their directions opposite to their orientations in $\vec{H}'$), and by the tails and heads of these edges. Then for any vertex p in $\vec{G}^*$, the relationship (2) is satisfied. Hence $\vec{G}^*$ consists of edge-disjoint directed circuits. The cost of $\vec{H}^*$ being less than that of $\vec{H}'$, $\vec{G}^*$ contains such a circuit $\vec{K}^*$ that the sum of the costs of its edges in $\vec{H}^*$ is less than the sum of the costs of the edges in $\vec{H}'$ corresponding to the rest of its edges (i.e. those with the original directions). In other words, let $\vec{K}'$ denote the subgraph of $\vec{G}'$ obtained from $\vec{K}^*$ by taking its edges originating from $\vec{H}'$ with their original direction (see figure 2.48 where the edges in $\vec{H}^*$ are continuous lines) and, further, let E_0^* and E_0' be the sets of the edges in $\vec{K}'$ included in $\vec{H}^*$ and $\vec{H}'$, respectively. Then

$$\varrho(E_0^*) < \varrho(E_0'). \tag{3}$$

If all edges of E_0' are simultaneously included in $\vec{H}$ then let us omit the edges of E_0' from $\vec{H}$ and let us add to $\vec{H}$ the edges of E_0^* along with their tails and heads. Let $\vec{H}_0$ be the new graph thus obtained from $\vec{H}$. Check that (1) and (2) are also valid for $\vec{H}_0$ (for inspection, the graph $\vec{K}'$ is shown in figure 2.48). Consequently, $\vec{H}_0$ contains k edge-disjoint $\overline{xy}$-paths. However, according to (3),

$$\varrho(\vec{H}_0) < \varrho(\vec{H}),$$

contrary to the fact that $\vec{H}$ is of minimal cost.

Now, we can suppose that there exists an edge in E_0' simultaneously included in $\vec{L}$. Let E_{01}' be the set of these edges of E_0' also contained in $\vec{L}$

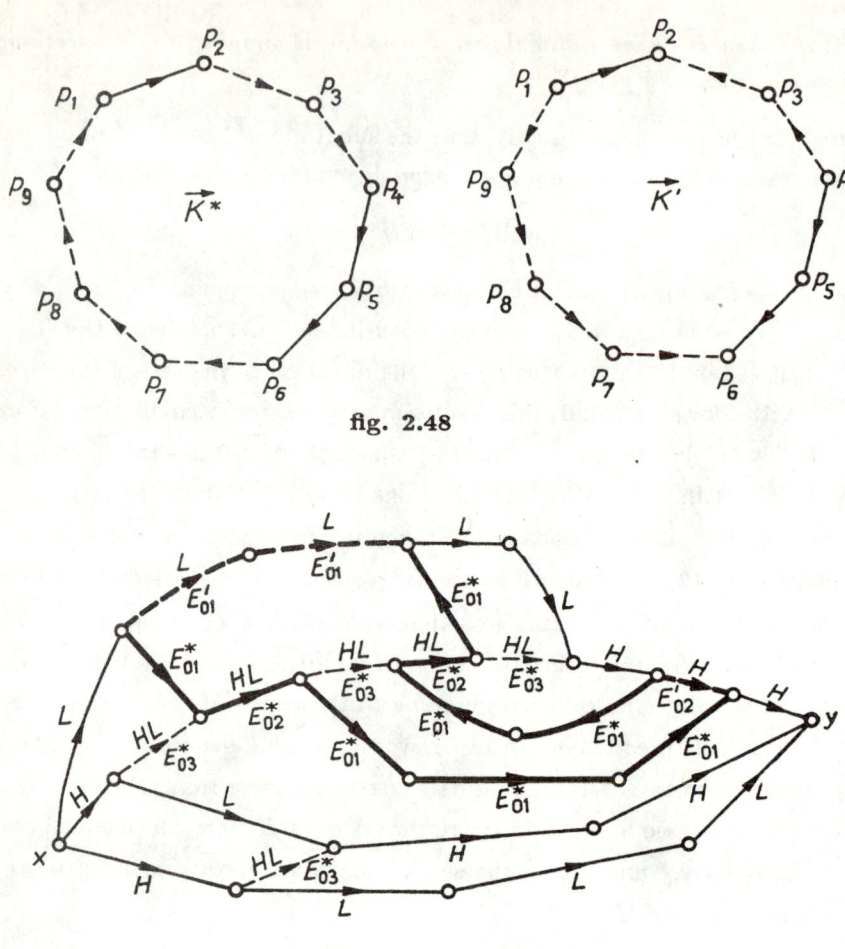

fig. 2.48

fig. 2.49

and E'_{02} be the set of edges of E'_0 not included in $\vec{L}$. (For example, $k = 2$ and the edges of $\vec{K'}$ are heavy in figure 2.49. The edges in $\vec{H}$ and $\vec{L}$ are denoted by H and L, respectively; the edges marked by HL are included in $\vec{L}$, naturally with opposite direction.) The edges in E'_{02} appear in $\vec{K}^*$ with opposite directions; let them form the set $\overline{E}'_{02}$ with these directions. There may be edges present in $\vec{H}$ among the edges in E_0^* which have also appeared in $\vec{L}$ with opposite directions (the heavy ones marked HL in the diagram), let these constitute the set E_{02}^* and let $E_{01}^* = E_0^* - E_{02}^*$. The edges of E_{02}^* appear in $\vec{L}$ with opposite directions; let them form the set $\overline{E}_{02}^*$. Let us further consider the edges in $\vec{L}$ appearing in $\vec{H}$ with opposite directions. From among these, let us omit those in $\overline{E}_{02}^{*}$ and let the set of the remaining ones be $\overline{E}_{03}^{*}$. Let us now construct the subgraph $\overrightarrow{L}^{*}$ of $\overrightarrow{G}'(\overrightarrow{H})$ from $\overrightarrow{L}$ as follows: let us delete the edges of $E_{01}' \cup \overline{E}_{02}^{*}$ from $\overrightarrow{L}$ and let us add the edges of $E_{01}^{*} \cup \overline{E}_{02}'$ to L. The vertices of $\overrightarrow{L}^{*}$ are the necessary tails and heads. (This $\overrightarrow{L}^{*}$ obtained from figure 2.49 is shown in figure 2.50.) The following statements can easily be verified: all edges in $\overrightarrow{L}^{*}$ are also in $\overrightarrow{G}'(\overrightarrow{H})$,

$$\varphi_{\text{out}}(x) - \varphi_{\text{in}}(x) = \varphi_{\text{in}}(y) - \varphi_{\text{out}}(y) = 1,$$

in $\overrightarrow{L}^{*}$, and

$$\varphi_{\text{out}}(p) = \varphi_{\text{in}}(p)$$

for any other vertex p. Hence, $\overrightarrow{L}^{*}$ includes at least one $\overrightarrow{xy}$-path, let $\overrightarrow{L}_{0}^{*}$ be one. If $\overrightarrow{L}_{0}^{*}$ does not exhaust the edges of $\overrightarrow{L}^{*}$ then some directed circuits $\overrightarrow{L}_{1}^{*}, \overrightarrow{L}_{2}^{*}, \ldots$, can be further found in it and, in the case $i \neq j$, $\overrightarrow{L}_{i}^{*}$ and $\overrightarrow{L}_{j}^{*}$ are edge-disjoint. Let us consider the costs. The inequality (3) can be written as follows:

$$\varrho(E_{01}^{*}) + \varrho(E_{02}^{*}) < \varrho(E_{01}') + \varrho(E_{02}'),$$

i.e.

$$\varrho(E_{01}') - \varrho(E_{02}^{*}) > \varrho(E_{01}^{*}) - \varrho(E_{02}'). \tag{4}$$

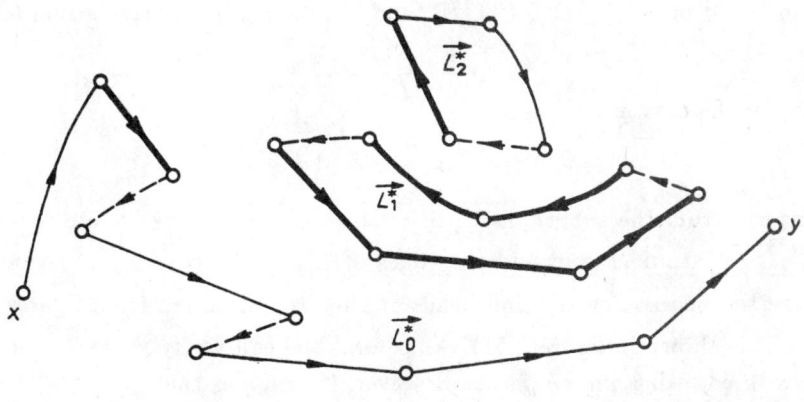

fig. 2.50

According to the relationship between the costs ϱ in $\overrightarrow{G}'$ and ϱ' in $\overrightarrow{G}'(\overrightarrow{H})$:

$$\varrho(E_{01}') = \varrho'(E_{01}'), \qquad -\varrho(E_{02}^{*}) = \varrho'(\overline{E}_{02}^{*}),$$
$$\varrho(E_{01}^{*}) = \varrho'(E_{01}^{*}), \qquad -\varrho(E_{02}') = \varrho'(\overline{E}_{02}').$$

Using these, (4) yields:
$$\varrho'(E'_{01}) + \varrho'(\overline{E}^*_{02}) > \varrho'(E^*_{01}) + \varrho(\overline{E}'_{02}),$$
and hence
$$\varrho'(\overrightarrow{L}) > \varrho'(\overrightarrow{L}^*). \tag{5}$$

Now, if $\overrightarrow{L}^*$ is the only $\overrightarrow{xy}$-path then (5) contradicts the fact that $\overrightarrow{L}$ is a minimal cost $\overrightarrow{xy}$-path in the graph $\overrightarrow{G}'(\overrightarrow{H})$. If $\overrightarrow{L}^*$ is not the only $\overrightarrow{xy}$-path then it will now be proved that

$$\varrho'(\overrightarrow{L}^*_i) \geq 0 \quad \text{if } i \neq 0. \tag{6}$$

Hence,
$$\varrho'(\overrightarrow{L}) > \varrho'(\overrightarrow{L}^*_0)$$
is obtained, again a contradiction.

Indirectly assume that
$$\varrho'(\overrightarrow{L}^*_i) < 0 \tag{7}$$

for some i. This will imply that $\overrightarrow{H}$ is not of minimal cost, a contradiction. Let E_{04} denote the set of edges of $\overrightarrow{L}$ also belonging to $\overrightarrow{H}'$ but not to E'_{01}. Now, the edges of $\overrightarrow{L}^*_i$ can only belong to the following sets: E^*_{01}, $\overline{E}'_{02}$, $\overline{E}^*_{03}$ and E_{04}. Let us denote the subsets formed by the elements of these sets and also included in $\overrightarrow{L}^*_i$ by E^*_{i1}, $\overline{E}'_{i2}$, $\overline{E}^*_{i3}$ and E_{i4}, respectively. According to (7):

$$\varrho'(E^*_{i1}) + \varrho'(\overline{E}'_{i2}) + \varrho'(\overline{E}^*_{i3}) + \varrho'(E_{i4}) < 0.$$

Then, for the cost ϱ:
$$\varrho(E^*_{i1}) + \varrho(E_{i4}) < \varrho(\overline{E}'_{i2}) + \varrho(\overline{E}^*_{i3}). \tag{8}$$

Let us construct the subgraph $\overrightarrow{H}_i$ of $\overrightarrow{G}'$ from $\overrightarrow{H}$ as follows: delete the edges of $\overline{E}'_{i2} \cup \overline{E}^*_{i3}$ from $\overrightarrow{H}$ and add the edges of $E^*_{i1} \cup E_{i4}$ to $\overrightarrow{H}$. The vertices of $\overrightarrow{H}_i$ are the necessary tails and heads. (This $\overrightarrow{H}_1$ obtained from figures 2.49 and 2.50 is shown in figure 2.51). As before, one can verify that $\overrightarrow{H}_i$ contains at least k edge-disjoint $\overrightarrow{xy}$-paths. However, (8) implies that

$$\varrho(\overrightarrow{H}_i) < \varrho(\overrightarrow{H}),$$

i.e. the aforementioned contradiction arises.

This concludes the proof of Theorem 46, and the problems arising in connection with the augmentation process have also been clarified. While the reasoning was made clearer by changing the original graph $\overrightarrow{G}$ to $\overrightarrow{G}'$, but the application of the procedure has been unquestionably complicated. However,

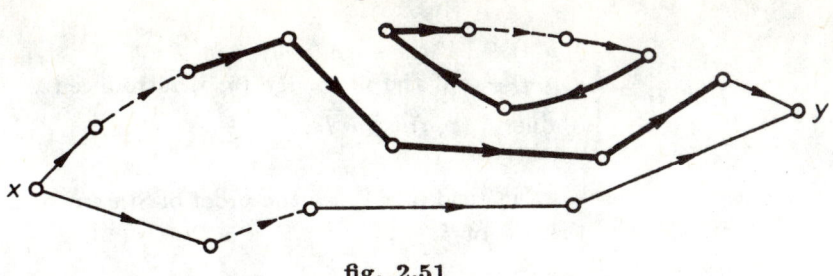

fig. 2.51

the augmentation can also be carried out in the original graph. It is not difficult to verify that our reasoning confirms the procedure below applicable to the graph $\vec{G}$:

47. Algorithm *for constructing minimal cost flows.*

The graph $\vec{G} = (P, E, \vec{\mathcal{G}})$ is given with a source x and a sink y. The positive, integer edge-capacity function κ and the non-negative cost function ϱ are defined over the set E. (The edges with zero capacity are deleted.)

(1) If there is no $\overrightarrow{xy}$-path in $\vec{G}$ then the only xy-flows are of zero value and cost, so the procedure has ended.

(2) Let us find a minimal cost $\overrightarrow{xy}$-path $\vec{L}$ in $\vec{G}$ (see Algorithm 43).
Let us define the function f over the set E as follows:

$$f(e) = \begin{cases} 1 & \text{if } e \text{ is an edge of } \vec{L} \\ 0 & \text{otherwise.} \end{cases}$$

(3) Let us construct the graph $\vec{G}(f) = (P, E', \vec{\mathcal{G}}')$ from $\vec{G}$ as follows.

If $f(e) = 0$, then $e \in E'$.

If $f(p, q) = \kappa(p, q)$

then $\begin{cases} (p,q) \notin E', \text{ but let us introduce a new edge } (q,p) \\ \text{and} \\ \text{let } (q,p) \in E'. \end{cases}$

If $0 < f(p, q) < \kappa(p, q)$

then $\begin{cases} (p,q) \in E', \text{ and} \\ \text{let us also introduce a new edge } (q,p) \in E'. \end{cases}$

Let the cost function ϱ' in $\vec{G}(f)$ be defined as

$$\varrho'(e) = \begin{cases} \varrho(e) & \text{if } e \in E, \\ -\varrho(e) & \text{if } e \notin E, \end{cases}$$

If there is no $\overrightarrow{xy}$-path in $\vec{G}(f)$, then jump to (5).

(4) Let us find a minimal cost $\overrightarrow{xy}$-path $\vec{L}$ in $\vec{G}(f)$.
Let us define the function f' over the set E as follows:

$$f'(p,q) = \begin{cases} f(p,q) + 1 \text{ if the edge } (p,q) \text{ is in } \overrightarrow{L}; \\ f(p,q) - 1 \text{ if the new edge } (q,p) \text{ introduced} \\ \qquad \text{due to } (p,q) \text{ is in } \overrightarrow{L}; \\ 0 \text{ otherwise.} \end{cases}$$

(One can also take ε_0 instead of ± 1, see the proof of Statement 26.)

Let f' take over the role of f.

Return to (3).

(5) Record the result: the function f yields a minimal cost xy-flow of maximal value in $\overrightarrow{G}$.

Remark. If the edge-capacity values are integers, then — just as at Algorithm 29 — the following upper bound can be given for the number of steps:

$$|P|^2 \cdot \sum_{e \in E} \kappa(e).$$

Here, the use of the shortest flow augmenting path is not enough to eliminate the effect of the edge-capacities. J Edmonds and R M Karp have given an algorithm [12] with its number of steps independent of the capacity values.

Algorithm 47 is based on Theorem 46. Results essentially equivalent to this theorem are due to W S Jewell, R G Busacker and P J Gowen, and R G Busacker and T L Saaty ([30], [31] and [32] Theorems 7–8). D R Fulkerson was the first to solve the problem by means of linear programming techniques [33]. It can be shown that the problem of finding a minimal cost flow is equivalent to the minimal cost transportation problem discussed in §2.8 (see [11], III. §4; in [11], §7 and §8 show the connection with other applications as well).

The majority of the problems treated in this chapter can also be solved by linear programming techniques (see in [11]). Minimax theorems connected with or related to several flow problems can be formulated. (Beside those above see, for example, [34] and [35].) It can be shown (although not very easily) that most of these theorems are reducible to a fundamental result, the so-called duality theorem of linear programming.

Many algorithms have been devised for a multitude of flow problems. Some of them can be found in chapter IX of the book of B Roy [36]. The books of T C Hu [37] and E Lawler [38] are also recommended in connection with this field.

2.10 Problems

48. A minimal vertex xy-cut in a graph G consists of the vertex z alone. Show that z is an articulation of G.

49. Let x_1, x_2 and x_3 be three vertices of a connected graph G. Assume that there are no two vertex-disjoint paths in the set H of $x_i x_j$-paths ($i, j = 1, 2, 3$; $i \neq j$) of G. Prove that there are two vertices of G (but usually not fewer) so that at least one of them is contained in any path of H as an inner vertex.

50. Let $x_1, x_2, \ldots, x_k$ be k vertices ($k \geq 4$) of a connected graph G. Assume that there are no two vertex-disjoint paths in the set H of $x_i x_j$-paths ($i, j = 1, 2, \ldots, k$; $i \neq j$) in G. In view of Theorem 1 and of the previous problem it is surprising that there is a vertex in G which is an inner vertex of any path of H. Prove this.

51. In a graph $G = (P, E, \mathcal{G})$, the minimal number of xy-cut edges is k. Let H denote the subgraph of G formed by k of its edge-disjoint xy-paths, and let P_x be the set of vertices accessible from x along a bypass path relative to H. Prove that if a minimal edge xy-cut in G induces the partition $\{P_1, Q_1\}$ of the set P with $x \in P_1$, then $P_x \subseteq P_1$.

52. The minimal edge xy-cuts W_1 and W_2 of a graph $G = (P, E, \mathcal{G})$ induce the partitions $\{P_1, Q_1\}$ and $\{P_2, Q_2\}$ of the set P, respectively, with $x \in P_i$ ($i = 1, 2$). Let $P_3 = P_1 \cap P_2$, $Q_3 = P - P_3$, $P_4 = P_1 \cup P_2$ and $Q_4 = P - P_4$. Show that both the set of $P_3 Q_3$-edges and the set of $P_4 Q_4$-edges constitute a minimal edge xy-cuts in G.

53. A *graph* G is called *n-connected* if it is connected, the number of its vertices is at least $n + 1$ and, deleting fewer than n vertices in any manner from G, the remaining graph remains connected. So, for graphs containing at least two vertices, the terms '1-connected' and 'connected' have the same meaning. Let, at $n \geq 2$, p and q be two arbitrary vertices of an n-connected graph G. Prove that by deleting the edges $\{p, q\}$ from G (with the vertices and the other edges retained), the graph obtained will be $(n - 1)$-connected.

54. Show that for any two vertices p and q of an n-connected graph G, there are n vertex-disjoint pq-paths in G.

55. Prove that for an arbitrary set $x, y, z_1, z_2, \ldots, z_{n-1}$ of vertices in an n-connected graph G, there is an xy-path in G containing all the $n-1$ vertices z_i.

56. Let x and y be two vertices of an n-connected graph G and let A and B be vertex xy-cuts, each of n elements. Let X denote the set of those vertices in $A \cup B$ which can be reached from x along a path P in G so that P does not contain any other vertex of $A \cup B$. Prove that $|X| = n$ (Lovász [39], 6.58).

57. Deduce Theorem 8 from Theorem 10.

58. Show that a graph G with at least $2n$ vertices ($n \geq 2$) is n-connected if and only if, for any disjoint n-element subsets A and B of its vertex set, there are n independent AB-paths in G (Dirac's theorem, [6]).

59. An $\overleftrightarrow{xy}$-path of a directed graph is a path which is either an $\overrightarrow{xy}$-path or a $\overrightarrow{yx}$-path. Any two $\overleftrightarrow{xy}$-paths of a graph $\overrightarrow{G}$ have a common edge. Is it true that $\overrightarrow{G}$ has an edge included in all $\overleftrightarrow{xy}$-paths of $\overrightarrow{G}$?

60. In a bipartite graph $G = G(A, B)$ the degree of all vertices is k ($k \geq 1$). Prove the existence of such a partition $\{E_1, E_2, \ldots, E_k\}$ of the edge set of G where any vertex of G is incident to exactly one edge of each set E_i.

61. Persons at a firm are qualified for several jobs and several persons take part in any job simultaneously. The firm needs certain data concerning all jobs of each person. The data can be collected in two basic ways: all employees get a questionnaire, or one questionnaire is set up at each job-site to be filled out by those working there. The two procedures can be combined. Devise a plan enabling the firm to collect the data with the aid of a minimal number of questionnaires.

62. Prove that if there are more than n such bands of a matrix $M = M_{n,n}$ that elements included in at least two of these bands are all zero, then $\det M = 0$.

63. Prove the following reverse of the statement in the previous problem: if all expansion terms (in the determinant) of a matrix $M = M_{n,n}$ are zero then there are more than n bands of M so that elements included in at least two of these bands are zero.

64. A firm intends to assign five of its employees to six kinds of jobs with minimising the sum of the percentage of faulty products by them. The percentages of faulty products are shown in the matrix below: the jth entry in the ith row is the percentage of faulty products by the ith employee in the jth job. Design the plan of assignment.

$$\begin{bmatrix} 2 & 15 & 100 & 11 & 100 & 2 \\ 3 & 100 & 100 & 6 & 4 & 100 \\ 100 & 6 & 4 & 5 & 18 & 2 \\ 3 & 9 & 100 & 7 & 100 & 100 \\ 100 & 12 & 5 & 100 & 5 & 100 \end{bmatrix}$$

65. In a graph $\overrightarrow{G} = (P, E, \overrightarrow{\mathcal{G}})$, $\overrightarrow{W}_1$ and $\overrightarrow{W}_2$ are edge $\overrightarrow{xy}$-cuts of minimal capacity for a given edge-capacity function. They induce the partitions $\{P_1, Q_1\}$ and $\{P_2, Q_2\}$ of P, respectively, with $x \in P_i$ ($i = 1, 2$). Let $P_3 = P_1 \cap P_2$, $Q_3 = P - P_3$, $P_4 = P_1 \cup P_2$ and $Q_4 = P - P_4$. Prove that both the set of $\overrightarrow{P_3 Q_3}$-edges and the set of $\overrightarrow{P_4 Q_4}$-edges constitute edge $\overrightarrow{xy}$-cuts of minimal capacity (cf. Problem 52).

66. In a graph $\overrightarrow{G} = (P, E, \overrightarrow{\mathcal{G}})$ let f be an xy-flow of maximal value limited by the edge-capacity function κ and let X be the set of vertices accessible along a flow relative to f. Let the edge $\overrightarrow{xy}$-cut $\overrightarrow{W}_1$ of minimal capacity in $\overrightarrow{G}$ induce the partition $\{X_1, Y_1\}$ of P with $x \in X_1$. Prove that $X \subseteq X_1$ (cf. Problem 51).

67. In a graph $\overrightarrow{G} = (P, E, \overrightarrow{\mathcal{G}})$, f is an xy-flow of maximal value limited by a given edge-capacity function and X is the set of vertices accessible along a flow relative to f. Show that X is independent of the choice of f (among the xy-flows of maximal value).

68. Let κ_1 be a positive edge-capacity function in a graph $\overrightarrow{G}_1 = (P, E_1, \overrightarrow{\mathcal{G}}_1)$, f_1 be an xy-flow of maximal value limited by κ_1, X_1 be the set of vertices accessible along a flow relative to f_1 and let $Y_1 = P - X_1$. The graph $\overrightarrow{G}_2 = (P, E_2, \overrightarrow{\mathcal{G}}_2)$ is derived from $\overrightarrow{G}_1$ by reversing the orientation of every edge of $\overrightarrow{G}_1$. The edge-capacity function κ_2 in $\overrightarrow{G}_2$ is defined as follows: if the edge (p, q) of E_2 has been obtained from the edge (q, p) of E_1 by reversing the orientation then

$$\kappa_2(p, q) = \kappa_1(q, p).$$

In the graph $\overrightarrow{G}_2$ let f_2 be a yx-flow of maximal value limited by κ_2, Y_2 be the set of vertices accessible along a flow relative to f_2 and let $X_2 = P - Y_2$. Prove that the edge $\overrightarrow{xy}$-cut of minimal capacity in $\overrightarrow{G}_1$ is unique if and only if the set of $\overrightarrow{X_1 Y_1}$-edges coincides with the set of $\overrightarrow{X_2 Y_2}$-edges.

69. The numbers written on the edges in figure 2.52 are the values of the edge-capacity function. Find an xy-flow of maximal value and an edge $\overrightarrow{xy}$-cut of minimal capacity in the graph.

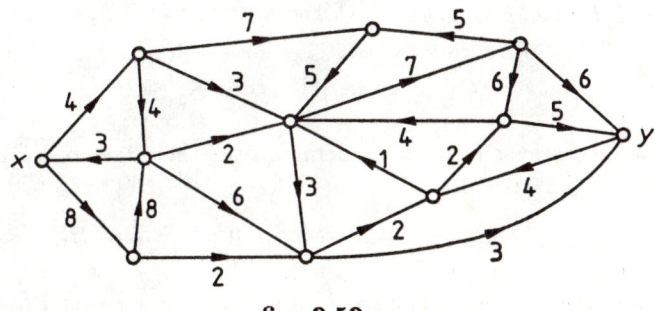

fig. 2.52

70. Let x and y be two fixed vertices of a graph $\overrightarrow{G}$ and $\kappa \geq 0$ the edge-capacity function in $\overrightarrow{G}$. Let $\overrightarrow{L}_1, \overrightarrow{L}_2, \ldots$ denote the $\overrightarrow{xy}$-paths in $\overrightarrow{G}$, let E_i be the set of edges in $\vec{L}_i$ $(i = 1, 2, \ldots)$ and $\vec{W}_1, \vec{W}_2, \ldots$ be the edge $\vec{xy}$-cuts in $\vec{G}$. Prove that

$$\max_i \min_{e \in E_i} \kappa(e) = \min_j \max_{e \in \vec{W}_j} \kappa(e).$$

(J Edmonds and D R Fulkerson; see [40].)

71. Solve the mobilisation problem sketched in figure 2.53. The numbers written on the edges are the capacities and those in parentheses are the crossing times. Determine the minimal time required for the concentration into y_1 and find the mobilisation plan executable in minimal time.

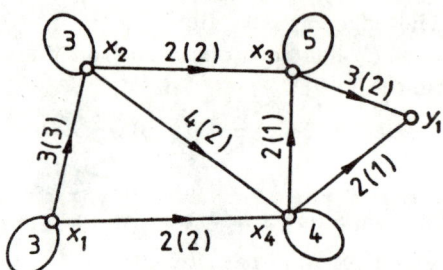

fig. 2.53

72. In a bipartite graph $\vec{G} = \vec{G}(A, B)$ let the non-negative, integer functions α, β and κ be defined over the subsets A and B of vertices and over the set of edges of $\vec{G}$, respectively. Prove that the requirements

$$0 \leq f(x, y) \leq \kappa(x, y) \quad \text{(for every edge } (x, y) \text{ in } \vec{G}\text{)},$$
$$f(x, B) \geq \alpha(x) \quad (x \in A),$$
$$f(A, y) \leq \beta(y) \quad (y \in B)$$

are feasible by an integer function f defined over the edges of $\vec{G}$ if and only if for any sets $A_1 \subseteq A$ and $B_1 \subseteq B$:

$$\alpha(A_1) \leq \beta(B_1) + \kappa(A_1, B_2) \qquad \text{if } B_2 = B - B_1.$$

73. The non-negative, integer functions κ_1, κ_2 and f defined over set E of edges of a graph $\vec{G} = (P, E, \vec{\mathcal{G}})$ satisfy

$$f(p, P) - f(P, p) = 0 \qquad \text{if } p \in P,$$
$$\kappa_1(p, q) \leq f(p, q) \leq \kappa_2(p, q) \qquad \text{if } (p, q) \in E.$$

Prove that for any partition $\{X, Y\}$ of P:

$$f(X, Y) = f(Y, X)$$
$$\kappa_1(X, Y) \leq \kappa_2(Y, X).$$

74. Let the non-negative integer function α be defined over the set of vertices of the complete graph G with n vertices. We intend to construct a graph $\vec{G}$ by directing the graph G with

$$\varphi_{\text{out}}(p) = \alpha(p)$$

for each vertex p. Prove that this is possible if and only if

$$\sum_{i=1}^{k} \alpha_i \leq \binom{k}{2} + k(n-k) \quad \text{if } k = 1, 2, \ldots, n-1 \text{ and } \sum_{i=1}^{n} \alpha_i = \binom{n}{2},$$

provided the subscripts determine a decreasing order, i.e. $\alpha_1 \geq \alpha_2 \geq \ldots \geq \alpha_n$.

75. Five families consist of 8, 8, 5, 5 and 4 persons, respectively. The families organise a joint car-trip. They have ten cars at their disposal. The capacities of the cars are 5, 5, 4, 4, 4, 3, 2, 1, 1, 1, persons, respectively, in addition to the luggage area. The thirty persons wish, in taking their places in the cars, to have no two members of the same family sit in the same car. The question is whether this arrangement is feasible and, if it is, devise a plan realising it.

76. Find an $\overrightarrow{xy}$-path of maximal value in the graph shown in figure 2.54. The indicated numbers are the values of the edges. How can the ideas of Algorithm 43 be used without changing the signs of the values of the edges?

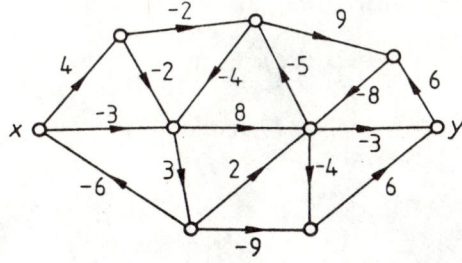

fig. 2.54

77. A transportation is to be executed from the supplies 4, 5, 3 and 9 to satisfy the demands 3, 3, 6, 2, 1 and 2. The cost matrix associated with the transportation is:

$$\begin{bmatrix} 5 & 3 & 7 & 3 & 8 & 5 \\ 5 & 6 & 12 & 5 & 7 & 11 \\ 2 & 8 & 3 & 4 & 8 & 2 \\ 9 & 6 & 10 & 5 & 10 & 9 \end{bmatrix}.$$

Find a minimum cost transportation plan satisfying the demands. For the 'manual calculations', speed up Algorithm 44 by subtracting, if possible, more than one from the rows or adding to the columns of the cost matrix.

78. Find a minimum cost xy-flow of maximal value in the graph shown in figure 2.55. The flow is limited by both a vertex- and an edge-capacity function. The numbers written in the vertices are the vertex-capacity values, the first numbers in the pairs written on the edges are the edge-capacity values and the second ones are the costs.

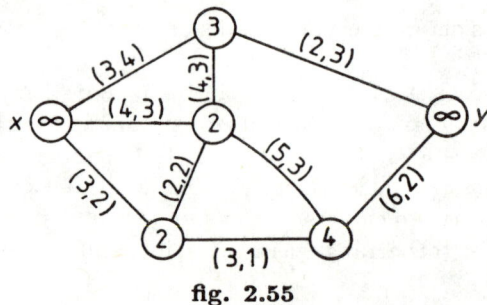

fig. 2.55

79. Let the positive, integer edge-capacity function κ and the non-negative cost function ϱ be defined over the set of edges of a graph $\vec{G}$ and let the integer function f limited by κ be a minimum cost xy-flow of value $\sigma > 0$ in $\vec{G}$. Prove that provided there is a minimum cost integer xy-flow of value σ limited by κ which differs from f then there is a directed circuit $\vec{K}$ in the graph $\vec{G}(f)$ defined in Algorithm 47 with

$$\varrho'(\vec{K}) = 0$$

where ϱ' is the cost function in $\vec{G}(f)$.

3

Graphs and Matrices

Certain properties of graphs can be investigated with the help of matrices in various ways. This has already been shown by the cost matrix introduced in §2.8 of the previous chapter, and the matrix shown in figure 2.11 in §2.3 directly indicates how a matrix is usually associated with a bipartite graph. The matrix defined below expresses the adjacency of the vertices of a graph in a similar fashion.

3.1 The adjacency matrix

The vertices of a graph G are $p_1, p_2, \ldots, p_c$ and the number of edges in G connecting p_i and p_j is a_{ij}. The square matrix $A = [a_{ij}]$ of order c is called the *adjacency matrix* of G or simply the matrix associated with G. In the case of a directed graph $\overrightarrow{G}$, the element a_{ij} of the adjacency matrix $A = [a_{ij}]$ denotes the number of edges with p_i as their tail and p_j as their head.

As an example, consider the graph $\overrightarrow{G}$ shown in figure 3.1. Let the adjacency matrices of G and $\overrightarrow{G}$ be $A = [a_{ij}]$ and $\overline{A} = [\overline{a}_{ij}]$, respectively. Then

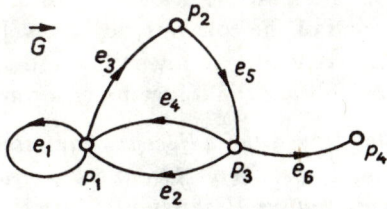

fig. 3.1

$$A = \begin{bmatrix} 1 & 1 & 2 & 0 \\ 1 & 0 & 1 & 0 \\ 2 & 1 & 0 & 1 \\ 0 & 0 & 1 & 0 \end{bmatrix} \quad \text{and} \quad \overline{A} = \begin{bmatrix} 1 & 1 & 0 & 0 \\ 0 & 0 & 1 & 0 \\ 2 & 0 & 0 & 1 \\ 0 & 0 & 0 & 0 \end{bmatrix}.$$

The adjacency matrix of G is clearly symmetric, and the degree of the vertex p_i is

$$\varphi(p) = 2a_{ii} + \sum_{\substack{j=1 \\ i \neq j}}^{c} a_{ij} \quad (i = 1, 2, \ldots, c),$$

and in case of $\overrightarrow{G}$, the indegree and outdegree of the vertex p_i are

$$\varphi_{\text{out}}(p_i) = \sum_{j=1}^{c} \overline{a}_{ij}$$

and

$$\varphi_{\text{in}}(p_i) = \sum_{j=1}^{c} \overline{a}_{ji} \quad (i = 1, 2, \ldots, c),$$

respectively. If the adjacency matrices of $\overrightarrow{G}$ and of G (obtained from $\overrightarrow{G}$ by disregarding the orientations) are $[\overline{a}_{ij}]$ and $[a_{ij}]$, respectively, then $a_{ij} = a_{ji} = \overline{a}_{ij} + \overline{a}_{ji}$ in the case $i \neq j$, otherwise $a_{ii} = \overline{a}_{ii}$.

The adjacency matrix clearly defines the graph up to isomorphism. Conversely, different adjacency matrices correspond to the same graph if the vertices of the graph are numbered differently. However, the matrices thus obtained differ only in as much as any of them can be constructed from any other by rearranging some rows and, in the same way, the corresponding columns. If two graphs are isomorphic, then renumbering the vertices of one to those of the other according to an isomorphic projection results in identical adjacency matrices. Accordingly, two graphs are isomorphic if and only if the adjacency matrix of one can be constructed from that of the other by identical rearrangement of rows and columns. The rows and columns of a matrix can be rearranged by multiplying it by a permutation matrix. It is well known that if a permutation matrix P has been obtained from the unit matrix E by rearrangement of rows then the multiplications PA and AP^* give the same rearrangement of the rows and of the columns, respectively, of A. Recall further that a permutation matrix is always invertible and its inverse coincides with its transpose. Hence we obtain the following theorem:

1. *The matrices A_1 and A_2 are the adjacency matrices of the graphs (or directed graphs) G_1 and G_2, respectively. G_1 and G_2 are isomorphic if and only if there is a permutation matrix P satisfying $A_1 = P^{-1} A_2 P$.*

The two theorems below can also be proved without difficulty.

2. *The graph or directed graph with the adjacency matrix A is disconnected or not strongly connected if and only if there exist a permutation matrix P, two square matrices A_{11} and A_{22} and two zero matrices N_{12} and N_{21} satisfying*

$$P^{-1}AP = \begin{bmatrix} A_{11} & N_{12} \\ N_{21} & A_{22} \end{bmatrix} \quad \text{and} \quad P^{-1}AP = \begin{bmatrix} A_{11} & N_{12} \\ A_{21} & A_{22} \end{bmatrix},$$

respectively.

3. *A graph with adjacency matrix A is bipartite if and only if there exist a permutation matrix P, a square matrix A_{12} and two zero matrices N_{11} and N_{22} satisfying*

$$P^{-1}AP = \begin{bmatrix} N_{11} & A_{12} \\ A_{12}^* & N_{22} \end{bmatrix}.$$

The cardinalities of the two independent subsets of vertices of the bipartite graph are clearly equal the number of rows and of columns in A_{12}, respectively. Since A_{12} contains every information about the entire adjacency matrix, the adjacency matrix of a bipartite graph is often considered to be A_{12} only. This was mentioned in the second introductory sentence of this chapter.

The further discussion is simplified by the following concepts: the sequence of the vertices $p_0, p_1, p_2, \ldots, p_m$ and of the edges $\{p_0, p_1\}, \{p_1, p_2\}, \{p_2, p_3\}, \ldots, \{p_{m-1}, p_m\}$ of a graph G together is called a $p_0 p_m$-*edge-sequence of length* m in G. Similarly, the sequence of the vertices $p_0, p_1, p_2, \ldots, p_m$ and of the edges $(p_0, p_1), (p_1, p_2), (p_2, p_3), \ldots, (p_{m-1}, p_m)$ of $\vec{G}$ together is called a *directed $p_0 p_m$-edge-sequence of length* m in a graph $\vec{G}$. Both vertices and edges may be repeated in an edge-sequence. If all the edges are different, the sequence is an edge-train and if so are all the vertices then it is a path or directed path. If $p_0 = p_m$, but no other vertex is repeated then the sequence is a circuit or a directed circuit (see [2]).

The *distance of the vertices q and p* in a graph G is the length of the shortest pq-path in G, provided such a path exists. If there is no pq-path in G, then the distance of p and q in G is said to be ∞. In a directed graph $\vec{G}$, the distance of the vertex q from the vertex p is similarly defined as the minimum of the length of the directed $\overrightarrow{pq}$-paths.

Let the adjacency matrices of the graphs G and $\vec{G}$ be $A = [a_{ij}]$ and $\overline{A} = [\overline{a}_{ij}]$, respectively. The entries of these matrices can also be interpreted in this way: a_{ij} is the number of $p_i p_j$-edge-sequences of length 1, and $\overline{a}_{ij}$ is the number of directed $p_i p_j$-edge-sequences of length 1. Let us consider the elements of $A^2 = [a_{ij}^{(2)}]$:

$$a_{ij}^{(2)} = a_{i1}a_{1j} + a_{i2}a_{2j} + \ldots + a_{ic}a_{cj}.$$

The first term in the sum is the number of $p_i p_j$-edge-sequences of length 2 through p_1 in G, the second term is the number of $p_i p_j$-edge-sequences of length 2 through p_2 in G etc. Therefore, $a_{ij}^{(2)}$ is the number of p_ip_j-edge-sequences of length 2 in G. A similar reasoning yields that the element $\overline{a}_{ij}^{(2)}$ of the matrix $\overline{A}^2$ is the number of directed p_ip_j-edge-sequences of length 2 in $\overrightarrow{G}$. Since $A^k = A^{k-1} \cdot A$, our reasoning can be extended:

4. *If the vertices of a graph or directed graph with adjacency matrix A are $p_1, p_2, \ldots, p_c$, then the element $a_{ij}^{(k)}$ of the matrix A^k is the number of p_ip_j-edge-sequences or directed p_ip_j-edge-sequences of length k in the graph or in the directed graph, respectively.*

The following three statements are obvious consequences of this theorem:

5. *A directed graph with adjacency matrix A contains no directed circuit if and only if there is a number k_0 so that A^k is the zero matrix for all positive integers $k \geq k_0$.*

6. *If $p_1, p_2, \ldots, p_c$ are the vertices of a connected graph with adjacency matrix $A = [a_{ij}]$ then the distance of the vertices p_i and p_j ($i \neq j$) is the lowest integer n with $a_{ij}^{(n)} \neq 0$ in A^n.*

7. *If $p_1, p_2, \ldots, p_c$ are the vertices of a simple graph with adjacency matrix $A = [a_{ij}]$ then the degree of p_i is $\varphi(p_i) = a_{ii}^{(2)}$ ($i = 1, 2, \ldots, c$).*

The adjacency matrix is suitable for investigating changes in the state of certain systems. Indeed, if the various states of a system are represented by the vertices of a graph and a directed edge designates the possibility of a direct transition from the state corresponding to its tail to the one corresponding to its head, then the above considerations allow the discovery of indirect transitions between the states. As an example, the solution of the so-called missionary–cannibal puzzle is presented [41]. Two missionaries and two cannibals reach the left bank of a river simultaneously and all of them intend to cross the river with the aid of the single available boat which carries no more than two persons. On no bank may the cannibals outnumber the missionaries lest the former jeopardise the latter (with the obvious exception of the case when the number of missionaries is zero). A state on the left bank of the river is defined by a pair of numbers: the first element is the number of missionaries and the second is that of the cannibals. The vertices representing permissible states are:

$$p_1 = (2,2), \quad p_2 = (2,1), \quad p_3 = (2,0), \quad p_4 = (1,1),$$
$$p_5 = (0,2), \quad p_6 = (0,1), \quad p_7 = (0,0).$$

The other states are obviously impermissible, since those would mean that the missionaries are in danger on one of the banks. Let the edge (p_i, p_j) of the graph $\overrightarrow{G}_1$ designate that the state p_j can be directly attained from the

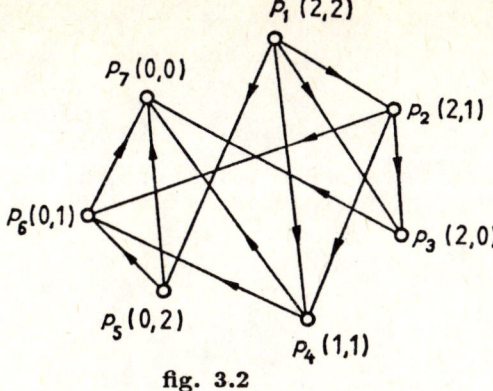

fig. 3.2

state p_i while the boat gets to the right bank (see figure 3.2). The adjacency matrix of this graph is:

$$A = [a_{ij}] = \begin{bmatrix} 0 & 1 & 1 & 1 & 1 & 0 & 0 \\ 0 & 0 & 1 & 1 & 0 & 1 & 0 \\ 0 & 0 & 0 & 0 & 0 & 0 & 1 \\ 0 & 0 & 0 & 0 & 0 & 1 & 1 \\ 0 & 0 & 0 & 0 & 0 & 1 & 1 \\ 0 & 0 & 0 & 0 & 0 & 0 & 1 \\ 0 & 0 & 0 & 0 & 0 & 0 & 0 \end{bmatrix}.$$

Consider the following. The state p_j can be attained from the state p_i by rowing from the left bank to the right one if and only if p_i is attainable from p_j by rowing from the right bank to the left one. So, if the direction of every edge of the graph $\vec{G}_1$ is reversed, the edges of the new graph $\vec{G}_2$ indicate the direct accessibility of states from other states when the boat gets from the right bank to the left one. However, the adjacency matrix of $\vec{G}_2$ is A^*. The element in the position (i,j) of the matrix AA^* (the jth element of its ith row) is

$$a_{i1}a_{j1} + a_{i2}a_{j2} + \ldots + a_{i7}a_{j7},$$

and this gives the number of ways the state p_j can be attained from the state p_i by a single back and forth trip of the boat. Observe that the element in the position (i,j) of the product $(AA^*)A$ gives the number of ways p_j can be reached from p_i by rowing back and forth and then over to the right bank. Continuing the procedure, the problem is clearly solvable provided there is an integer m so that the position $(1,7)$ of the matrix

$$(AA^*)^m A$$

is non-zero, i.e. then p_7 can be attained from p_1 by a boat trip which ends in rowing to the right bank, and namely in $2m+1$ steps: by rowing m times back and forth and finally rowing to the right bank. It is obviously desirable to find the lowest number m with this property. All the non-zero entries of the matrices

$$AA^*, \quad (AA^*)A, \quad (AA^*)^2, \quad (AA^*)^2A, \quad \ldots$$

can be replaced by one if the number of ways of solving the problem is of no concern. Hence the calculations are:

$$AA^* = \begin{bmatrix} 1 & 1 & 0 & 0 & 0 & 0 & 0 \\ 1 & 1 & 0 & 1 & 1 & 0 & 0 \\ 0 & 0 & 1 & 1 & 1 & 1 & 0 \\ 0 & 1 & 1 & 1 & 1 & 1 & 0 \\ 0 & 1 & 1 & 1 & 1 & 1 & 0 \\ 0 & 0 & 1 & 1 & 1 & 1 & 0 \\ 0 & 0 & 0 & 0 & 0 & 0 & 0 \end{bmatrix}, \quad (AA^*)A = \begin{bmatrix} 0 & 1 & 1 & 1 & 1 & 1 & 0 \\ 0 & 1 & 1 & 1 & 1 & 1 & 1 \\ 0 & 0 & 0 & 0 & 0 & 1 & 1 \\ 0 & 0 & 1 & 1 & 0 & 1 & 1 \\ 0 & 0 & 1 & 1 & 0 & 1 & 1 \\ 0 & 0 & 0 & 0 & 0 & 1 & 1 \\ 0 & 0 & 0 & 0 & 0 & 0 & 0 \end{bmatrix},$$

$$(AA^*)^2 = \begin{bmatrix} 1 & 1 & 0 & 1 & 1 & 0 & 0 \\ 1 & 1 & 1 & 1 & 1 & 1 & 0 \\ 0 & 1 & 1 & 1 & 1 & 1 & 0 \\ 1 & 1 & 1 & 1 & 1 & 1 & 0 \\ 1 & 1 & 1 & 1 & 1 & 1 & 0 \\ 0 & 1 & 1 & 1 & 1 & 1 & 0 \\ 0 & 0 & 0 & 0 & 0 & 0 & 0 \end{bmatrix}, \quad (AA^*)^2A = \begin{bmatrix} 0 & 1 & 1 & 1 & 1 & 1 & 1 \\ 0 & 1 & 1 & 1 & 1 & 1 & 1 \\ 0 & 0 & 1 & 1 & 0 & 1 & 1 \\ 0 & 1 & 1 & 1 & 1 & 1 & 1 \\ 0 & 1 & 1 & 1 & 1 & 1 & 1 \\ 0 & 0 & 1 & 1 & 0 & 1 & 1 \\ 0 & 0 & 0 & 0 & 0 & 0 & 0 \end{bmatrix}.$$

So, the puzzle can be solved in five steps. The course of the solution can be obtained by proceeding backwards from the last matrix. To this end, let us start from the entry 1 in the position $(1,7)$ of the matrix $(AA^*)^2A$ and let us check why this is non-zero. This number 1 has been obtained by multiplying the first row of $(AA^*)^2$ by the last column of A. Which element of the first row of $(AA^*)^2$ yields a non-zero result when multiplied by the corresponding element of the last column of A? One possibility is the element of $(AA^*)^2$ in the position $(1,4)$ since the element in the position $(4,7)$ of A is non-zero. The second possibility is offered by the position $(1,5)$. Let us select the former one. This means that the last step of the solution is (p_4, p_7). The following question is how can an entry 1 appear in position $(1,4)$ of $(AA^*)^2$. On inspection, this turns out to result from the elements in the position $(1,6)$ of the matrix $(AA^*)A$ and in the position $(6,4)$ of A^*. This means that the penultimate step of the solution is (p_6, p_4). Now, the origin of the entry 1 in the position $(1,6)$ of the matrix $(AA^*)A$ is sought. This results from the positions $(1,2)$ of AA^* and $(2,6)$ of A. Thus the third from the last step of the solution is (p_2, p_6). The fourth and fifth steps from the last one can be similarly found: (p_3, p_2) and (p_1, p_3). So, the steps of the solution expressed with the aid of the states on the left bank are:

$$(2,2); \quad (2,0); \quad (2,1); \quad (0,1); \quad (1,1); \quad (0,0).$$

The above reasoning and the details of the course of the solution suggest the way of solving a similar puzzle if the number of missionaries and cannibals is increased or if further restrictions are made, e.g. that only a single cannibal or a single missionary is able to row.

3.2 The incidence matrix

The *edge matrix* $B = [b_{ij}]$ below or, in other words, the *incidence matrix* expresses how the edges of the graph are incident with its vertices. If the vertices of the graph G are $p_1, p_2, \ldots, p_c$ and its edges are $e_1, e_2, \ldots, e_e$ then B is a $c \times e$ matrix, i.e. one consisting of c rows and e columns and

$$b_{ij} = \begin{cases} 1 & \text{if } e_j \text{ is not a loop and is incident to the vertex } p_i, \\ 0 & \text{if } e_j \text{ is a loop or is not incident to the vertex } p_i. \end{cases}$$

In the case of a directed graph with the above constituents, the incidence matrix $B = [b_{ij}]$ satisfies

$$b_{ij} = \begin{cases} 1 & \text{if } e_j \text{ is not a loop and its tail is } p_i, \\ -1 & \text{if } e_j \text{ is not a loop and its head is } p_i \\ 0 & \text{if } e_j \text{ is a loop or is not incident with the vertex } p_i. \end{cases}$$

As an example, let the incidence matrices of the graphs G and $\overrightarrow{G}$ shown in figure 3.1 be B_1 and B_2, respectively. Then

$$B_1 = \begin{bmatrix} 0 & 1 & 1 & 1 & 0 & 0 \\ 0 & 0 & 1 & 0 & 1 & 0 \\ 0 & 1 & 0 & 1 & 1 & 1 \\ 0 & 0 & 0 & 0 & 0 & 1 \end{bmatrix} \quad \text{and} \quad B_2 = \begin{bmatrix} 0 & -1 & 1 & -1 & 0 & 0 \\ 0 & 0 & -1 & 0 & 1 & 0 \\ 0 & 1 & 0 & 1 & -1 & 1 \\ 0 & 0 & 0 & 0 & 0 & -1 \end{bmatrix}.$$

The matrix B expressing the incidence conditions clearly does not carry any information on the incidence of loops. Therefore it is sometimes customary to express the incidence of the loop e_j with the vertex p_i by the choice $b_{ij} = 1$ or, with the number of incidences also indicated, by $b_{ij} = 2$. An argument in favour of the above variant is the observation of O. Veblen [42]. In the linear

+	0	1
0	0	1
1	1	0

·	0	1
0	0	0
1	0	1

fig. 3.3

space over the mod 2 field (i.e. in the field formed by the modulus 2 residue classes and by the operations shown in figure 3.3) interesting relationships can be discovered between the undirected graphs and the associated incidence matrices (and also the circuit and cutset matrices, see the later treatment). (For linear space, see in [43]). Now, a loop is incident with a single vertex but twice, i.e. the sum of incidences with the relevant vertex is $1 + 1$ which is congruent to zero mod 2. In case of directed graphs, the linear space over the mod 2 field proves to be too tight. In view of the application to so-called linear electrical networks, the calculations are better carried out in the linear space over the field of real numbers. Then the sum of incidences at the endpoints of a loop is $1 - 1$ which is again zero in the field of the real numbers. It is easy to verify that, aside from loops, the incidence matrix determines the graph uniquely, i.e. the following theorem, similar to Theorem 1, can be stated:

8. *Two graphs or two directed graphs without loops are isomorphic if and only if the incidence matrix of one of them can be constructed from that of the other by the interchange of rows and columns.*

So, let us consider the rows and columns of the incidence matrix B of a non-directed graph G with c vertices and e edges in mod 2 sense (i.e. as vectors in the linear space over the mod 2 field). An obvious question is the maximal number of linearly independent vectors among these, i.e. the rank $r(B)$ of the matrix B. The rank of this matrix will be shown to equal the rank $\varrho(G)$ of the graph G which can be expressed with the number of components k of the graph G (see [2], page 42) as

$$\varrho(G) = c - k \quad \text{(mod 2)}.$$

(This rank is a non-negative integer. In such cases mod 2 refers to the linear space.) If the row vectors of B are $\mathbf{b}_1^*, \mathbf{b}_2^*, \ldots, \mathbf{b}_c^*$ and $\mathbf{0}^*$ is a zero vector of e coordinates, then

$$\mathbf{b}_1^* + \mathbf{b}_2^* + \ldots + \mathbf{b}_c^* = \mathbf{0}^* \quad \text{(mod 2)}$$

since the corresponding coordinates of the vectors $\mathbf{b}_i^*$ are either all zero or there are exactly two ones among them, but

$$1 + 1 = 0 \quad \text{(mod 2)}.$$

Hence, the following theorem has been verified:

9. *The row vectors of the incidence matrix of any graph are linearly dependent* (mod 2).

In the case of a connected graph, all but one row vectors of the incidence matrix are linearly independent. This is stated in the following theorem:

10. *Any $c - 1$ of the row vectors of the incidence matrix of any connected graph containing $c \geq 2$ vertices are linearly independent* (mod 2).

Proof. Let B be the incidence matrix of a connected graph G containing c vertices. Indirectly, assume that B has $c-1$ linearly dependent row vectors, i.e. $c-1$ row vectors having a non-trivial linear combination resulting in the zero vector. Since in a mod 2 linear combination any coefficient is either 0 or 1, this means that B has less than c row vectors whose sum (i.e. linear combination with all coefficients 1) is the zero vector. Let such a system be denoted by $\mathbf{b}_1^*, \mathbf{b}_2^*, \ldots, \mathbf{b}_d^*$ ($d < c$). Therefore:

$$\mathbf{b}_1^* + \mathbf{b}_2^* + \ldots + \mathbf{b}_d^* = \mathbf{0}^*. \tag{1}$$

The vectors $\mathbf{b}_i^*$ correspond to d vertices of G. In view of the connectedness of G, there is an edge e of G with one of its endpoints being one of these d vertices and the other not among these (see [2], 1.25). Therefore, in the column of B corresponding to e, the vectors $\mathbf{b}_i^*$ contain exactly one entry equal to 1, so the sum of their coordinates corresponding to this column cannot be zero, a contradiction to (1). This completes the proof of Theorem 10.

The two previous theorems directly imply the following one:

11. *If B is the incidence matrix of a connected graph G containing c vertices then*

$$r(B) = \varrho(G) = c - 1 \quad (\mathrm{mod}\ 2).$$

This theorem and Theorem 10 imply that if an arbitrary row is deleted from the incidence matrix of a connected graph G containing c vertices then a matrix B_0 is obtained with

$$r(B_0) = \varrho(G) = c - 1. \tag{2}$$

Such a matrix B_0 is called a *reduced incidence matrix* (or *reduced edge matrix*) of the connected graph G.

Let the components of a disconnected graph G containing c vertices be denoted by $G_1, G_2, \ldots, G_k$ and the number of vertices in G_i by c_i. Then

$$c = \sum_{i=1}^{k} c_i.$$

Let us construct the incidence matrix B of the graph G so that the rows and columns of B be arranged in the order of the components. Then B can be written as the hypermatrix partitioned as follows:

$$B = \begin{bmatrix} B_1 & N_{12} & \ldots & N_{1k} \\ N_{21} & B_2 & \ldots & N_{2k} \\ \ldots & \ldots & \ldots & \ldots \\ N_{k1} & N_{k2} & \ldots & B_k \end{bmatrix} = \begin{bmatrix} P_1^* \\ P_2^* \\ \vdots \\ P_k^* \end{bmatrix}.$$

Here, B_i is the incidence matrix of the graph G_i and all N_{ij} are zero matrices. The rank of the hyper-row $P_i^* = [N_{i1} \ldots B_i \ldots N_{ik}]$ of the hyper-matrix B

equals the rank of B_i since the matrix P_i^* has been obtained by augmenting B_i by zero vectors. However, in view of Theorem 11,

$$r(B_i) = \varrho(G_i) = c_i - 1.$$

Therefore, the maximal number of linearly independent row vectors in P_i^* is $c_i - 1$. Consequently the maximal number s of linearly independent row vectors of B satisfies

$$s \leq \sum_{i=1}^{k}(c_i - 1) = c - k,$$

i.e.

$$r(B) \leq c - k. \tag{3}$$

According to (2), the rank of any reduced incidence matrix of G_i is $c_i - 1$. So, an arbitrary reduced incidence matrix of G_i possesses a non-singular minor matrix B_{0i} of order $(c_i - 1)$, i.e. one satisfying

$$\det B_{0i} \neq 0 \quad (i = 1, 2, \ldots, k). \tag{4}$$

Accordingly, the following matrix can be obtained from B by rearranging its rows and columns.

$$B' = \begin{bmatrix} B_{01} & N'_{12} & \ldots & N'_{1k} & X_1 \\ N'_{21} & B_{02} & \ldots & N'_{2k} & X_2 \\ \ldots & \ldots & \ldots & \ldots & \ldots \\ N'_{k1} & N'_{k2} & \ldots & B_{0k} & X_k \\ X_{k+1} & X_{k+2} & \ldots & X_{2k} & X_{2k+1} \end{bmatrix},$$

X_i has been used to denote the remaining matrices irrelevant for our purposes. The minor

$$B'' = \begin{bmatrix} B_{01} & N'_{12} & \ldots & N'_{1k} \\ N'_{21} & B_{02} & \ldots & N'_{2k} \\ \ldots & \ldots & \ldots & \ldots \\ N'_{k1} & N'_{k2} & \ldots & B_{0k} \end{bmatrix}$$

of the hyper-matrix B' is a square matrix of order $c - k$ and is non-singular because, using the Laplace expansion of the determinant and taking (4) into account:

$$\det B'' = \det B_{01} \cdot \det B_{02} \cdot \ldots \cdot \det B_{0k} \neq 0.$$

Consequently,

$$r(B) = r(B') \geq c - k. \tag{5}$$

Comparing (3) and (5) yields

$$r(B) = c - k.$$

Since any incidence matrix of the graph G can be constructed from B by rearrangement of rows and columns (these operations do not change the rank), the following theorem has been obtained:

12. *If B is the incidence matrix of a graph G containing c vertices and consisting of k components then*

$$r(B) = \varrho(G) = c - k \qquad (\text{mod } 2).$$

In this way we also obtain that if a row corresponding to a vertex of each component is deleted from B then a matrix B_0 is obtained with

$$r(B_0) = \varrho(G) = c - k.$$

Such a matrix B_0 is called a *reduced incidence matrix* (or *reduced edge matrix*) of the graph G. The fact that the rank of the reduced incidence matrix is $c-k$ also implies that B_0 has a non-singular square minor of order $(c-k)$. It will emerge in what follows that a square minor of B_0 possesses this property if and only if the edges corresponding to its columns are the edges of a spanning forest of the graph (for the definitions of spanning tree and spanning forest see [2] pages 39 and 42). Our assertion is based on the following theorem:

13. *The column vectors of any reduced incidence matrix corresponding to edges forming a circuit of the graph are linearly dependent* (mod 2).

Proof. If certain column vectors of the incidence matrix are linearly dependent then, of course, so are those of the reduced matrix. It will, however, be shown that if the edges corresponding to the column vectors $\mathbf{e}_1, \mathbf{e}_2, \ldots, \mathbf{e}_m$ are the edges of a circuit K of the graph, then their sum is zero (mod 2). Let p_i be the vertex of the graph corresponding to the ith row of the incidence matrix. If p_i is not a vertex of the circuit K then K has no edge incident with p_i which also implies that the ith coordinate of all vectors $\mathbf{e}_j$ is zero. If p_i is a vertex of K then K has exactly two edges incident with p_i which also implies that among the vectors $\mathbf{e}_j$, exactly two have 1 as their ith entry and that of the others is zero. Consequently,

$$\mathbf{e}_1 + \mathbf{e}_2 + \ldots + \mathbf{e}_m = 0 \qquad (\text{mod } 2)$$

which implies that the vectors $\mathbf{e}_j$ are linearly dependent. This completes the proof of Theorem 13.

14. *A $(c-1)$-order square minor of a reduced incidence matrix of a connected graph containing c vertices is non-singular* (mod 2) *if and only if the edges corresponding to its columns are the edges of a spanning tree of the graph.*

Proof. Let B_1 be a $(c-1)$-order square minor of a reduced incidence matrix of the connected graph containing c vertices.

To prove the sufficiency of the condition, assume that the edges corresponding to the columns of B_1 constitute a spanning tree F of the graph. Consequently, B_1 is a reduced incidence matrix of a connected graph F containing c vertices and thus, according to (2), $r(B_1) = c - 1$ which also implies that $\det B_1 \neq 0$.

To prove the necessity of the condition, assume that B_1 is non-singular. Then the column vectors of B_1 are linearly independent and so, according to Theorem 12, the edges corresponding to the columns of B_1 constitute a subgraph without any circuit. Now, this subgraph can be proved (see [2], 2.26) to be a spanning tree of the graph and hence the statement of Theorem 14 follows.

The reasoning can be extended to disconnected graphs, too. Then it is useful to arrange the rows and columns of the reduced incidence matrix of the graph according to its components. Employing the Laplace expansion of determinants and using the above theorem, the following result can be verified:

15. *A $(c - k)$-order square minor of a reduced incidence matrix of a graph containing c vertices and consisting of k components is non-singular (mod 2) if and only if the edges corresponding to its columns are the edges of a spanning forest of the graph.*

Note that if the incidence matrix of a non-directed graph is considered in the linear vector space over the field of real numbers then the rank of this matrix is not necessarily equal to the rank of the graph. For example, if B_1 is the incidence matrix corresponding to the graph G shown in figure 3.1 then

$$r(B_1) = 4 > \varrho(G) = 3$$

since deleting the first two columns of B_1 the resulting submatrix B_{12} has

$$\det B_{12} = \det \begin{bmatrix} 1 & 1 & 0 & 0 \\ 1 & 0 & 1 & 0 \\ 0 & 1 & 1 & 1 \\ 0 & 0 & 0 & 1 \end{bmatrix} = -2.$$

We have seen already that the incidence matrix of a directed graph can better be studied in the linear space over the field of real numbers (we always use this space unless (mod 2) is explicitly written). In spite of this, the theorems stated for non-directed graphs as well as their proofs will be seen to carry over to directed graphs virtually without change. The *spanning tree*, *spanning forest* and *rank* of a directed graph $\vec{G}$ are defined as the same for the graph G.

If a column vector of the incidence matrix B of a connected directed graph containing c vertices corresponds to a loop, then it is a zero vector, otherwise exactly one of its coordinates is 1 and one is -1. Consequently, the sum of the row vectors of B is a zero vector. However, any $c - 1$ of the row vectors of B are linearly independent. This can be verified similarly to the proof of Theorem 10. (1) should be now replaced by

$$\lambda_1 \mathbf{b}_1^* + \lambda_1 \mathbf{b}_2^* + \ldots + \lambda_d \mathbf{b}_d^* = \mathbf{0}^*$$

where all λ_i are real and none of them is zero. (Instead of [2] 1.25, the corresponding theorem [2] 3.8 is recalled here.) Therefore, Theorem 11 is valid for directed graphs as well. The reasoning following theorem is also applicable to directed graphs. This yields the following correspondent of Theorem 12:

16. *If B is the incidence matrix of a directed graph $\vec{G}$ containing c vertices and consisting of k components then*

$$r(B) = \varrho(\vec{G}) = c - k.$$

If a row corresponding to a vertex of each component is deleted from B then a matrix B_0 is obtained with

$$r(B_0) = \varrho(\vec{G}) = c - k.$$

Such a matrix B_0 is called a *reduced incidence matrix* (or *reduced edge matrix*) of the graph $\vec{G}$.

Theorem 13 is also valid for directed graphs but if its proof should be applied to directed graphs, the following modification is necessary: let us traverse K along its edges, let $\lambda_i = 1$ if the direction of the edge corresponding to e_i is identical to the orientation of the walk, and let $\lambda_i = -1$ otherwise. Then

$$\lambda_1 \mathbf{e}_1 + \lambda_2 \mathbf{e}_2 + \ldots + \lambda_m \mathbf{e}_m = \mathbf{0},$$

since, if p_i is not a vertex of the circuit K, then the ith coordinate of all vectors $\mathbf{e}_j$ is zero, and if p_i is a vertex of the circuit K, then K has exactly two edges incident with p_i which also implies that exactly two of the vectors $\mathbf{e}_j$ have non-zero ith coordinate, say $\mathbf{e}_a$ and $\mathbf{e}_b$. But the ith coordinates of $\mathbf{e}_a$ and $\mathbf{e}_b$ coincide if and only if $\lambda_a \neq \lambda_b$.

Theorems 14 and 15 as well as their proofs are valid for directed graphs without change, so the following theorem can be stated:

17. *A $(c-k)$-order square minor of a reduced incidence matrix of a directed graph containing c vertices and consisting of k components is non-singular if and only if the edges corresponding to its columns are the edges of a spanning forest of the graph.*

The determinant of a non-singular square minor of a matrix is necessarily 1 if operations are performed mod 2 ; but in the linear space over the field of real numbers it can be any non-zero number, even if all entries of the matrix are 0 or 1. Therefore, it is interesting that if the incidence matrix of a directed graph is considered then the determinant of any non-singular square minor is 1 or −1, i.e. the following theorem of Poincaré [44] is valid:

18. *If a matrix M is a non-singular square minor of an incidence matrix of a directed graph, then*

$$|\det M| = 1.$$

Proof. There can be no more than two non-zero elements in any column of M, one of value 1 and one of (-1). Not all columns contain two non-zero elements, otherwise the sum of the row vectors of M is a zero vector implying that M is singular. Nor can M have a column consisting of zeros only. Consequently, M has a column with only one non-zero entry. Let us expand $\det M$ along such a column. The minor appearing in this expansion is again non-singular, and must therefore have a column with a single non-zero element. The determinant of this minor is expanded along such a column, and so on. $\det M$ is finally obtained as the product of values 1 and (-1) which proves the theorem.

3.3 The circuit matrix

The *circuit matrix* $K = [k_{ij}]$ indicates the inclusion of the edges of the graph G in circuits. Its rows correspond to the circuits and its columns to the edges of the graph. If the circuits and edges of G are $k_1, k_2, \ldots, k_q$ and $e_1, e_2, \ldots, e_e$, respectively, then the matrix K has q rows and e columns, and

$$k_{ij} = \begin{cases} 1 & \text{if } e_j \text{ is an edge of the circuit } k_i, \\ 0 & \text{if } e_j \text{ is not an edge of the circuit } k_i. \end{cases}$$

In case of a directed graph $\vec{G}$, k_i still denotes the ith circuit of the non-directed graph G, and e_j is the jth edge of the graph $\vec{G}$. Let us further establish an orientation along the edges of each circuit of G. Then, the circuit matrix $K = [k_{ij}]$ of the graph $\vec{G}$ is defined as follows:

$$k_{ij} = \begin{cases} 1 & \text{if } e_j \text{ is an edge of the circuit } k_i \text{ and its direction coincides with the orientation of } k_i, \\ -1 & \text{if } e_j \text{ is an edge of the circuit } k_i \text{ and its direction is opposite to the orientation of } k_i, \\ 0 & \text{if } e_j \text{ is not an edge of the circuit } k_i. \end{cases}$$

As an example, consider the graph $\vec{G}$ shown in figure 3.4. The circuit matrices of G and $\vec{G}$ are:

$$K_1 = \begin{bmatrix} 1 & 0 & 0 & 0 & 0 & 1 & 1 & 0 & 0 \\ 0 & 1 & 0 & 0 & 0 & 1 & 1 & 0 & 0 \\ 0 & 0 & 1 & 0 & 0 & 1 & 0 & 1 & 0 \\ 0 & 0 & 0 & 1 & 0 & 0 & 1 & 1 & 0 \\ 0 & 0 & 0 & 0 & 1 & 0 & 0 & 0 & 0 \\ 1 & 1 & 0 & 0 & 0 & 0 & 0 & 0 & 0 \\ 1 & 0 & 1 & 0 & 0 & 0 & 1 & 1 & 0 \\ 1 & 0 & 0 & 1 & 0 & 1 & 0 & 1 & 0 \\ 0 & 1 & 1 & 0 & 0 & 0 & 1 & 1 & 0 \\ 0 & 1 & 0 & 1 & 0 & 1 & 0 & 1 & 0 \\ 0 & 0 & 1 & 1 & 0 & 1 & 1 & 0 & 0 \\ 1 & 0 & 1 & 1 & 0 & 0 & 0 & 0 & 0 \\ 0 & 1 & 1 & 1 & 0 & 0 & 0 & 0 & 0 \end{bmatrix},$$

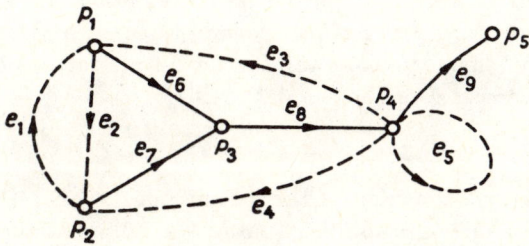

fig. 3.4

$$K_2 = \begin{bmatrix} 1 & 0 & 0 & 0 & 0 & 1 & -1 & 0 & 0 \\ 0 & 1 & 0 & 0 & 0 & -1 & 1 & 0 & 0 \\ 0 & 0 & 1 & 0 & 0 & 1 & 0 & 1 & 0 \\ 0 & 0 & 0 & 1 & 0 & 0 & 1 & 1 & 0 \\ 0 & 0 & 0 & 0 & 1 & 0 & 0 & 0 & 0 \\ 1 & 1 & 0 & 0 & 0 & 0 & 0 & 0 & 0 \\ -1 & 0 & 1 & 0 & 0 & 0 & 1 & 1 & 0 \\ 1 & 0 & 0 & 1 & 0 & 1 & 0 & 1 & 0 \\ 0 & 1 & 1 & 0 & 0 & 0 & 1 & 1 & 0 \\ 0 & -1 & 0 & 1 & 0 & 1 & 0 & 1 & 0 \\ 0 & 0 & 1 & -1 & 0 & 1 & -1 & 0 & 0 \\ 1 & 0 & -1 & 1 & 0 & 0 & 0 & 0 & 0 \\ 0 & 1 & 1 & -1 & 0 & 0 & 0 & 0 & 0 \end{bmatrix}.$$

The edges drawn by continuous lines in the diagram constitute the edges of a spanning tree of the graph. The subscripts of the circuits and edges have been arranged according to the following rule:

(6) A spanning tree or, in case of a disconnected graph, a spanning forest F is selected in the graph, and the chords of the graph with respect to F receive the first subscripts. The fundamental circuits induced by F receive the same subscripts as the chords defining these circuits (for the definition of fundamental circuits and for a related theorem see [2] page 40). In the case of a directed graph, the orientation of each fundamental circuit is chosen to coincide with the direction of the chord included in it.

Note that the terms *fundamental circuit* and *fundamental system of circuits* are sometimes shortened to *f-circuit* and *f-system of circuits*.

Theorem 2.17 of [2] implies that if the circuit matrix K' of a graph $\vec{G}$ or G, containing e edges and c vertices and consisting of k components, is given in accordance with the rule above, then the $(e-c+k)$-order upper left square minor of K' is a unit matrix. It is also obvious that any circuit matrix K of the graph can be constructed from K' by rearranging its rows and columns.

This leads to the following relation between the *cyclomatic number* $\mu(\vec{G})$ of the graph, in other words, its *nullity* (defined to equal $\mu(G)$, see [2] page 42) and the rank of K (this formula is valid in both linear spaces defined over the field of real numbers or over the mod 2 field):

$$r(K) \geq e - c + k = \mu(\vec{G}) = \mu(G). \tag{7}$$

We now claim that the rank of the circuit matrix, in (mod 2) sense in the case of a non-directed graph, is equal to the nullity of the graph.

Let us consider the circuit matrices K_1 and K_2 of the graphs G and $\vec{G}$, respectively, shown in figure 3.4 and let us construct the incidence matrices B_1 and B_2 of the graphs G and $\vec{G}$, respectively, using the same subscripts for the edges:

$$B_1 = \begin{bmatrix} 1 & 1 & 1 & 0 & 0 & 1 & 0 & 0 & 0 \\ 1 & 1 & 0 & 1 & 0 & 0 & 1 & 0 & 0 \\ 0 & 0 & 0 & 0 & 0 & 1 & 1 & 1 & 0 \\ 0 & 0 & 1 & 1 & 0 & 0 & 0 & 1 & 1 \\ 0 & 0 & 0 & 0 & 0 & 0 & 0 & 0 & 1 \end{bmatrix},$$

$$B_2 = \begin{bmatrix} -1 & 1 & -1 & 0 & 0 & 1 & 0 & 0 & 0 \\ 1 & -1 & 0 & -1 & 0 & 0 & 1 & 0 & 0 \\ 0 & 0 & 0 & 0 & 0 & -1 & -1 & 1 & 0 \\ 0 & 0 & 1 & 1 & 0 & 0 & 0 & -1 & 1 \\ 0 & 0 & 0 & 0 & 0 & 0 & 0 & 0 & -1 \end{bmatrix}.$$

Observe that the scalar product of any row of K_i with any row of B_i, in mod 2 sense in the case of G, is zero. This is always true, i.e. we shall prove the following theorem which shows the close connection between the circuit matrix and the incidence matrix:

19. *If the columns of the incidence matrix B and the circuit matrix K of a graph G or $\vec{G}$ are arranged in the same order of the edges, then (in (mod 2) sense in the case of G):*

$$BK^* = N_1 \quad \text{and} \quad KB^* = N_2$$

where N_1 and N_2 are zero matrices of appropriate dimensions.

Proof. We prove the undirected case only. The proof is applicable to $\vec{G}$ with a slight modification. Since the two equalities follow from each other by transposition, it suffices to prove the first one. Let the ith row of B be $\mathbf{b}_i^*$ and the jth column of K^* be $\mathbf{k}_j$ and let p_i be the vertex corresponding to $\mathbf{b}_i^*$ and K_j be the circuit corresponding to $\mathbf{k}_j$. If p_i is not a vertex of K_j then none of the edges incident to p_i are included in K_j which implies that if some coordinate of $\mathbf{b}_i^*$ is non-zero then the same coordinate of $\mathbf{k}_j$ is certainly zero and hence:

$$\mathbf{b}_i^* \cdot \mathbf{k}_j = 0. \tag{8}$$

If p_i is a vertex of K_j and the length of K_j is at least 2 then exactly two of the edges incident with p_i are included in K_j. Hence, among the non-zero coordinates of $\mathbf{b}_i^*$ there are exactly two so that the same coordinates of $\mathbf{k}_j$ are also 1. Thus, taking $1 + 1 = 0$ (mod 2) into account, (8) proves to hold in this case, too. If p_i is a vertex of K_j and the length of K_j is one, then K_j contains a single edge which is a loop. Then only a single coordinate of $\mathbf{k}_j$ is 1, but the corresponding coordinate of $\mathbf{b}_i^*$ is 0, i.e. (8) is still true. This proves the undirected version of the theorem.

Let B denote an incidence matrix and K a circuit matrix of a graph G or $\vec{G}$ containing e edges, c vertices and consisting of k components. Then the number of both the columns of B and the rows of K^* is e. According to Theorems 12 and 16 (in mod 2 sense in the case of G):

$$r(B) = c - k.$$

In view of this, of Theorem 19 and of Sylvester's theorem — stating that if A and B are matrices of size $m \times n$ and $n \times p$, respectively, and if $A \cdot B$ is a zero matrix then $r(A) + r(B) \leq n$ — the following is obtained (in mod 2 sense in the case of G):

$$r(K) = r(K^*) \leq e - c + k.$$

Comparing this with (7), we obtain

20. *If K is the circuit matrix of a graph G or $\vec{G}$ containing e edges and c vertices and consisting of k components then, in* mod 2 *sense in the case of G,*

$$r(K) = \mu(G) = \mu(\vec{G}) = e - c + k.$$

According to Theorem 20, the maximal number of linearly independent row vectors of K is $e - c + k$. A minor K_0 of the matrix K consisting of $e - c + k$ of its linearly independent row vectors is called a *reduced circuit matrix* of the graph G or $\vec{G}$. If a forest of G or $\vec{G}$ is selected and rule (6) is followed then the minor K_f consisting of the first $e - c + k$ rows of K can be partitioned as follows:

$$K_f = [E \quad K_{12}],$$

where E is a unit matrix. Such circuit matrices K_f are called *fundamental circuit matrices*, in short, *f-circuit matrices* of the graph G or $\vec{G}$. Evidently,

$$r(K_f) = r(K_0) = \mu(G) = \mu(\vec{G}) = e - c + k,$$

and any f-circuit matrix is also a reduced circuit matrix. The following theorem is the 'analogue' of Theorems 15 and 17:

21. *An $(e-c+k)$-order square minor of a reduced circuit matrix of a graph G or $\vec{G}$ containing e edges and c vertices and consisting of k components is non-singular (in* mod 2 *sense in the case of G) if and only if the edges corresponding to its columns constitute a system of chords of G.*

Proof. We give the detailed proof for the undirected case only. Let K_{11} denote an $(e-c+k)$-order square minor of a reduced circuit matrix of G. Let us rearrange the columns of the reduced circuit matrix accordingly and so let

$$K_0 = [K_{11} \quad K_{12}]. \tag{9}$$

In order to prove the sufficiency, assume that the edges corresponding to the columns of K_{11} to constitute a system of chords of G. Let

$$K_f = [E \quad K'_{12}],$$

be the f-circuit matrix of G defined by this system of chords with its columns arranged in the same order of the edges as in K_0. The rows of K_f are present in any circuit matrix K of G (if the edges follow in the same order as in K_f) and they constitute there a maximal linearly independent system. Consequently, the row vectors of K and hence also of K_0 can be expressed as linear combinations of the row vectors of K_f. Let us express the row vectors of K_0 as the linear combinations of the row vectors of K_f. Denote by D the matrix formed by the coefficients of these linear combinations. Then, in a concise form

$$K_0 = D \cdot K_f. \tag{10}$$

Since the row vectors of K_0 are linearly independent, K_0 has a non-singular, $(e - c + k)$-order square minor K_0'. Let K_f' denote the minor of K_f formed by the columns corresponding to the columns of K_0'. Then, according to (10):

$$K_0' = D \cdot K_f'.$$

Thus, by the theorem concerning the determinant of the product of two matrices,

$$\det K_0' = \det D \cdot \det K_f',$$

and since $\det K_0' \neq 0$, D cannot be singular, i.e.

$$\det D \neq 0. \tag{11}$$

Writing equation (10) in detail:

$$[K_{11} \quad K_{12}] = D \cdot [E \quad K_{12}'] = [D \quad D \cdot K_{12}'].$$

Consequently,

$$K_{11} = D,$$

and so, in view of (11), K_{11} cannot be singular.

In order to prove the necessity, assume that the matrix K_{11} appearing in (9) is non-singular, i.e.

$$\det K_{11} \neq 0. \tag{12}$$

We must prove that the edges corresponding to the columns of K_{11} constitute a system of chords of G, i.e. the edges corresponding to the columns of K_{12} constitute the edges of a spanning forest of G. To this end, it suffices to show that the edges corresponding to the columns of K_{12} form no circuit of G (since the number of columns in K_{12} is $c - k$, and then the edges corresponding to the columns of K_{12} constitute the edges of a spanning forest of G; see the solutions of [2] 2.4 and 2.26). Assume, indirectly, that G has a circuit C and all of its edges appear among those corresponding to the columns of K_{12}. Let us augment the matrix K_0 by the row vector corresponding to C:

$$K' = \begin{bmatrix} K_{11} & K_{12} \\ N & K_{22} \end{bmatrix}$$

where N is a zero matrix consisting of a single row. The matrix $[N \quad K_{22}]$ cannot be included in K_0 since, in view of (12), K_{11} cannot contain a row consisting of zeros only. However, this row vector is present in the circuit matrix K of G with its columns arranged in the order of the columns of K_0.

K_{22} includes at least one non-zero element, by the definition of C. Let us augment the matrix

$$\begin{bmatrix} K_{11} \\ N \end{bmatrix}$$

by a column of K' containing such an element. The matrix obtained

$$K'' = \begin{bmatrix} K_{11} & X \\ N & 1 \end{bmatrix}$$

is an $(e - c + k + 1)$-order square minor of K. Expanding its determinant along the last row:

$$\det K'' = \det K_{11} \neq 0$$

which yields

$$r(K) \geq e - c + k + 1 > e - c + k.$$

This is, however, impossible in view of Theorem 20. This completes the proof of Theorem 21.

3.4 Cutsets and the cutset matrix

In this section, just as in the two previous ones, the mathematical model of linear electrical networks is being developed. Parallel and series connections are sometimes called duals of each other. This means that currents in one network and voltages in the other one satisfy similar equations. The recognition of dualities may facilitate the investigations and may also simplify the description of the results. The dualities are present in the mathematical model, too, and often it is the model that enables their easy recognition. Now, the duals of the circuits defined by circuit-currents are sought in the graph model. The edges in a circuit are 'connected in series'. Let us connect the same edges in parallel, too. This is illustrated in figure 3.5. The two graphs can be drawn one over the other so that the corresponding edges intersect. This is shown in figure 3.6 where the edges of one graph have been drawn by broken lines and the vertices of the other have been shown as full circles. Observe that if a circuit of the first graph is traversed, the other is simultaneously 'cut' into two along its intersected edges. This cutting will now be made precise.

The deletion of some set of edges of a graph means that the elements of the set of edges are deleted but not their endpoints. If a graph does have edges and not only loops, then it naturally decomposes, i.e. its number of components increases, if we delete its set of edges. Now, consider a minimal set of edges whose deletion increases the number of components of the graph by one. Since the deletion of a set of edges does not affect the vertices, a set of

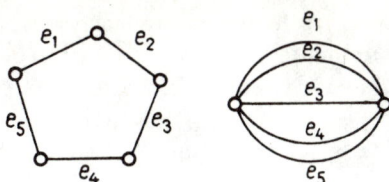

fig. 3.5

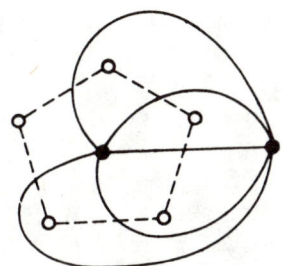

fig. 3.6

edges is required whose deletion decreases by one the rank $c - k$ of the graph containing c vertices and consisting of k components.

Let V be a subset of the set of edges of a graph G. V is said to be a *cutset* of G if the deletion of V decreases the rank of G and V is minimal with respect to this property, i.e. the rank of G fails to decrease if we delete any proper subset of V.

The definition implies that the deletion of a cutset increases the number of components of the graph by exactly one. The deletion of a loop does not affect the number of components of the graph. Hence, owing to its minimality, a cutset cannot contain a loop. The minimality also implies that deleting all but one of the edges of a cutset does not change the number of components. The deletion of a single edge cannot increase the number of components by no more than one since, on reinsertion, an edge can connect at most two components.

A single edge e cannot constitute a cutset if it is contained in a circuit of the graph. If, however, e is not included in any circuit then the deletion of e is easily seen to increase the number of components of the graph. Hence, in view of the minimality, a cutset of more than one edge cannot contain any edge not included in some circuit of the graph. So, the following theorem holds:

22. *An edge of the graph is a cutset of the graph if and only if it is not included in any circuit of the graph. All edges of any cutset of at least two edges are included in some circuit of the graph.*

Consider the graph shown in figure 1.8 and check whether a set of edges belonging to different components or blocks can constitute a cutset. May the set of edges incident with a single vertex be a cutset? Is there a circuit in the graph including an even or, on the other hand, an odd number of the edges of some cutset? The statements below formulate the fact that our observations are valid in general.

23. *If V is a cutset of a graph then all edges of V belong to the same block — and hence to the same component — of the graph.*

Proof. Indirectly, assume that the element $e_i = \{p_i, q_i\}$ of a cutset V of a graph G are included in the block T_i of the graph ($i = 1, 2$). Recall the properties of blocks (Theorems 11 and 12 of Chapter 1). The minimality of the cutsets implies that the rank of G — i.e. the number of its components — remains unchanged if we delete the edges of V from G with the exception of e_1. Hence there is a p_2q_2-path in G containing no edge of V. This path, along with e_2, yields a circuit K_2 with all of its edges in T_2. Similarly, if the edges of V except e_2 are deleted then there is a circuit K_1 in T_1 with e_1 being its only edge in V. It follows from the structure of blocks that K_1 and K_2 share no edge. Now, if the number of components of G fails to increase if we delete the edges of V with the exception of e_1, then neither can it increase on deleting e_1, since p_1 and q_1 stay accessible from each other along the undeleted arc of K_1. This contradicts the fact that V is a cutset of G. Hence Theorem 23 has been proved.

This theorem permits us to restrict the study of cutsets to connected graphs only. The following theorem is proved first:

24. *The set V is formed by the edges incident to the vertex p of a graph G. V is a cutset of G if and only if p is not an articulation.*

Proof. By the above result, G can be assumed to be connected. Theorems 14 and 23 of Chapter 1 prove the necessity of the condition of Theorem 24. In order to prove the sufficiency, assume the vertex p not to be an articulation of G. The deletion of V obviously decreases the rank of G. Now we prove that deleting a proper subset of V, G stays connected, i.e. its rank does not decrease. Let us assume that the edge $e = \{p, q\}$ in V has not been deleted. The definition of blocks and Theorem 14 of Chapter 1 imply that for any edge $\{p, r\}$ in V, there is a circle K in G containing both the edges e and $\{p, r\}$. The circuit K cannot contain more than two edges of V. So, G stays connected after the partial deletion, because the vertices p and r are accessible from each other along the undeleted arc of K. This completes the proof of Theorem 24.

This reasoning also implies that a proper subset of edges incident with the same vertex p of a graph and simultaneously belonging to the same block cannot form a cutset of the graph, whether p is an articulation or not. If V is a set of edges (without loops) incident with an articulation p and simultaneously belonging to the same block, then V is a cutset of the graph. Indeed, the properties of blocks (Theorem 17 of Chapter 1) imply that the deletion of V increases the number of components of the graph. So, either V or a proper subset of it is a cutset of the graph. The latter, however, is impossible in view of the foregoing. This completes the proof of the following theorem:

25. *If p is an articulation of a graph G, then, disregarding loops, the edges incident with p and belonging to the same block constitute a cutset of G.*

The following theorems indicate the close connection between the cutsets and spanning forests of the graph:

26. *Let us delete a set E of edges from a graph G. The number of components of G increases if and only if E includes at least one edge of any spanning forest of G.*

Proof. The discussion can again be restricted to connected graphs since E contains at least one edge of any spanning forest if and only if G has a component K so that E contains at least one edge of any of its spanning trees.

The condition of the theorem — namely that E includes at least one edge of any spanning tree of G — is sufficient because G cannot stay connected after the deletion of E since any connected graph has a spanning tree (see [2] 2.32).

The condition is necessary because if it is not satisfied, i.e. if E fails to include any edge of some spanning tree F of G, then the graph stays connected after deleting E (since F is not affected). This concludes the proof of Theorem 26.

Hence, and from the definition of a cutset, we directly obtain:

27. *A subset of the edges of a graph G is a cutset of G if and only if it includes at least one edge from any spanning forest of G and it is minimal with respect to this property.*

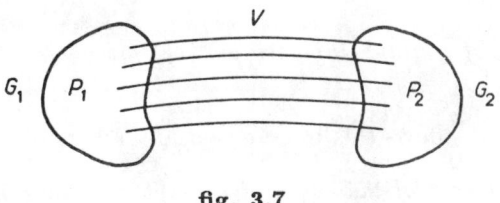

fig. 3.7

Let G be a connected graph. We saw that G falls into exactly two components by deleting a cutset V. Let these components be denoted by G_1 and G_2 and let the sets of vertices of G, G_1 and G_2 be P, P_1 and P_2, respectively. Then $P = P_1 \cup P_2$. Now, the edges of V are by the P_1P_2-edges of G. This fact helps us to imagine the cutsets (see figure 3.7); and vice versa: if the set P of vertices of a graph G can be partitioned into two disjoint parts P_1 and P_2 with $P = P_1 \cup P_2$ and with the subgraphs of G induced by both P_1 and P_2 (see [2] page 127) being connected, then the set V of the $P_1 P_2$-edges constitute a cutset of G. Indeed, the deletion of V increases the number of components of G, but the deletion of $V - \{e\} = V'$ fails to do that since, on augmenting a spanning tree of each of the two induced subgraphs by e, a spanning tree of G is obtained which is not affected by the deletion of V'. Any edge e of any spanning forest L of a graph G containing c vertices and consisting of k components define such a partition of the set of vertices of any component K containing e as follows: let F denote the component of L which is the spanning tree of K. The edge e is a cutset of F and also of L and so the vertices of F, i.e. the vertices of K, can be uniquely decomposed into two sets in the aforementioned way with the help of e. Therefore, this cutset of K defined via e — which is also a cutset of G — contains only e from among the edges of F, its further edges are chords of K with respect to F and hence also of G with respect to L. Since L has $c - k$ edges, this has defined $c - k$ cutsets of G. These cutsets are the *fundamental cutsets*, in short, *f-cutsets* of the graph G with respect to L and they together give the *fundamental system of cutsets*, i.e. *f-system of cutsets* of G with respect to L.

The f-system of cutsets defined by the spanning forest L is a dual concept of the f-system of circuits also defined by L as shown by the two statements below. Each element of the f-system of cutsets defined by L includes exactly one edge belonging to L and each element of the f-system of circuits defined by L (see [2] 2.17) includes exactly one edge not belonging to L.

We saw that all edges of a cutset of the graph belong to the same block of the graph. A similar statement holds for circuits. We also saw that a cutset V always 'bisects' a single component as shown in figure 3.7. Now, let us consider a circuit K of the graph and let us traverse it along its edges. Then, either no edge of V has been crossed or we have proceeded from P_1 to P_2 exactly as many times as vice versa. Therefore, the following theorem is valid:

28. *Any cutset of a graph includes an even number of edges from any circuit of the graph.*

The theorem below shows this to be essentially reversible:

29. *A non-empty set of the edges of a graph including an even number of edges from any circuit of the graph is the union of pairwise edge-disjoint cutsets of the graph.*

Proof. The proof can obviously be restricted to connected graphs. Let V be the relevant set of edges of G. V will be shown to cut the graph. Indeed, if V fails to cut G then let L be a spanning tree of the graph obtained from

G by deleting V and, further, let $e \in V$. Consider the fundamental circuit of e relative to L. This contains a single edge of V which is a contradiction.

So, V includes a cutset V_1. Then $V - V_1$ also includes an even number of edges from any circuit of G. So, repeating the deletion process, the assertion of Theorem 29 follows. If the concepts in the above proof are replaced by their respective duals, the following dual of Theorem 29 can be proved:

30. *A non-empty set of the edges of a graph including an even number of edges from any cutset of the graph is the union of the edge-sets of pairwise edge-disjoint circuits of the graph.*

This theorem is noted to be in close connection with the routes in graphs along so-called Eulerian lines (cf. [2] 3.5).

The *cutsets* of a directed graph $\vec{G}$ are the cutsets of G with orientations, as suggested by figure 3.7 (e.g. from P_1 toward P_2).

Thereupon, the rows of the *cutset matrix* $V = [v_{ij}]$ of a graph G are associated with the cutsets of the graph and the columns of V with the edges of the graph. More precisely, if the cutsets and edges of G are $v_1, v_2, \ldots, v_s$ and $e_1, e_2, \ldots, e_e$, respectively, then V is an $s \times e$ matrix and

$$v_{ij} = \begin{cases} 1 & \text{if } v_i \text{ includes the edge } e_j, \\ 0 & \text{if } v_i \text{ does not include the edge } e_j. \end{cases}$$

In the case of a directed graph, with the above notation and with the orientations of the cutsets, the cutset matrix $V = [v_{ij}]$ is defined as

$$v_{ij} = \begin{cases} 1 & \text{if } v_i \text{ includes } e_j \text{ and their directions coincide}, \\ -1 & \text{if } v_i \text{ includes } e_j \text{ and their directions are opposite} \\ 0 & \text{if } v_i \text{ does not contain } e_j. \end{cases}$$

As an example, let us consider the graph $\vec{G}$ shown in figure 3.4. The cutset matrices of G and $\vec{G}$ are:

$$V_1 = \begin{bmatrix} 1 & 1 & 1 & 0 & 0 & 1 & 0 & 0 & 0 \\ 1 & 1 & 0 & 1 & 0 & 0 & 1 & 0 & 0 \\ 0 & 0 & 1 & 1 & 0 & 0 & 0 & 1 & 0 \\ 0 & 0 & 0 & 0 & 0 & 0 & 0 & 0 & 1 \\ 0 & 0 & 1 & 1 & 0 & 1 & 1 & 0 & 0 \\ 1 & 1 & 1 & 0 & 0 & 0 & 1 & 1 & 0 \\ 1 & 1 & 0 & 1 & 0 & 1 & 0 & 1 & 0 \\ 0 & 0 & 0 & 0 & 0 & 1 & 1 & 1 & 0 \end{bmatrix}$$

and

$$V_2 = \begin{bmatrix} -1 & 1 & -1 & 0 & 0 & 1 & 0 & 0 & 0 \\ 1 & -1 & 0 & -1 & 0 & 0 & 1 & 0 & 0 \\ 0 & 0 & -1 & -1 & 0 & 0 & 0 & 1 & 0 \\ 0 & 0 & 0 & 0 & 0 & 0 & 0 & 0 & 1 \\ 0 & 0 & -1 & -1 & 0 & 1 & 1 & 0 & 0 \\ 1 & -1 & 1 & 0 & 0 & 0 & 1 & -1 & 0 \\ 1 & -1 & 0 & -1 & 0 & -1 & 0 & 1 & 0 \\ 0 & 0 & 0 & 0 & 0 & -1 & -1 & 1 & 0 \end{bmatrix}.$$

The selection of the subscripts and the orientation of the cutsets of $\vec{G}$ correspond to the following rule:

(13) Let us select a spanning tree or, in case of a disconnected graph, a spanning forest F. Number the edges of the graph from 1 to e, starting with the chords of F. Then number the cutsets, starting with the f-cutsets defined by F, in the same order as the corresponding chords. In the case of a directed graph, orient the f-cutsets to coincide with the directions of the edges in F included in them.

The properties of f-cutsets imply that if the cutset matrix V' of a graph $\vec{G}$ or G, containing c vertices and consisting of k components, is written in accordance with rule (13), then the $(c-k)$-order upper right square minor of V' is a unit matrix. It is also obvious that any cutset matrix V of the graph can be constructed from V' by rearranging its rows and columns. This leads to the following relation both in the linear spaces over the field of reals and in those over the mod 2 field:

$$r(V) \geq c - k = \varrho(\vec{G}) = \varrho(G). \tag{14}$$

We have already seen the circuit matrices K_1 and K_2 of the graphs shown in figure 3.4. Observe that the scalar product of any row of the matrix V_i with any row of K_i yields zero (in mod 2 sense in case of G). This is true in general, i.e. the following theorem holds:

31. *If the columns of the cutset matrix V and the circuit matrix K of a graph G or $\vec{G}$ are arranged in the same order of the edges, then (in mod 2 sense in the case of G):*

$$V \cdot K^* = N_1 \quad \text{and} \quad K \cdot V^* = N_2$$

where N_1 and N_2 are zero matrices of appropriate dimensions.

Proof. We have to prove only one of the equalities since the other then follows by transposition.

In case of a non-directed graph simply refer to Theorem 28 and recall that the sum of an even number of ones is 0 in the mod 2 field.

In the case of a directed graph, consider the scalar product of the row vectors

$$[a_1, a_2, \ldots, a_e]^* \quad \text{and} \quad [b_1, b_2, \ldots, b_e]^*$$

in K and V corresponding to an arbitrary circuit and cutset, respectively:

$$\sum_{i=1}^{e} a_i \cdot b_i.$$

Besides Theorem 28, observe that if the circuit is traversed, say, in accordance with its orientation then this walk proceeds from P_1 toward P_2 along as many edges as it does in the opposite direction. Therefore the orientation of the cutset is identical to that of the circuit as many times as it is not. Among the even number of non-zero terms in the above scalar product, if the signs of a_i and b_i coincide (the first case above) then their product is 1. In the same number of instances, the signs of a_i and b_i are opposite and so their product is -1. Therefore, the above scalar product is zero. This completes the proof of Theorem 31.

This theorem, Theorem 20, (14) and Sylvester's theorem imply the following theorem:

32. *If V is the cutset matrix of a graph G or $\vec{G}$ containing c vertices and consisting of k components then, in* mod 2 *sense in the case of G:*

$$r(V) = \varrho(G) = \varrho(\vec{G}) = c - k.$$

The minors V_0 of the matrix V consisting of $c-k$ of its linearly independent row vectors are called *reduced cutset matrices* of the graph G or $\vec{G}$ (with linear independence meant in mod 2 sense in the case of G). If rule (13) is followed then the minor V_f consisting of the first $c - k$ rows of V can be partitioned as follows:

$$V_f = [V_{11} \quad E]$$

where E is a unit matrix. Such cutset matrices V_f are called *fundamental cutset matrices*, in short, *f-cutset matrices* of the graph G or $\vec{G}$. Evidently, any f-cutset matrix is also a reduced cutset matrix and

$$r(V_f) = r(V_0) = \varrho(G) = \varrho(\vec{G}) = c - k$$

(in mod 2 sense in the case of G).

Accordingly, the rank of the cutset matrix of a graph is equal to the rank of the incidence matrix of the graph. These two matrices will later be shown to provide the same information for investigating networks.

Let us consider again the incidence matrices B_i and cutset matrices V_i of the graphs G and $\vec{G}$ shown in figure 3.4 ($i = 1, 2$). Observe that the rows 1,

2, 3 and 5 of B_1 coincide with the rows 1, 2, 8 and 4 of V_1, respectively, in this order and that the row vector 4 of B_1 is the sum of the row vectors 3 and 4 of V_1. The vertex corresponding to the latter row of B_1 is an articulation of the graph. A similar phenomenon can be observed between B_2 and V_2 and it can be further seen that any row vector of B_i or V_i can be expressed as the linear combination of the row vectors of the other. The following theorem shows the general validity of these observations:

33. *If the columns of the incidence matrix and the cutset matrix of a graph G or $\vec{G}$ are arranged in the same order of the edges, then the row vectors of any of these matrices can be expressed as linear combinations of the row vectors of the other matrix (in mod 2 sense in the case of G).*

Proof. (Follow the reasoning in mod 2 sense when referring to G.) Theorems 24 and 25 imply that the row vectors of the incidence matrix can be expressed as linear combinations of the row vectors of the cutset matrix. Hence, and by the definition of a reduced cutset matrix, the row vectors of a reduced incidence matrix B_0 of a graph or directed graph containing c vertices and consisting of k components can be expressed as linear combinations of the row vectors of its reduced cutset matrix V_0. Let D denote the $(c-k)$-order square matrix formed by the coefficients appearing in this expansion, i.e. let

$$B_0 = D \cdot V_0. \tag{15}$$

Since $r(B_0) = c - k$, B_0 has a $(c-k)$-order non-singular square minor B_0':

$$\det B_0' \neq 0. \tag{16}$$

Let V_0' be the minor of V_0 selected in accordance with the columns of B_0'. Then

$$B_0' = D \cdot V_0'. \tag{17}$$

Hence, in view of the theorem concerning the determinant of the product of matrices,

$$\det B_0' = \det D \cdot \det V_0'.$$

Now, (16) implies that D is non-singular and so V_0 can be expressed from (15) as

$$V_0 = D^{-1} \cdot B_0. \tag{18}$$

This completes the proof of Theorem 33.

Since D is non-singular, the above reasoning also implies that V_0' is non-singular if and only if B_0' is non-singular. The following theorem is a consequence of this and of Theorems 15 and 17:

34. *A $(c-k)$-order square minor of a reduced cutset matrix of a graph or directed graph containing c vertices and consisting of k components is non-*

singular if and only if the edges corresponding to its columns constitute the edges of a spanning forest of the graph.

3.5 Interrelations between the matrices of graphs

Theorems 19, 31 and 33 indicate interrelations among the incidence matrix, the circuit matrix and the cutset matrix. These interrelations will now be made clearer using vector spaces associated with graphs. First, however, the following reverse of Theorem 31 is proved:

35. *Let K be a circuit matrix of a graph G or $\vec{G}$ containing e edges, let the number of coordinates be e in a non-zero column vector $\mathbf{w}$ formed by values 0 and 1 in the case of G and of values 0, 1 and -1 in the case of $\vec{G}$. Finally let $K\mathbf{w}$ be a zero vector (in* mod 2 *sense in the case of G). Then the edges associated with the non-zero coordinates of $\mathbf{w}$ form the union of pairwise edge-disjoint cutsets of G.*

Proof. The theorem is a consequence of Theorem 29. To show this for G, observe that if the components of $\mathbf{w}$ are associated with the edges of G in the same order as in K, then the one-entries in $\mathbf{w}$ correspond to a subset of the set of edges of G. Now, the fact that $K\mathbf{w}$ is a zero vector is equivalent to the requirement that the set of edges corresponding to $\mathbf{w}$ includes an even number of edges from any circuit of G.

In the case of $\vec{G}$, observe that $K\mathbf{w}$ can be a zero vector only if among the non-zero terms in the scalar product of $\mathbf{w}$ with any row of K there are as many elements in $\mathbf{w}$ and K with the same signs as there are with opposite signs. Therefore, there is an even number of such terms. Thus, the set of edges in G corresponding to $\mathbf{w}$ again includes an even number of edges from any circuit of G. This completes the proof of Theorem 35.

In order to make the mentioned interrelations clearer, let us consider the 'weighted' subgraphs of G. That is, associate an element of a field T with each edge of G as a 'weight', and weighting by the zero element of T is interpreted as deleting the edge. If G has e edges (with a fixed order), then any weighted subgraph of G can be described by the ordered set of the corresponding e edge-weights and, conversely, any ordered set of e elements of T determines a weighted subgraph of G. Therefore, the weighted subgraphs of G can be identified with the ordered sets of e elements of T. So, these form an e-dimensional linear vector space; let us denote it by $\mathcal{R}_G^e$.

Now let T be the mod 2 field. Then the edges are weighted by 0 or 1 only, with the obvious interpretation of the relevant edge having been deleted or retained when constructing the subgraph. Then, the sum of two weighted subgraphs G_1 and G_2 (i.e. the sum of two sets of e elements 0 or 1) yields a subgraph consisting of those edges of G_1 and G_2 which are not present in the other subgraph. The vectors containing a single element 1, i.e. the subgraphs of G with one edge, form a basis.

For a graph G containing e edges and c vertices and consisting of k components, let a reduced incidence matrix be B_0, a reduced circuit matrix be K_0 and a reduced cutset matrix be V_0. Since

$$r(B_0) = r(V_0) = c - k \quad \text{and} \quad r(K_0) = e - c + k,$$

the vectors obtainable as linear combinations of the row vectors of the matrices B_0, V_0 and K_0 form the $(c-k)$-dimensional subspaces $\mathcal{R}_B^{c-k}$, $\mathcal{R}_V^{c-k}$ and the $(e-c+k)$-dimensional subspace $\mathcal{R}_K^{e-c+k}$ of $\mathcal{R}_G^e$ and, according to Theorem 32,

$$\mathcal{R}_B^{c-k} = \mathcal{R}_V^{c-k}. \tag{19}$$

The significance of the vectors of these subspaces is shown by the two theorems below:

36. *The subspace $\mathcal{R}_K^{c-k}$ of the linear vector space $\mathcal{R}_G^e$ of the subgraphs of a graph G containing e edges, c vertices and consisting of k components is constituted by 2^{c-k} vectors all of which, except the zero vector, are unions of edge-disjoint cutsets of G.*

Proof. The number of the vectors in the subspace stems from the fact that each of these vectors is represented by a sequence of length $c - k$ formed by elements of two possible values. The rest of the theorem is due to Theorem 35, since if an arbitrary base of a subspace is orthogonal to an arbitrary base of another subspace (which is here the case by Theorem 31) then the two subspaces are orthogonal.

37. *The subspace $\mathcal{R}_K^{e-c+k}$ of the linear vector space $\mathcal{R}_G^e$ of the subgraphs of a graph G containing e edges, c vertices and consisting of k components is constituted by 2^{e-c+k} vectors all of which, except the zero vector, are unions of edge-disjoint circuits of G.*

Proof. The number of the vectors in the subspace stems from the fact that each of these vectors is represented by a sequence of length $e - c + k$ formed by elements of two possible values. The rest of the theorem can be verified by the reasoning below.

Any vector of the subspace $\mathcal{R}_K^{e-c+k}$ can be obtained as the sum of certain rows of the matrix K_0, each representing a circuit (keep those edges of the circuits which appear in an odd number of terms in the sum). The degree of every vertex of a circuit is two. If the degree of every vertex of a graph obtained by a certain number of such additions is even then this property is retained after the addition of a further circuit K (the degree of a vertex p also belonging to K, with original degree $2n$ becomes $2n + 2$, $2n$ or $2n - 2$

depending on whether the number of edges incident with p and included in K is 0, 1 or 2). Now, if all vertices of a graph are of even degree then its edges constitute the union of the sets of edges of edge-disjoint circuits (see [2] 3.5). This completes the proof of Theorem 37.

Let us now consider a graph $\vec{G}$ containing c vertices, e edges and consisting of k components, and let us weight its edges by real numbers. Let B_0 be a reduced incidence matrix, V_0 a reduced cutset matrix and K_0 a reduced circuit matrix of $\vec{G}$, and let $\mathcal{R}_B^{c-k}$, $\mathcal{R}_V^{c-k}$ and $\mathcal{R}_K^{e-c+k}$ be the subspaces of the vector space $\mathcal{R}_{\vec{G}}^e$ over the field of real numbers formed by the linear combinations of the rows of B_0, V_0 and K_0, respectively. According to Theorem 33, (19) is still valid now. In the linear vector space over the field of real numbers, orthogonality implies linear independence. Hence, and since the sum of $c-k$ and $e-c+k$ is e, the following consequence of Theorem 31 is obtained:

38. $\mathcal{R}_{\vec{G}}^e$ *is the linear vector space of the weighted subgraphs of a graph* $\vec{G}$ *containing e edges and c vertices and consisting of k components. The weights are real numbers. The two subspaces* $\mathcal{R}_B^{c-k}$ *(or* $\mathcal{R}_V^{c-k}$*) and* $\mathcal{R}_K^{e-c+k}$ *of* $\mathcal{R}_{\vec{G}}^e$ *are two orthogonal components of* $\mathcal{R}_{\vec{G}}^e$ *and their direct sum is* $\mathcal{R}_{\vec{G}}^e$*, i.e.:*

$$\mathcal{R}_B^{c-k} \oplus \mathcal{R}_K^{e-c+k} = \mathcal{R}_V^{c-k} \oplus \mathcal{R}_K^{e-c+k} = \mathcal{R}_{\vec{G}}^e.$$

This is the mathematical background for a relation between the equations expressing Kirchhoff's voltage and current laws below.

Remark. The orthogonality of the subspaces corresponding to those in Theorem 38 has already played a role in proving Theorem 36. However, it has not been possible there to quote the independence of the vectors of the two subspaces since, for example, a vector containing an even number of values 1 is orthogonal to itself without being independent. Further, a set of edges forming a circuit of a non-directed graph G may simultaneously be one of its cutsets. The simplest example of this is the set of all edges of G if G is a circuit of length 2. In spite of this, the relevant subspaces which, incidentally, coincide are orthogonal there, too.

We saw that the incidence matrix essentially defines the graph and thus it defines the circuit matrix, too. The theorem below indicates the way an f-circuit matrix of the graph can be constructed from the reduced incidence matrix.

39. *If $B_0 = [B_{11} \quad B_{12}]$ is a reduced incidence matrix of the graph with B_{12} a non-singular square matrix, then the matrix*

$$K_f = [E \quad B_{11}^* \cdot B_{12}^{-1*}]$$

(mod 2) *is an f-circuit matrix of the graph where E is a unity matrix of appropriate size.*

Proof. According to Theorem 15, the edges corresponding to the columns of B_{12} constitute the edges of a spanning forest of the graph G, so the edges corresponding to the columns of B_{11} form a system of chords of G. Let K_f be an f-circuit matrix of G written in accordance with rule (6). Then K_f can be partitioned as follows:

$$K_f = [E \quad K_{12}]$$

where E is a unit matrix. Theorem 19 implies that

$$B_0 \cdot K_f^* = N$$

where N is a zero matrix. Expanding the product

$$B_{11} + B_{12} \cdot K_{12}^* = N.$$

In view of the 'subtraction rule' valid in the mod 2 field:

$$B_{12} \cdot K_{12}^* = B_{11}.$$

Multiplying this by B_{12}^{-1} from the left,

$$K_{12}^* = B_{12}^{-1} \cdot B_{11}$$

is obtained, and hence, by transposition:

$$K_{12} = B_{11}^* \cdot B_{12}^{-1*}$$

which proves Theorem 39.

As an example, let us apply this theorem to the graph G of figure 3.4. An incidence matrix B_1 of G has been written just before Theorem 19 was stated. Let B_0 be the reduced incidence matrix of G obtained from B_1 by deleting its last row. Then:

$$B_{11} = \begin{bmatrix} 1 & 1 & 1 & 0 & 0 \\ 1 & 1 & 0 & 1 & 0 \\ 0 & 0 & 0 & 0 & 0 \\ 0 & 0 & 1 & 1 & 0 \end{bmatrix}, \quad B_{12} = \begin{bmatrix} 1 & 0 & 0 & 0 \\ 0 & 1 & 0 & 0 \\ 1 & 1 & 1 & 0 \\ 0 & 0 & 1 & 1 \end{bmatrix},$$

$$B_{11}^* = \begin{bmatrix} 1 & 1 & 0 & 0 \\ 1 & 1 & 0 & 0 \\ 1 & 0 & 0 & 1 \\ 0 & 1 & 0 & 1 \\ 0 & 0 & 0 & 0 \end{bmatrix}, \quad B_{12}^{-1*} = \begin{bmatrix} 1 & 0 & 1 & 1 \\ 0 & 1 & 1 & 1 \\ 0 & 0 & 1 & 1 \\ 0 & 0 & 0 & 1 \end{bmatrix},$$

$$B_{11}^* \cdot B_{12}^{-1*} = \begin{bmatrix} 1 & 1 & 0 & 0 \\ 1 & 1 & 0 & 0 \\ 1 & 0 & 1 & 0 \\ 0 & 1 & 1 & 0 \\ 0 & 0 & 0 & 0 \end{bmatrix}.$$

Consequently:

$$K_f = [E \quad B_{11}^* \cdot B_{12}^{-1*}] = \begin{bmatrix} 1 & 0 & 0 & 0 & 0 & 1 & 1 & 0 & 0 \\ 0 & 1 & 0 & 0 & 0 & 1 & 1 & 0 & 0 \\ 0 & 0 & 1 & 0 & 0 & 1 & 0 & 1 & 0 \\ 0 & 0 & 0 & 1 & 0 & 0 & 1 & 1 & 0 \\ 0 & 0 & 0 & 0 & 1 & 0 & 0 & 0 & 0 \end{bmatrix}.$$

This matrix, which is defined by the spanning tree of G indicated in figure 3.4, coincides with the minor formed by the first five rows of K_1 already written in connection with that diagram.

The following theorem points out how an f-cutset matrix of a graph G can be constructed from its arbitrary f-circuit matrix and vice versa, as well as one of its f-cutset matrices from any of its reduced incidence matrices:

40. *If $B_0 = [B_{11} \quad B_{12}]$ is a reduced incidence matrix of a graph with B_{12} a non-singular square matrix and further, if*

$$K_f = [E_1 \quad K_{12}] \quad \text{and} \quad V_f = [V_{11} \quad E_2],$$

(where E_1, E_2 are unit matrices) are an f-circuit matrix and an f-cutset matrix of G, respectively, with their columns written in the same order as the edges defined by B_0, then

$$V_{11}^* = K_{12} \quad \text{and} \quad V_f = B_{12}^{-1} \cdot B_0 \quad (\text{mod } 2).$$

Proof. According to Theorem 39:

$$K_f = [E_1 \quad K_{12}] = [E_1 \quad B_{11}^* \cdot B_{12}^{-1*}]. \tag{20}$$

Theorem 31 implies that

$$K_f \cdot V_f^* = N$$

where N is a zero matrix of appropriate size. In more detail:

$$[E_1 \quad B_{11}^* \cdot B_{12}^{-1*}] \cdot \begin{bmatrix} V_{11}^* \\ E_2 \end{bmatrix} = N,$$

i.e.

$$V_{11}^* + B_{11}^* \cdot B_{12}^{-1*} = N.$$

Hence, expressing V_{11}^*:

$$V_{11}^* = B_{11}^* \cdot B_{12}^{-1*}$$

which, in view of (20), proves the first formula in the theorem. Now,

$$V_f = [B_{12}^{-1*} \cdot B_{11} \quad E_2],$$

and so

$$B_{12}^{-1} \cdot B_0 = B_{12}^{-1} [B_{11} \quad B_{12}] = [B_{12}^{-1} \cdot B_{11} \quad E_2].$$

This implies the second formula.

As an exercise, let us carry out the calculations for the graph G shown in figure 3.4. The matrix K_f has just been written, and the matrix V_f coincides with the minor formed by the first four rows of V_1 written just before the introduction of rule (13). The first formula of the theorem is easily verified, and to obtain the second one, follow the calculations below:

$$B_{12}^{-1} \cdot B_0 = \begin{bmatrix} 1 & 0 & 0 & 0 \\ 0 & 1 & 0 & 0 \\ 1 & 1 & 1 & 1 \\ 1 & 1 & 1 & 1 \end{bmatrix} \cdot \begin{bmatrix} 1 & 1 & 1 & 0 & 0 & 1 & 0 & 0 & 0 \\ 1 & 1 & 0 & 1 & 0 & 0 & 1 & 0 & 0 \\ 0 & 0 & 0 & 0 & 0 & 1 & 1 & 1 & 0 \\ 0 & 0 & 1 & 1 & 0 & 0 & 0 & 1 & 1 \end{bmatrix}$$

$$= \begin{bmatrix} 1 & 1 & 1 & 0 & 0 & 1 & 0 & 0 & 0 \\ 1 & 1 & 0 & 1 & 0 & 0 & 1 & 0 & 0 \\ 0 & 0 & 1 & 1 & 0 & 0 & 0 & 1 & 0 \\ 0 & 0 & 0 & 0 & 0 & 0 & 0 & 0 & 1 \end{bmatrix} = V_f.$$

The two last theorems and their proofs are 'essentially' valid for directed graphs, too. The only difference is that, subtraction is to be carried out there in the field of real numbers. This results in differences in signs only. More precisely, the following theorem is obtained:

41. *If $B_0 = [B_{11} \quad B_{12}]$ is a reduced incidence matrix of a directed graph $\vec{G}$ with B_{12} a non-singular square matrix and further, if*

$$K_f = [E_1 \quad K_{12}] \quad \text{and} \quad V_f = [V_{11} \quad E_2],$$

(where E_1, E_2 are unit matrices) are an f-circuit matrix and an f-cutset matrix of $\vec{G}$, respectively, with their columns written in the same order as the edges defined by B_0, then

$$K_{12} = -B_{11}^* \cdot B_{12}^{-1*}, \qquad V_{11}^* = -K_{12} \quad \text{and} \quad V_f = B_{12}^{-1} \cdot B_0.$$

As an example, let us check the results of the theorem on figure 3.4. Deleting the last two rows of B_2 leads to B_0, and the matrices K_f and V_f are formed by the first five rows of K_2 and the first four rows of V_2, respectively. The second formula in the theorem is immediately obvious. To obtain the third and first ones, let us follow the following calculations:

$$B_{12}^{-1} = \begin{bmatrix} 1 & 0 & 0 & 0 \\ 0 & 1 & 0 & 0 \\ -1 & -1 & 1 & 0 \\ 0 & 0 & -1 & 1 \end{bmatrix}^{-1} = \begin{bmatrix} 1 & 0 & 0 & 0 \\ 0 & 1 & 0 & 0 \\ 1 & 1 & 1 & 0 \\ 1 & 1 & 1 & 1 \end{bmatrix}.$$

$$B_{12}^{-1} \cdot B_0 = \begin{bmatrix} 1 & 0 & 0 & 0 \\ 0 & 1 & 0 & 0 \\ 1 & 1 & 1 & 0 \\ 1 & 1 & 1 & 1 \end{bmatrix} \cdot \begin{bmatrix} -1 & 1 & -1 & 0 & 0 & 1 & 0 & 0 & 0 \\ 1 & -1 & 0 & -1 & 0 & 0 & 1 & 0 & 0 \\ 0 & 0 & 0 & 0 & 0 & -1 & -1 & 1 & 0 \\ 0 & 0 & 1 & 1 & 0 & 0 & 0 & -1 & 1 \end{bmatrix}$$

$$= \begin{bmatrix} -1 & 1 & -1 & 0 & 0 & 1 & 0 & 0 & 0 \\ 1 & -1 & 0 & -1 & 0 & 0 & 1 & 0 & 0 \\ 0 & 0 & -1 & -1 & 0 & 0 & 0 & 1 & 0 \\ 0 & 0 & 0 & 0 & 0 & 0 & 0 & 0 & 1 \end{bmatrix} = V_f,$$

$$B_{11}^* \cdot B_{12}^{-1*} = \begin{bmatrix} -1 & 1 & 0 & 0 \\ 1 & -1 & 0 & 0 \\ -1 & 0 & 0 & 1 \\ 0 & -1 & 0 & 1 \\ 0 & 0 & 0 & 0 \end{bmatrix} \cdot \begin{bmatrix} 1 & 0 & 1 & 1 \\ 0 & 1 & 1 & 1 \\ 0 & 0 & 1 & 1 \\ 0 & 0 & 0 & 1 \end{bmatrix}$$

$$= \begin{bmatrix} -1 & 1 & 0 & 0 \\ 1 & -1 & 0 & 0 \\ -1 & 0 & -1 & 0 \\ 0 & -1 & -1 & 0 \\ 0 & 0 & 0 & 0 \end{bmatrix} = -K_{12}.$$

The following theorem establishes a connection between the incidence matrix of a directed graph $\vec{G}$ and the adjacency matrix of G:

42. *Let B be the incidence matrix of a directed graph $\vec{G}$, A be the adjacency matrix of the graph G and D be a diagonal matrix where the ith element is equal to the number of edges incident with the ith vertex of G. Then*

$$BB^* = D - A.$$

Corollary. *BB^* is independent of the orientation of the edges of $\vec{G}$.*

Proof. The corollary is obvious.

Let $\mathbf{b}_i^*$ denote the ith row vector of B. Then, the (i,j) entry of BB^* is $\mathbf{b}_i^* \mathbf{b}_j$. If $i \neq j$ then the non-zero elements in the ith and jth rows of B are in those columns where the corresponding edges join the ith and the jth vertex of G. In $\vec{G}$, one of these vertices is a tail and the other is a head, and so the two non-zero terms below each other are a $(+1)$ and a (-1), their product being (-1). So, in case $i \neq j$, $\mathbf{b}_i^* \mathbf{b}_j$ is the number of edges in G connecting the vertices i and j multiplied by (-1). This appears at the right-hand side as well.

The product $\mathbf{b}_i^* \mathbf{b}_i$, however, yields the number of non-loop edges incident to vertex i, and this is equal to the element of $D - A$ in position (i,i). This concludes the proof of Theorem 42.

3.6 The spectrum of graphs. The complexity

Among the above matrices associated with graphs, the adjacency matrix is a square matrix. Its characteristic polynomial is also called the *characteristic polynomial of the graph*. The *characteristic matrix, characteristic equation*, an *eigenvalue* or an *eigenvector of the graph* are defined similarly. The *spectrum of a graph* is defined as the spectrum of its adjacency matrix, that is, another matrix of two rows, the first of which contains the eigenvalues of the matrix and the second of which is formed by the multiplicities of the eigenvalues. For the sake of uniformity, only simple graphs are considered here and their eigenvectors are understood as right eigenvectors.

Let us consider the characteristic polynomial $p(\lambda) = p(G, \lambda)$ of a simple graph G with n vertices, with the convention that the coefficient of the highest order term is positive, i.e.

$$p(G, \lambda) = p(\lambda) = \det(\lambda E - A)$$

$$= (-1)^n \det \begin{bmatrix} a_{11} - \lambda & a_{12} & \cdots & a_{1n} \\ a_{21} & a_{22} - \lambda & \cdots & a_{2n} \\ \cdots & \cdots & \cdots & \cdots \\ a_{n1} & a_{n2} & \cdots & a_{nn} - \lambda \end{bmatrix},$$

where $A = [a_{ij}]$ is the adjacency matrix of G and E is a unit matrix. Here, $a_{ii} = 0$ for any i and the value of any a_{ij} is either 0 or 1, since G is a simple graph. The determinant is the sum of its expansion terms:

$$p(\lambda) = (-1)^n \sum_{i=1}^{n!} (-1)^I b_{1i_1} b_{2i_2} \ldots b_{ni_n}$$

where

$$b_{ii} = a_{ii} - \lambda,$$
$$b_{ij} = a_{ij} \quad \text{if } i \neq j$$

and I is the number of inversions in the ith permutation of the second subscripts. Let us examine the coefficients in the polynomial form

$$p(\lambda) = \lambda^n + c_1 \lambda^{n-1} + c_2 \lambda^{n-2} + \ldots + c_{n-1} \lambda + c_n.$$

It is easy to see that $c_1 = 0$, since c_1 is the trace of the matrix A, i.e.

$$c_1 = \sum_{i=1}^{n} a_{ii}.$$

To examine the further coefficients, recall that $a_{ii} = 0$ for all i. Let us now find the coefficient of λ^{n-2}. To this end, the expansion terms with $n-2$ factors $(-\lambda)$ from the main diagonal must be considered. Let d be such a term. The two factors in d different from $(-\lambda)$ are terms of A chosen from the two rows and columns with the corresponding $(-\lambda)$ in the main diagonal missing. It is not hard to verify that c_2 is the sum of the determinants of the leading second-order minors of the matrix A. (A leading minor is one with its main diagonal included in the main diagonal of A.) Of these minors, the only non-singular one can be

$$\begin{bmatrix} 0 & 1 \\ 1 & 0 \end{bmatrix}.$$

This corresponds to an edge of G, and

$$\det \begin{bmatrix} 0 & 1 \\ 1 & 0 \end{bmatrix} = -1.$$

Similar reasoning yields the following for $(-c_3)$: it is the sum of the determinants of the leading third-order minors of the matrix A. It is easy to verify that only

$$\begin{bmatrix} 0 & 1 & 1 \\ 1 & 0 & 1 \\ 1 & 1 & 0 \end{bmatrix}$$

can be non-singular among these minors. This corresponds to a triangle of G, and

$$\det \begin{bmatrix} 0 & 1 & 1 \\ 1 & 0 & 1 \\ 1 & 1 & 0 \end{bmatrix} = 2.$$

So, the following theorem has been proved:

43. *If the characteristic polynomial of a simple graph G is $\lambda^n + c_1 \lambda^{n-1} + c_2 \lambda^{n-2} + \ldots + c_{n-1} \lambda + c_n$ then*

$$c_1 = 0,$$

$$-c_2 = \text{the number of edges of } G \text{ and}$$

$$-\frac{c_3}{2} = \text{the number of triangles in } G.$$

The question is obvious: do the rest of the coefficients have such simple meaning? Already the examination of c_4 indicates that the answer is negative. c_4 is seen to be the sum of the determinants of the leading fourth-order minors of the matrix A. Every such minor is obviously the adjacency matrix of a

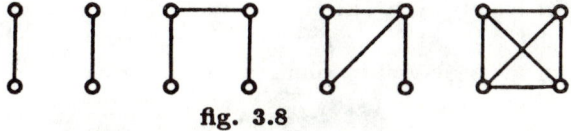

fig. 3.8

subgraph of G induced by four vertices. One can easily see that figure 3.8 shows all simple graphs of four vertices with their adjacency matrix non-singular. Among these, the determinant of the complete graph with four vertices is -3 and that of all others is 1. We have already seen that the eigenvectors and eigenvalues of a simple graph G, with adjacency matrix A and containing n vertices, are given by the solutions of the equations

$$A\mathbf{x} = \lambda \mathbf{x} \quad \text{and} \quad \det(A - \lambda E) = 0.$$

To explain the combinatorial significance of an eigenvalue and an eigenvector, let $\mathbf{a}_i^*$ denote the ith row vector of the matrix A, i.e. the row vector corresponding to vertex p_i of the graph G. Let an arbitrary eigenvector belonging to an eigenvalue λ of G be

$$\mathbf{x} = [x_1, x_2, \ldots, x_n]^*.$$

According to the eigenvector equation:

$$\mathbf{a}_i^* \mathbf{x} = \lambda x_i \qquad (i = 1, 2, \ldots, n),$$

this implies

44. *Place the ith coordinate of an eigenvector $\mathbf{x}$ to the vertex p_i. Then the sum of the numbers at the neighbours of p_i equals the number at p_i multiplied by λ ($i = 1, 2, \ldots, n$).*

One can easily guess that the eigenvector belonging to the eigenvalue k of a k-regular simple graph is a vector with all of its coordinates equal. As the following theorem states, a graph cannot have any eigenvalue with its absolute value greater than k.

45. *If the simple graph G is k-regular, then*

1. *k is an eigenvalue of G;*
2. *the multiplicity of the eigenvalue k is 1, provided G is connected;*
3. *$|\lambda| \leq k$ for any eigenvalue λ of G.*

Proof. Statement 1 is obvious since $[1, 1, \ldots, 1]^*$ is an eigenvector. In order to prove Statement 2, it suffices to show that the coordinates of any eigenvector belonging to k are equal, i.e. there are no two linearly independent vectors among the eigenvectors belonging to k. Indeed, it is well known that the multiple roots of the characteristic equation of a symmetric matrix involve several linearly independent eigenvectors. So, let

$$\mathbf{x} = [x_1, x_2, \ldots, x_n]^*$$

be an eigenvector belonging to the eigenvalue k of the adjacency matrix A of a graph G, and let x_i be the coordinate with maximum absolute value. The jth coordinate of the eigenvector equation

$$A\mathbf{x} = k\mathbf{x}$$

is:

$$s_j^* \mathbf{x} = k x_j, \tag{21}$$

i.e., placing the ith coordinate to the ith vertex as in Statement 44,

$$\Sigma' x_i = k x_j \tag{22}$$

where Σ' extends over the subscripts of the k neighbours of the jth vertex. $|x_j|$ being maximal, each of these k coordinates is x_j. Similarly, the coordinates placed to the second neighbours of the jth vertex are also x_j. Since G is connected, the procedure extends to all vertices with application.

To prove Statement 3, let λ be an arbitrary eigenvalue of the graph G, and $\mathbf{y}$ be an eigenvector belonging to λ, i.e.

$$A\mathbf{y} = \lambda \mathbf{y}.$$

Let y_i be the coordinate of the vector $\mathbf{y}$ with the highest absolute value. As in (21) and (22),

$$\Sigma' y_i = \lambda y_j, \tag{23}$$

hence

$$|\lambda||y_j| = |\Sigma' y_i| \leq \Sigma'|y_i| \leq k|y_j|,$$

and so, in view of $|y_j| \neq 0$,

$$|\lambda| \leq k.$$

This completes the proof of Theorem 45.

Remark. One can also easily prove that the multiplicity of the greatest eigenvalue of a connected, non-regular graph is 1 (see [46]).

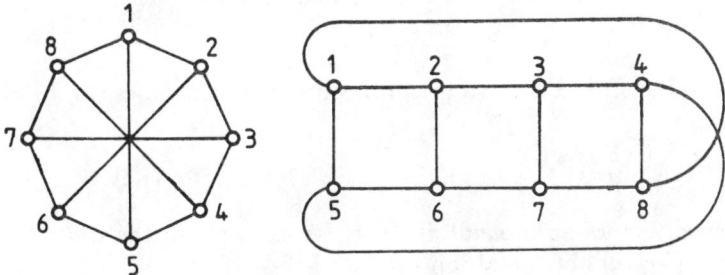

fig. 3.9

As an example, let us calculate the eigenvalues of the *Möbius ladder* with $2h$ vertices ($h \geq 3$). This 3-regular graph can be constructed by connecting the opposite vertices of a regular $2h$-gon by edges. Its name is justified by the following construction: let us draw a ladder with h steps on a strip and twist one of its ends to form a Möbius strip. This results in a Möbius ladder containing $2h$ edges. In figure 3.9, the Möbius ladder in the case $h = 4$ is shown in diagrams corresponding to both ways of its derivation. Its characteristic matrix is

$$\left[\begin{array}{cccc:cccc} -\lambda & 1 & 0 & 0 & 1 & 0 & 0 & 1 \\ 1 & -\lambda & 1 & 0 & 0 & 1 & 0 & 0 \\ 0 & 1 & -\lambda & 1 & 0 & 0 & 1 & 0 \\ 0 & 0 & 1 & -\lambda & 1 & 0 & 0 & 1 \\ \hdashline 1 & 0 & 0 & 1 & -\lambda & 1 & 0 & 0 \\ 0 & 1 & 0 & 0 & 1 & -\lambda & 1 & 0 \\ 0 & 0 & 1 & 0 & 0 & 1 & -\lambda & 1 \\ 1 & 0 & 0 & 1 & 0 & 0 & 1 & -\lambda \end{array}\right].$$

The dotted lines serve to emphasise its structure, hence the characteristic matrix of a general Möbius ladder containing $2h$ vertices can be easily given. However, instead of doing this, the eigenvalues can be determined by means of the following heuristic procedure, too. Let us assign numbers to the vertices of the Möbius ladder of $2h$ vertices as shown in figure 3.9. Let A be the adjacency matrix of the graph, ε a unit root of order $2h$, i.e. one of the complex numbers $\sqrt[2h]{1}$ and

$$\mathbf{x}^* = [\varepsilon, \varepsilon^2, \ldots, \varepsilon^{2h}].$$

We try to find an eigenvalue in this form. Let us write the kth coordinate of $A\mathbf{x}$ ($\bar{\varepsilon}$ =the conjugate of ε):

$$\varepsilon^{k-1} + \varepsilon^{k+1} + \varepsilon^{k+h} = \left(\varepsilon + \frac{1}{\varepsilon}\right)\varepsilon^k + \varepsilon^{k+h} = \left(\varepsilon + \bar{\varepsilon} + \varepsilon^h\right)\varepsilon^k.$$

Obviously, $\varepsilon^h = +1$ or (-1) and $\varepsilon + \bar{\varepsilon}$ is real. Consequently,

$$\varepsilon + \bar{\varepsilon} + \varepsilon^h$$

is an eigenvalue of A. Using all the unit roots, all eigenvalues can be obtained. These are:

$$2\cos\frac{\pi j}{h} + (-1)^j \qquad (j = 0, 1, 2, \ldots, 2h-1).$$

The $2h$ eigenvectors are not all distinct, for example in the cases $h = 3$ and $h = 4$, the spectrum of the Möbius ladder is:

$$\begin{bmatrix} -3 & 0 & 3 \\ 1 & 4 & 1 \end{bmatrix} \quad \text{and} \quad \begin{bmatrix} -\sqrt{2}-1 & -1 & 1 & \sqrt{2}-1 & 3 \\ 2 & 1 & 2 & 2 & 1 \end{bmatrix},$$

respectively.

Knowing the spectrum of a regular graph, let us determine the spectrum of its complement. Prior to Theorem 45, it has been established that k is the eigenvalue of maximal absolute value of a k-regular simple graph G containing n vertices, and that its corresponding eigenvector is an n-dimensional vector $\mathbf{j}$ with all the entries equal to one. Consequently, $n-1-k$ is the eigenvalue of maximal absolute value of the similarly regular complement $\overline{G}$ of the graph G, and j is a corresponding eigenvector. Let an orthogonal system of the eigenvectors of G (such a system exists since the adjacency matrix A is symmetric) be:

$$\mathbf{j} = \mathbf{v}_1, \mathbf{v}_2, \ldots, \mathbf{v}_n,$$

and let these belong to the eigenvalues

$$\lambda_1 \geq \lambda_2 \geq \ldots \geq \lambda_n$$

of A, respectively. We claim that these vectors are eigenvectors of $\overline{G}$, too, i.e. of the adjacency matrix $\overline{A}$ of the graph $\overline{G}$ as well. (This has been already verified for $\mathbf{v}_1$.) Let E be the unity matrix of size n and J the n-order square matrix with all entries equal to one. Then, obviously:

$$\overline{A} = J - A - E.$$

For the verification of our assertion in case $i \geq 2$:

$$\overline{A}\mathbf{v}_i = (J - A - E)\mathbf{v}_i = J\mathbf{v}_i - A\mathbf{v}_i - E\mathbf{v}_i.$$

Owing to the stipulated orthogonality, $\mathbf{j}^*\mathbf{v}_i = 0$, and so, $J\mathbf{v}_i$ is a zero matrix, and

$$A\mathbf{v}_i = \lambda_i \mathbf{v}_i.$$

Consequently, in the case $i \geq 2$,

$$\overline{A}\mathbf{v}_i = (-\lambda_i - 1)\mathbf{v}_i.$$

Hence we obtain the following theorem:

46. *If the eigenvalues of a k-regular simple graph containing n vertices are $\lambda_1 \geq \lambda_2 \geq \ldots \geq \lambda_n$, then the eigenvalues of its complement are:*

$$n - 1 - \lambda_1 \quad \text{and} \quad -\lambda_i - 1 \quad \text{for every} \quad 2 \leq i \leq n.$$

Remarks. 1. The spectrum of a regular graph also implies the multiplicities of the eigenvalues of its complement, i.e. the entire spectrum of the complement. For example, it is easy to verify that the spectrum of a graph consisting of e vertex-disjoint edges and their endpoints is

$$\begin{bmatrix} -1 & 1 \\ e & e \end{bmatrix},$$

and hence, the spectrum of its complement can be immediately written:

$$\begin{bmatrix} -2 & 0 & 2e - 2 \\ e - 1 & e & 1 \end{bmatrix}.$$

This graph also shows that, in case of disconnected regular graphs, the greatest eigenvalue may also be a multiple one (cf. Statement 2 of Theorem 45). The spectrum of a non-regular graph does not in general determine that of its complement (cf. figure 4.19).

2. According to Statement 3 of Theorem 45 and to Theorem 46, the absolute values of the eigenvalues of a k-regular graph containing n vertices and those of its complement are bounded from above by k and by $n-1-k$, respectively. In other words, if two graphs are given, the spectrum of the one with more edges can 'stretch further'. The following theorem states a similar result concerning the relation between a graph and its subgraph:

47. *If the greatest eigenvalues of a simple graph and of its subgraph are λ_m and λ'_m, respectively, then:*

$$0 \leq \lambda'_m \leq \lambda_m.$$

Proof. Since the sum of the eigenvalues is zero, the maximal eigenvalue of any simple graph is non-negative. Let G and G' denote a simple graph and its subgraph, respectively, and let their respective adjacency matrices be A and A'. G' can be assumed to include all vertices of G since otherwise, extending G' by the missing vertices as isolated ones, the set of the eigenvalues of G' is clearly augmented by zeros only. According to the theorem of Frobenius and Perron (see [45], Theorems 9.1.3 and 9.1.6, or [46]), any matrix with non-negative elements has an eigenvector of non-negative coordinates only, corresponding to its largest eigenvalue. For A' and λ'_m, let $\mathbf{v}$ be such a normalised eigenvector, i.e. $|\mathbf{v}| = 1$. Then, using that $A' \leq A$ for any of their entries,

$$\lambda'_m = \mathbf{v}^* \cdot (\lambda'_m \mathbf{v}) = \mathbf{v}^* \cdot A' \mathbf{v} \leq \mathbf{v}^* \cdot A \mathbf{v}.$$

Further, it is well known (see [46]) that the normalised eigenvector with non-negative coordinates corresponding to the maximal eigenvalue maximises the quadratic form

$$\mathbf{x}^* \cdot A \mathbf{x}$$

over the unit sphere, and so:

$$\lambda_m = \max_{|\mathbf{x}|=1} \mathbf{x}^* \cdot A\mathbf{x} \geq \mathbf{v}^* \cdot A\mathbf{v}.$$

This completes the proof of Theorem 47.

The *edge graph* $L(G)$ of a simple graph G is defined as follows: $L(G)$ is a simple graph, its vertices correspond to the edges of G and two vertices of $L(G)$ are adjacent if and only if the edges of G corresponding to the two vertices share an endpoint. The relationship between the spectrum of a connected, regular graph and that of its edge graph will be investigated. Some remarks are made first.

Let us examine the product B^*B constructed from the incidence matrix B of a simple graph G. The main diagonal of this product is formed by values 2 and its entry (i,j) in the case $i \neq j$ is 1 or 0 depending on whether the vertices i and j of the graph G share an endpoint, i.e. the latter equals the element of the adjacency matrix of $L(G)$ in the position (i,j). The following has been proved:

48. *If the incidence matrix of a simple graph with e edges is B and the adjacency matrix of its edge graph is A_L, then*

$$B^*B = A_L + 2 \cdot E_e$$

where E_e is the unit matrix of order e.

Let us now consider the product BB^* constructed from the incidence matrix B of a k-regular, simple graph G. Each element in the main diagonal of this product is k and its entry (i,j) in the case $i \neq j$ is 1 or 0 depending on whether the vertices i and j of the graph G are adjacent. Hence, the following statement has been proved:

49. *If the incidence matrix of a k-regular, simple graph with n vertices is B and its adjacency matrix is A then*

$$BB^* = A + k \cdot E_n$$

where E_n is the unit matrix of order n.

The corollary of the following theorem indicates the aforementioned relation between the spectrum of a connected, regular graph and that of its edge graph (see H Sachs [47]):

50. *If a k-regular, connected, simple graph G contains n vertices and $e = \frac{1}{2} n \cdot k$ edges, then*

$$p(L(G), \lambda) = (\lambda + 2)^{e-n} p(G, \lambda + 2 - k).$$

Corollary. Also using Theorem 45, if the spectrum of G is

$$\begin{bmatrix} k & \lambda_1 & \lambda_2 & \ldots & \lambda_s \\ 1 & m_1 & m_2 & \ldots & m_s \end{bmatrix}$$

then the spectrum of $L(G)$ is

$$\begin{bmatrix} 2k-2 & k-2+\lambda_1 & k-2+\lambda_2 & \ldots & k-2+\lambda_s & -2 \\ 1 & m_1 & m_2 & \ldots & m_s & e-n \end{bmatrix}.$$

The proof of Theorem 50. The notations used in Statements 48 and 49 will be retained. Consider the two matrices below and then their products (N is the zero matrix):

$$U = \begin{bmatrix} E_n & B \\ B^* & \lambda E_e \end{bmatrix}, \quad V = \begin{bmatrix} \lambda E_n & -B \\ N & E_e \end{bmatrix};$$

$$UV = \begin{bmatrix} \lambda E_n & N \\ \lambda B^* & \lambda E_e - B^*B \end{bmatrix}, \quad VU = \begin{bmatrix} \lambda E_n - BB^* & N \\ B^* & \lambda E_e \end{bmatrix}.$$

Now,

$$\det UV = \det VU$$

yields the following formula:

$$\lambda^n \det(\lambda E_e - B^*B) = \lambda^e \det(\lambda E_n - BB^*).$$

Employing this formula and Statements 48 and 49 we make the following calculations:

$$\begin{aligned} p(L(G), \lambda) &= \det(\lambda E_e - A_L) = \det\left((\lambda + 2)E_e - B^*B\right) \\ &= (\lambda + 2)^{e-n} \det\left((\lambda + 2)E_n - BB^*\right) \\ &= (\lambda + 2)^{e-n} \det\left((\lambda + 2 - k)E_n - A\right) \\ &= (\lambda + 2)^{e-n} p(G, \lambda + 2 - k). \end{aligned}$$

This completes the proof of Theorem 50.

We have studied the spectrum of a Möbius ladder containing $2h$ vertices. The special cases $h = 3$ and $h = 4$ were explicitly presented after the calculations. The spectrum of the former is clearly symmetric to the origin, while that of the latter is asymmetric. This could be proved in the general case as well: the spectrum of the Möbius ladder, i.e. its eigenvalues along with the corresponding multiplicities, is symmetric to the origin if and only if h is odd. Moreover, in that case (and only then) its maximal eigenvalue multiplied by (-1) appears among its eigenvalues. The Möbius ladder is clearly a bipartite graph if and only if h is odd. The following two theorems indicate that the observed relation between these phenomena is even more general (see L Lovász and J Pelikán [48]).

51. *A simple graph with maximal eigenvalue Λ is bipartite if and only if $(-\Lambda)$ is also one of its eigenvalues.*

52. *A simple graph is bipartite if and only if its spectrum is symmetric to the origin.*

To prove Theorem 51, at first we verify the sufficiency of the condition. To this end, let $A = [a_{ij}]$ be the adjacency matrix of the connected, simple graph

G, let Λ be its highest eigenvalue and let $\mathbf{v} = [v_1, \ldots, v_n]^*$ be its eigenvector belonging to the eigenvalue $(-\Lambda)$. Let

$$\mathbf{v}' = [|v_1|, |v_2|, \ldots, |v_n|]^* \quad \text{and} \quad \mathbf{v}'_0 = \frac{\mathbf{v}'}{|\mathbf{v}'|}.$$

Then, on the one hand,

$$|(A\mathbf{v})^* \cdot \mathbf{v}| = |-\Lambda \mathbf{v}^* \cdot \mathbf{v}| = \Lambda \mathbf{v}^* \cdot \mathbf{v}.$$

On the other hand, we use the fact that any normalised eigenvector without non-negative coordinates, corresponding to the maximal eigenvalue maximises the quadratic form

$$\mathbf{x}^* \cdot A\mathbf{x} \tag{24}$$

over the unit sphere, i.e. with $|\mathbf{x}| = 1$ (see [46]):

$$|(A\mathbf{v})^* \cdot \mathbf{v}| = \left| \sum_i \sum_j a_{ij} v_i v_j \right| \leq \sum_i \sum_j a_{ij} |v_i| \cdot |v_j|$$

$$= (A\mathbf{v}')^* \cdot \mathbf{v}' = \frac{\mathbf{v}'^* A \mathbf{v}'}{|\mathbf{v}'|^2} (\mathbf{v}'^* \cdot \mathbf{v}')$$

$$= (\mathbf{v}_0'^* \cdot A\mathbf{v}_0') \cdot (\mathbf{v}'^* \cdot \mathbf{v}') \leq \Lambda (\mathbf{v}'^* \cdot \mathbf{v}') = \Lambda(\mathbf{v}^* \cdot \mathbf{v}).$$

(In the penultimate step, we used the reasoning, presented for (24), in the form $(\mathbf{v}_0'^* \cdot A\mathbf{v}_0') \leq \Lambda$, while in the fourth step we used $A^* = A$.) Comparing this with the previous formula, we see that here equality holds at both places. Consequently, $\mathbf{v}'$ is also an eigenvector of G belonging to Λ. Now, the definition of $\mathbf{v}$ implies that:

$$-\Lambda(\mathbf{v}^* \cdot \mathbf{v}) = \mathbf{v}^* \cdot (-\Lambda \mathbf{v}) = \mathbf{v}^* \cdot A\mathbf{v} = \sum_i \sum_j a_{ij} v_i v_j.$$

Comparing this with the previous formulae, we obtain:

$$-\sum_i \sum_j a_{ij} |v_i||v_j| = \sum_i \sum_j a_{ij} v_i v_j.$$

G being connected, $\mathbf{v}'$ has no zero coordinate in view of the theorem of Frobenius and Perron (see [46]). So, the last formula implies that the signs of v_i and v_j are opposite, provided $a_{ij} \neq 0$. Therefore, if the ith vertex of G is assigned to the sets P_1 or P_2 depending on whether $v_i > 0$ or $v_i < 0$, a bipartite graph $G = G(P_1, P_2)$ is obtained since no two vertices of P_1 or P_2 can be adjacent.

In order to prove the necessity in Theorem 51, let the graph $G(P_1, P_2)$ be a bipartite graph and $\mathbf{v} = [v_1, v_2, \ldots, v_n]^*$ be an eigenvector of this graph corresponding to its highest eigenvalue Λ. The vector $\mathbf{v}' = [v_1', v_2', \ldots, v_n']^*$ is obtained from $\mathbf{v}$ as follows:

$$v_i' = \begin{cases} v_i & \text{if the } i\text{th vertex} \in P_1, \\ -v_i & \text{if the } i\text{th vertex} \in P_2. \end{cases}$$

The visualisation of the eigenvalues and eigenvectors as described in Statement 44 implies that $\mathbf{v}'$ is an eigenvector belonging to the eigenvalue $(-\Lambda)$ of the graph.

This idea is helpful in verifying the necessity of the condition in Theorem 52 with the following additional considerations: the above reasoning did not make use either of the connectedness of the graph or of the maximality of the eigenvalue Λ. So, associating $\mathbf{v}'$ to $\mathbf{v}$ also indicates how to associate an eigenvector, belonging to an arbitrary eigenvalue λ, to an eigenvector belonging to the eigenvalue $(-\lambda)$. One can easily see that this assigns linearly independent vectors to linearly dependent ones and hence (taking into account that the eigenvectors corresponding to different eigenvalues are linearly independent) we also obtain that the multiplicities of λ and of $-\lambda$ coincide.

It only remains to prove that if the spectrum of G is symmetric to the origin then G is a bipartite graph. If G is connected, this statement follows from Theorem 51. Now let G be a disconnected graph with a spectrum symmetric to the origin. One can easily see that if λ appears in the spectra of the components of a graph with multiplicities $m_1, m_2, \ldots, m_r$, then the multiplicity of λ in the spectrum of the graph is $\sum_{i=1}^{r} m_i$. So, it suffices to verify that the components of G are bipartite. Let Λ be the maximal eigenvalue of the graph G, and G_1 be the component of G with $(-\Lambda)$ being its eigenvalue. Since the maximal eigenvalue is the one with the highest absolute value among the eigenvalues (see [46]), the highest eigenvalue of G_1 is not less than $|-\Lambda| = \Lambda$, and since no eigenvalue can be greater, this eigenvalue is Λ. So, in view of Theorem 51, G_1 is a bipartite graph. Let us delete G_1 from G. The spectrum of the graph obtained is again symmetric to the origin. Repeating the above line of reasoning, the result is that G has a bipartite component beside G_1, and so on. The final result is that all components of G are bipartite, and this completes both the proof of Theorem 51 and that of Theorem 52.

According to Theorem 45, the greatest eigenvalue of a k-regular simple graph is k. This result is a special case of the following theorem (see [39], §11.14.a):

53. *If the greatest eigenvalue of a simple graph G is Λ, and the lowest and highest degrees in G are d and D, respectively, then*

$$\max\left(d, \sqrt{D}\right) \leq \Lambda \leq D.$$

Further, if G is connected and $\Lambda = D$ then G is D-regular.

Proof. Let the vertices of G be $p_1, p_2, \ldots, p_n$, let the adjacency matrix of G be A, and let $\mathbf{j}$ be the n-dimensional column vector with all of its coordinates 1. Then, the ith coordinate of $A\mathbf{j}$ yields the degree of the ith vertex of G. Therefore:

$$A\mathbf{j} \geq d\mathbf{j},$$

and so,
$$\mathbf{j}^* \cdot A\mathbf{j} \geq dn.$$
Hence, in view of the remark for (24):
$$\Lambda = \max_{\mathbf{x} \neq 0} \frac{\mathbf{x}^* A\mathbf{x}}{\mathbf{x}^* \mathbf{x}} \geq \frac{\mathbf{j}^* A\mathbf{j}}{\mathbf{j}^* \mathbf{j}} \geq d. \qquad (25)$$

Now, let G' be the subgraph of G formed by a vertex of G with degree D, by the edges incident with it and by the other endpoints of the latter. One can verify by a simple calculation (see Problem 107) that the characteristic equation of G' is
$$\lambda^{D+1} - D\lambda^{D-1} = 0,$$
and so, the maximal eigenvalue of G' is $\sqrt{D}$. So, according to Theorem 47:
$$\sqrt{D} \leq \Lambda.$$

Evidently,
$$A\mathbf{j} \leq D\mathbf{j}.$$
Let $\mathbf{v} = [v_1 \cdot v_2, \ldots, v_n]^*$ be an eigenvector of G belonging to Λ with
$$\max_i v_i = v_m = 1.$$
Then:
$$\Lambda \mathbf{v} = A\mathbf{v} \leq A\mathbf{j} \leq D\mathbf{j},$$
and so, considering each coordinate:
$$\Lambda \leq D.$$

Now, assume that
$$\Lambda = D.$$
Then, the mth coordinate of both $A\mathbf{v}$ and $A\mathbf{j}$ is D, i.e., if E denotes the set of the edges of G, then
$$\sum_{\{p_m, p_i\} \in E} v_i = D.$$
In view of $v_i \leq 1$ (for any i), this can hold only if every term of the sum is 1. Then the same reasoning shows that the v_i-s corresponding to the second neighbours of p_m are 1, etc. Owing to the connectedness of G, all v_i can be attained, i.e.
$$\mathbf{v} = \mathbf{j}.$$
So,
$$A\mathbf{j} = \Lambda \mathbf{j} = D\mathbf{j},$$

i.e. G is D-regular. This completes the proof of Theorem 53.

If the degrees of G are $\varphi(p_i)$ then the *average degree* of G is defined as

$$\tilde{\varphi} = \frac{1}{n} \sum_{i=1}^{n} \varphi(p_i).$$

It is easy to check that

$$\frac{\mathbf{j}^* A \mathbf{j}}{\mathbf{j}^* \mathbf{j}} = \frac{\sum_{i=1}^{n} \varphi(p_i)}{n} = \tilde{\varphi},$$

so, in view of (25):

54. *If the average degree of a simple graph is $\tilde{\varphi}$ and its highest eigenvalue is Λ then*

$$\tilde{\varphi} \leq \Lambda.$$

To obtain a further lower bound of the highest eigenvalue (see H S Wilf [49]), let us define the *chromatic number* $\chi(G)$ of a graph G: it is the smallest possible number k with a partition $\{P_1, P_2, \ldots, P_k\}$ of the set of vertices of G so that each of the sets P_i is independent in G.

55. *If the chromatic number of a simple graph G is $\chi(G)$ and its largest eigenvalue is Λ then:*

$$\chi(G) \leq \Lambda + 1.$$

Proof. We use induction for the number of vertices. If G consists of a single vertex then $\Lambda = 0$ and $\chi(G) = 1$. Assume the statement of the theorem to be true for any simple graph containing n vertices. Let G be a simple graph containing $(n+1)$ vertices, Λ be the greatest eigenvalue of G, p be a vertex of G with minimal degree d, and

$$k = [\Lambda] + 1.$$

In view of Theorem 53, $d \leq \Lambda$, and so

$$d \leq k - 1. \tag{26}$$

Let G' be the graph obtained from G by deleting p and the edges incident to p. Let Λ' denote the greatest eigenvalue of G'. By induction and by Theorem 47:

$$\chi(G') \leq \Lambda' + 1 \leq \Lambda + 1.$$

Hence:

$$\chi(G') \leq k,$$

i.e. the set of vertices of G' can be partitioned into k independent sets and, according to (26), p can be included in one of these still keeping the independence, so:

$$\chi(G) \leq k \leq \Lambda + 1.$$

This proves Theorem 55.

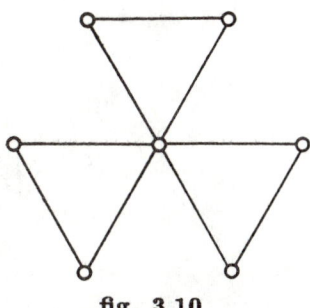

fig. 3.10

Finally, we present one of the numerous interesting applications of eigenvalue techniques, in connection with the proof of the so-called 'friendship theorem'. According to this theorem, if in a company any two persons have exactly one common friend then necessarily there is one person who is a friend of everyone, and the others can be partitioned in pairs being friends, and there is no further friendship (the friendships are assumed to be mutual). Then, the number n of people in the company is, naturally, odd. The case $n = 7$ is shown in figure 3.10, the edges indicate friendships. Clearly, these graphs can be drawn in a way reminiscent of windmills. Let a graph be called a *windmill* if its vertices are $p_0, p_1, q_1, p_2, q_2, \ldots, p_k, q_k$, and its edges are $\{p_0, p_i\}$, $\{p_0, q_i\}$ and $\{p_i, q_i\}$ for any i. Strictly formulated, the friendship theorem is (see P Erdős, A Rényi and V T Sós [50] and H S Wilf [51]):

56. *A simple graph is a windmill if and only if any two of its vertices share exactly one neighbour.*

Proof. The necessity of the condition of the theorem is obvious. So, assume that the condition is fulfilled for a graph G. Keep in mind that G cannot contain a circuit of length 4. If there is a vertex p in G adjacent to all of the others then let us consider the common neighbours of this vertex and of each of the others. This gives that G is a windmill. Any other case will now be shown to be impossible.

For the indirect reasoning, let us assume that the neighbours of a vertex p of G are $p_1, p_2, \ldots, p_k$ and that G has other vertices, too. G will be shown to be regular in this case. We can see in this case as well that the vertices p_i can be set in pairs with the elements of each pair adjacent to each other. Each of these vertices also shares exactly one neighbour with p. Consequently, each of the vertices not mentioned is adjacent with some p_i. Let the set Q_i of vertices be formed by those neighbours of the vertex p_i which have not yet been enumerated. It can be assumed that Q_i is not empty and that $q_1 \in Q_1$ (figure 3.11). The vertex q_1 shares a neighbour with each vertex p_i. So, q_1 has one neighbour in each Q_i except in Q_2, but not more than one, since a quadrangle would result otherwise. So, q_1 has k neighbours, i.e.

$$\varphi(p) = \varphi(q_1) = k.$$

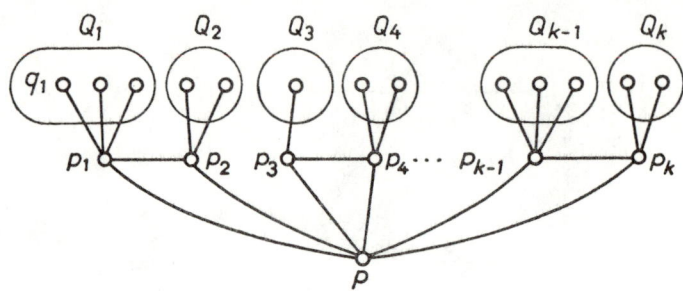

fig. 3.11

We have also obtained that any two non-adjacent vertices of G are of degree k. Now, if the complement $\overline{G}$ of G is connected then any two vertices in $\overline{G}$ can be connected by a path. This already implies the regularity of G. If, however, $\overline{G}$ is disconnected, then its set of vertices has a partition $\{P_1, P_2\}$ with all P_1P_2-edges existing in G. Since there can be no quadrangle in G, one of the sets P_i can have no more than one element. This element is adjacent with all vertices of G, contrary to our indirect assumption.

So, G can be assumed to be a k-regular graph. Let n denote its number of vertices and let A be its adjacency matrix. It is easy to see that

$$A^2 = J + (k-1)E$$

where all the elements of J are unity and E is a suitable unity matrix. Simple arithmetic yields that the spectrum of A^2 is

$$\begin{bmatrix} n+k-1 & k-1 \\ 1 & n-1 \end{bmatrix},$$

and so, the eigenvalues of A, and hence of G, cannot differ from the numbers $\pm\sqrt{n+k-1}$ and $\pm\sqrt{k-1}$. The graph G is evidently connected and, since it includes triangles, it cannot be bipartite. In view of Theorems 45 and 51, $-\sqrt{n+k-1}$ cannot be an eigenvalue of G, but $\sqrt{n+k-1}$ can, and namely its highest eigenvalue, i.e.

$$\sqrt{n+k-1} = k. \tag{27}$$

Let the multiplicity of the eigenvalue $\sqrt{k-1}$ be m. Then the multiplicity of $-\sqrt{k-1}$ is $n-1-m$. The sum of the eigenvalues of G is zero:

$$k - (n - 1 - 2m)\sqrt{k-1} = 0. \tag{28}$$

Therefore, $\sqrt{k-1}$ is rational, and consequently integer, say

$$\sqrt{k-1} = r.$$

Then:

$$k = r^2 + 1, \tag{29}$$

and so, in view of (28):

$$r^2 + 1 = (n - 1 - 2m)r.$$

Since r is a divisor of the right-hand side, it is also a divisor of the left. Consequently, r is a divisor of 1, too, i.e.:

$$r = 1.$$

So, by (29),

$$k = 2,$$

and by (27)

$$n = 3,$$

i.e. G is a single triangle which contradicts our indirect assumption.

This completes the proof of the friendship theorem.

Imagine that the edges of a connected graph G are distinctly labelled, say by distinct numbers. If the number of the vertices of G is n, then each of its spanning trees contains $n-1$ edges. Two spanning trees of G are called different if their edge-sets are different. The number $\kappa(G)$ of the different spanning trees of a connected graph G is called the *complexity* of G. Obviously, $\kappa(G) = 1$ if and only if G is a tree. We shall see that the complexity of graphs can easily be computed with the aid of the incidence matrix, and the complexity of regular graphs with the aid of the spectrum, too. This assertion is based on the following theorem:

57. *If B is the incidence matrix of a directed graph $\overrightarrow{G}$ and if $Q = BB^*$ then*

$$\operatorname{adj} Q = cJ$$

where J is a matrix of suitable dimensions consisting of entries 1 and c is a constant.

Proof. Let the number of vertices in G be n. The theorem concerning the determinant of the product of two matrices implies that $r(Q) = r(B)$. If G is disconnected, then $r(Q) < n - 1$ by Theorem 16 and so adj Q is a zero matrix, i.e. the statement of the theorem is true with $c = 0$. If, however, G is connected then $r(Q) = n - 1$. Consequently,

$$Q \cdot \text{adj } Q = E \cdot \det Q = N \tag{30}$$

where E is a suitable unity matrix and N is an appropriate zero matrix. It suffices to verify that each column of adj Q is a multiple of the vector $\mathbf{j} = [1, 1, \ldots, 1]^*$, i.e.

$$\text{adj } Q = [c_1 \mathbf{j}, c_2 \mathbf{j}, \ldots, c_n \mathbf{j}],$$

since Theorem 42 indicates that Q, and hence adj Q, is symmetric which implies that all the constants c_i are equal. To prove our assertion let

$$Q = [q_{ij}], \qquad \text{adj } Q = [Q_{ij}],$$

and consider the ith column of the matrix N appearing in (30):

$$\begin{aligned} q_{11}Q_{1i} + q_{12}Q_{2i} + \ldots + q_{1n}Q_{ni} &= 0 \\ q_{21}Q_{1i} + q_{22}Q_{2i} + \ldots + q_{2n}Q_{ni} &= 0 \\ &\ldots \\ q_{n1}Q_{1i} + q_{n2}Q_{2i} + \ldots + q_{nn}Q_{ni} &= 0. \end{aligned} \tag{31}$$

Theorem 42 implies that the sum of each row of the matrix

$$Q = BB^* = D - A$$

is zero. Hence, the choice

$$Q_{1i} = Q_{2i} = \ldots = Q_{ni} = 1$$

satisfies (31). But the number of linearly independent solutions of this set of equations is:

$$n - r(Q) = n - (n - 1) = 1$$

which implies that the ith column of adj Q is a multiple of $\mathbf{j}$. This completes the proof of Theorem 57.

The following, so-called matrix-tree theorem states that the constant c in Theorem 57 is the complexity of the graph, if the graph is connected (see G Kirchhoff, [52]).

58. *If B is the incidence matrix of a connected directed graph $\vec{G}$, $Q = BB^*$, and all elements of the matrix J of suitable dimensions are 1, then*

$$\text{adj } Q = \kappa(G) \cdot J.$$

Proof. In view of Theorem 57, it suffices to prove that an arbitrary element of adj Q equals $\kappa(G)$. The entry Q_{nn} in the lower right corner of the matrix adj Q is selected for this purpose. This can be constructed from B as follows: let us delete the last row of B, hence obtain a reduced incidence matrix B_0

of the graph $\vec{G}$. Since the elements in the last row and column of Q are all obtained as scalar products with the last row of B, we obtain

$$Q_{nn} = \det B_0 B_0^*.$$

Let us apply the so-called Binet–Cauchy theorem. Then

$$\det B_0 B_0^* = \sum \det B_i \cdot \det B_i^*,$$

and the summation here extends to all $(n-1)$-order square minors B_i of B_0. However, according to Theorems 17 and 18, $\det B_i = \pm 1$, i.e. $\det B_i \cdot \det B_i^* = 1$ if the edges corresponding to the columns of B_i constitute the edges of a spanning tree of G, and otherwise $\det B_i = 0$. Therefore:

$$Q_{nn} = \kappa(G),$$

and this concludes the proof of Theorem 58.

Let us apply this theorem to a complete graph G with n vertices. It is easy to verify that now,

$$Q = D - A = n \cdot E - J \tag{32}$$

where E is the n-order unity matrix. So, the $(n-1)$-order determinant is

$$\kappa(G) = Q_{nn} = \det \begin{bmatrix} n-1 & -1 & -1 & \ldots & -1 \\ -1 & n-1 & -1 & \ldots & -1 \\ \ldots & \ldots & \ldots & \ldots & \ldots \\ -1 & -1 & -1 & \ldots & n-1 \end{bmatrix}.$$

Let us subtract the last row from all the others and then add all columns to the last one. Then

$$\kappa(G) = Q_{nn} = \det \begin{bmatrix} n & 0 & 0 & \ldots & 0 & 0 \\ 0 & n & 0 & \ldots & 0 & 0 \\ \ldots & \ldots & \ldots & \ldots & \ldots & \ldots \\ 0 & 0 & 0 & \ldots & n & 0 \\ -1 & -1 & -1 & \ldots & -1 & 1 \end{bmatrix} = n^{n-2}.$$

So, the following corollary of Theorem 58 (see A Cayley [53]) has been obtained:

59. *If G is a complete graph with n vertices, then*

$$\kappa(G) = n^{n-2}.$$

The following formula concerning the complexity will be derived with the aid of Theorem 58 (see H N V Temperley [54]):

60. *If B is the incidence matrix of a connected directed graph $\vec{G}$ containing n vertices, $Q = BB^*$ and all elements of the matrix J of suitable dimensions are 1 then*

$$\kappa(G) = n^{-2} \det(J + Q).$$

Proof. Obviously,
$$n \cdot J = J^2, \tag{33}$$
and, by Theorem 42:
$$JQ = N \tag{34}$$
where N is the n-order zero matrix. These two formulas help to follow the calculations below where E is the n-order unity matrix:
$$(nE - J)(J + Q) = nJ + nQ - J^2 - JQ = nQ,$$
$$\operatorname{adj}(J + Q) \cdot \operatorname{adj}(nE - J) = \operatorname{adj} nQ. \tag{35}$$
Simple arithmetic yields:
$$\operatorname{adj} nQ = n^{n-1} \operatorname{adj} Q. \tag{36}$$
According to (32), $nE - J$ is nothing but the matrix Q with respect to the complete graph with n vertices. So, using Theorems 58 and 59:
$$\operatorname{adj}(nE - J) = n^{n-2} J. \tag{37}$$
Consequently, the following is obtained from (35) in view of the formulas (36), (37) and Theorem 58:
$$\operatorname{adj}(J + Q) \cdot J = n \cdot \kappa(G) \cdot J.$$
Multiplying both sides by the matrix $J + Q$, applying the relationships (33) and (34), and in view of the fact that $M \cdot \operatorname{adj} M = \det M \cdot E$. Provided M is a square matrix:
$$\det(J + Q) \cdot J = n^2 \kappa(G) \cdot J,$$
and hence
$$\det(J + Q) = n^2 \kappa(G)$$
follows, which completes the proof of Theorem 60.

Now, the complexity of regular graphs will be computed with the aid of the spectrum. To this end, some preliminary remarks concerning the spectrum are made. First of all, let us compute the spectrum of the n-order, square matrix J with all of its entries being 1. The rank of J is 1, so 0 is its eigenvalue with multiplicity $(n-1)$. Obviously, the number n is an eigenvalue, too, with the all one vector as eigenvector. So, the spectrum of J is:
$$\begin{bmatrix} n & 0 \\ 1 & n-1 \end{bmatrix}. \tag{38}$$

Proof. Obviously,
$$n \cdot J = J^2, \tag{33}$$

and, by Theorem 42:

$$JQ = N \qquad (34)$$

where N is the n-order zero matrix. These two formulas help to follow the calculations below where E is the n-order unity matrix:

$$(nE - J)(J + Q) = nJ + nQ - J^2 - JQ = nQ,$$

$$\operatorname{adj} (J + Q) \cdot \operatorname{adj} (nE - J) = \operatorname{adj} nQ. \qquad (35)$$

Simple arithmetic yields:

$$\operatorname{adj} nQ = n^{n-1} \operatorname{adj} Q. \qquad (36)$$

According to (32), $nE - J$ is nothing but the matrix Q with respect to the complete graph with n vertices. So, using Theorems 58 and 59:

$$\operatorname{adj} (nE - J) = n^{n-2} J. \qquad (37)$$

Consequently, the following is obtained from (35) in view of the formulas (36), (37) and Theorem 58:

$$\operatorname{adj} (J + Q) \cdot J = n \cdot \kappa(G) \cdot J.$$

Multiplying both sides by the matrix $J + Q$, applying the relationships (33) and (34), and in view of the fact that $M \cdot \operatorname{adj} M = \det M \cdot E$. Provided M is a square matrix:

$$\det(J + Q) \cdot J = n^2 \kappa(G) \cdot J,$$

and hence

$$\det(J + Q) = n^2 \kappa(G)$$

follows, which completes the proof of Theorem 60.

Now, the complexity of regular graphs will be computed with the aid of the spectrum. To this end, some preliminary remarks concerning the spectrum are made. First of all, let us compute the spectrum of the n-order, square matrix J with all of its entries being 1. The rank of J is 1, so 0 is its eigenvalue with multiplicity $(n-1)$. Obviously, the number n is an eigenvalue, too, with the all one vector as eigenvector. So, the spectrum of J is:

$$\begin{bmatrix} n & 0 \\ 1 & n-1 \end{bmatrix}. \qquad (38)$$

It is easy to verify that if λ is an eigenvalue of a matrix M then $(-\lambda)$ is an eigenvalue of $(-M)$, and that if E is a unit matrix and k is a constant then any vector is an eigenvector of kE, belonging to its single eigenvalue k. It can also be checked that if two matrices A_1 and A_2 of the same order have a common eigenvector belonging to the eigenvalues λ_1 and λ_2 of the two matrices respectively, then $\lambda_1 + \lambda_2$ is an eigenvalue of $A_1 + A_2$.

Thus, the promised theorem can be formulated as follows, taking Statements (1) and (2) of Theorem 45 also into account (see H Hutschenreuther, [55]):

61. *If G is a connected, k-regular simple graph with n vertices, its characteristic polynomial is $p(G,\lambda)$, and its spectrum is*

$$\begin{bmatrix} k & \lambda_1 & \ldots & \lambda_s \\ 1 & m_1 & \ldots & m_s \end{bmatrix}$$

then

$$\kappa(G) = n^{-1} \prod_{r=1}^{s} (k - \lambda_r)^{m_r} = n^{-1} p'(G, k)$$

where $p'(G, \lambda)$ denotes the derivative of the characteristic polynomial.

Proof. Using the notations of the previous theorem and Theorem 42,

$$J + Q = J + kE - A.$$

By the proof of Theorem 46, the spectrum of $J - E - A$ is:

$$\begin{bmatrix} n-1-k & -\lambda_1-1 & -\lambda_2-1 & \ldots & -\lambda_s-1 \\ 1 & m_1 & m_2 & \ldots & m_s \end{bmatrix},$$

and, obviously, the addition of $(k+1)E$ 'shifts' the spectrum by $(k+1)$. Hereupon, the first equality in the theorem follows from Theorem 60 taking into account the well-known identity stating that $\det(J + Q)$ equals the product of the eigenvalues of the matrix $J + Q$.

To verify the second equality in the theorem, consider the factor form of $p(G, \lambda)$ using its roots:

$$p(G, \lambda) = (\lambda - k)(\lambda - \lambda_1)^{m_1} \ldots (\lambda - \lambda_s)^{m_s}.$$

If this is differentiated and λ is replaced by k therein, the product

$$\prod_{r=1}^{s} (k - \lambda_r)^{m_r}$$

is obtained. Thus, Theorem 61 has been proved.

Let us apply this theorem to the edge graph $L(G)$ of a k-regular, connected simple graph G containing n vertices and e edges. Verify and keep in mind that $2e = nk$ and that $L(G)$ is a $(2k-2)$-regular graph. According to Theorem 61:

$$\kappa(L(G)) = e^{-1} p'(L(G), 2k - 2),) \qquad (39)$$
$$\kappa(G) = n^{-1} p'(G, k).$$

Let us differentiate the polynomial

$$p(L(G), \lambda) = (\lambda + 2)^{e-n} p(G, \lambda + 2 - k)$$

appearing in Theorem 50, let us substitute λ by $(2k-2)$ and let us take Statement (1) of Theorem 45 into account. Then:

$$p'(L(G), 2k-2) = (2k)^{e-n} p'(G, k).$$

Comparing this with the equality (39) and using the substitution

$$n = \frac{2e}{k},$$

the following is obtained (A K Kel'mans, see [56], Theorem 7.24):

$$\kappa(L(G)) = 2^{e-n+1} k^{e-n+1} \kappa(G).$$

If, particularly, G is a complete graph with n vertices then

$$k = n-1, \qquad e = \frac{n(n-1)}{2},$$

and so, using Theorem 59:

$$\kappa(L(G)) = 2^{\frac{1}{2}(n^2-3n+2)} (n-1)^{\frac{1}{2}(n^2-3n+2)} n^{n-2}.$$

An upper bound for the complexity of a regular graph can be obtained as follows with the aid of Theorem 61. Let G be a connected, k-regular, simple graph containing n vertices. All eigenvalues of G are known to be reals. Consequently, with the notation $\lambda_1 = k$ and in view of Theorem 61:

$$\kappa(G) = n^{-1} \prod_{i=2}^{n} (k - \lambda_i)$$

where the eigenvalues λ_i are not necessarily distinct. The sum S of the factors $k - \lambda_i$ is:

$$S = (n-1)k - \sum_{i=2}^{n} \lambda_i = nk - \left(k + \sum_{i=2}^{n} \lambda_i\right) = nk - \sum_{i=1}^{n} \lambda_i.$$

According to Theorem 43, the coefficient of the $(n-1)$-order term of the characteristic polynomial is zero, and its absolute value coincides with the sum of the roots of the characteristic equation. Consequently:

$$S = nk,$$

i.e. it is constant, so the product of the factors $k - \lambda_i$ is maximal if and only if

$$k - \lambda_i = \frac{nk}{n-1} \qquad (i = 2, 3, \ldots, n).$$

So, the following theorem has been proved (A K Kel'mans, see [56], Theorem 7.27):

62. *For a connected, k-regular, simple graph G with n vertices:*

$$\kappa(G) \leq \frac{1}{n} \left(\frac{nk}{n-1}\right)^{n-1}.$$

Remark. It is easy to verify that equality holds here if G is a complete graph with n vertices. It can also be proved (see [57] Theorem 6.5) that equality holds in no other case.

Finally, the reader may consult the book [56] treating the theory and applications of the spectra of graphs.

3.7 Linear electrical networks

In what follows, the computation of the parameters of linear electrical networks, i.e. their analysis will be presented as an application of the incidence, circuit and cutset matrices. The detailed formulation of the appropriate model, the elaborate investigations employ a rather extensive apparatus, so only a separate volume could present all of them (see [58], [59]). Therefore, no lengthy proofs will be presented, rather the problems of constructing the model and the calculations involving the matrices will be detailed.

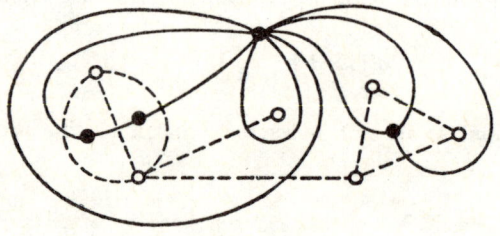

fig. 3.12

First, the duality mentioned in §3.4 will be discussed again, from the point of view of the graph model of networks. We shall examine how the network elements, joined to the network at two points each, can be interconnected with the properties of duality exhibited. We have mentioned that the series and parallel connections of the elements are sometimes called dual to each other. The series and parallel connections of the same elements have been illustrated in figure 3.6. Here, the 'corresponding' edges intersect. The traversal of a circuit of one has been seen to result in cutting the other along its intersected edges. This seems to be the case in more complex situations, too, as for example in figure 3.12 which also illustrates how the correspondents of edges of different types can be drawn in the other graph. Observe, too, that both graphs divide the page of this book into regions by means of their edges and, according to the diagram, the regions of both graphs can be associated with the vertices of the other.

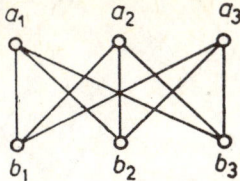

fig. 3.13

On the basis of all this, an intuitive way of constructing the dual of a graph seems to be drawing a diagram of the graph on the Euclidean plane and then the 'solid graph' is drawn over this 'dotted graph' as above. Unfortunately, already the first step of the instruction may result in breakdown since not all graphs are planar. As an example, we show that the bipartite graph G shown in figure 3.13 is not planar. This problem is sometimes formulated in interrogative form as follows: is it possible to stake out a path from each of three houses to each of three wells with no two of these nine paths intersecting? G will be proved not to be planar and so the answer is negative. For the indirect reasoning, let us assume G to be a planar graph and let G_0 denote a planar diagram of G in the plane S with no edges intersecting. Let K be a circuit of length 4 in G_0, for example the one formed by the edges a_1, b_1, a_2, b_2. According to the curve theorem of Jordan, K divides the points of S excluding the points of K into two disjoint regions: the interior and the exterior of K. Let us assume the vertex a_3 to be in the interior of K. (The reasoning is similar if it is in the exterior of K.) Since the edges of G_0 do not intersect, both the edge $\{b_1, a_3\}$ and the edge $\{a_3, b_2\}$ run in the interior of K. The path constituted by these two edges divides the interior of K into two regions. In figure 3.14, these regions have been denoted by T_1 and T_2, respectively, while T_3 denotes the exterior of K. Since the edges of G_0 do not intersect, all three edges $\{a_i, b_3\}$ are in the same region T_j together with b_3. This is, however, impossible since a_i possesses a neighbourhood not contained within T_i ($i = 1, 2, 3$). Therefore, G_0 does not exist, i.e. G is not a planar graph.

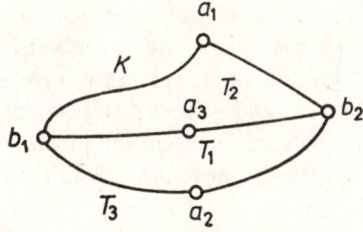

fig. 3.14

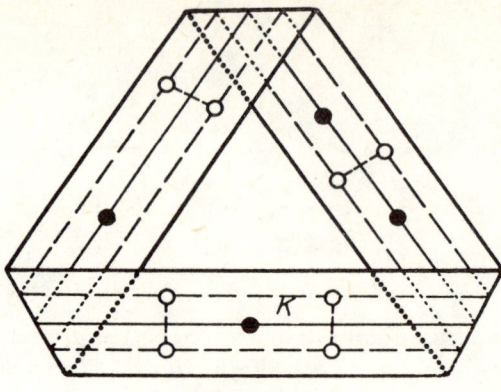

fig. 3.15

In order to construct the dual graph, the graph itself must be first realised on some surface to facilitate the drawing of the dual on this surface. A surface containing the diagram of a graph in space always exists since the graph can be imagined to be 'patched' into a surface. It is, however, possible that the dual so constructed fails to satisfy the requirements for the dual, for example that the duals of the edges of any circuit of the graph constitute a cutset in the dual. This case is illustrated by the broken line diagram on the Möbius strip in figure 3.15 or the Möbius ladder shown in figure 3.9. (The invisible sections of the edges are indicated by dotted lines.) As about the solid graph, only the circuit K of the four edges corresponding to the steps of the ladder have been drawn. Now, whether the graphs in figure 3.9 are regarded or the strip is cut along K, the steps of the ladder obviously fail to form a cutset of the Möbius ladder. Such a possibility can happen in case of non-planar graphs. If, however, the dual were to be defined for planar graphs only, it would still be doubtful whether it manifests the duality features of electrical networks, too. What do we mean by this? The determination of the currents and voltages in the branches of electrical networks is facilitated by the equations derived by means of Kirchhoff's laws. Kirchhoff's voltage law is applied to circuits. Among the equations thus obtained, the circuits of an f-system of circuits will be seen to provide a maximal linearly independent set and the number of circuits in such an f-system has been shown in §3.3 to equal the nullity of the graph. Kirchhoff's current law is applied to cutsets. Among the equations thus obtained, it is an f-system of cutsets that provides a maximal linearly independent set and it is a direct result of the statements proved in §3.4 that the number of cutsets in such an f-system equals the rank of the graph. On the basis of these, it is an entirely reasonable requirement to demand that the rank of a graph equals the nullity of its dual and its nullity equals the rank of its dual. This also suggests that duality should associate circuits to cutsets and vice versa. Therefore, let two graphs be the *duals* of each other, i.e.

the two graphs are *dual*, if a one-to-one correspondence can be established between the edges of the two graphs so that the set of edges of any circuit of one of them corresponds to a cutset in the other graph, and the correspondents of any of its cutsets constitute the set of edges of a circuit in the other graph. It immediately follows from this definition that if two graphs are dual, then a circuit matrix of any one of them is a cutset matrix of the other. In view of the results of §3.3 and §3.4, the following theorems can immediately be obtained:

63. *If G_1 and G_2 are dual graphs, then*

$$\varrho(G_1) = \mu(G_2) \quad \text{and} \quad \varrho(G_2) = \mu(G_1).$$

64. *If two graphs are dual then the one-to-one correspondence between their edges assigns any f-system of circuits of any one of them to an f-system of cutsets in the other, and vice versa.*

65. *If two graphs are dual then the one-to-one correspondence between their edges assigns the edges of any spanning tree of any one of them to the edges of a set of chords in the other, and vice versa.*

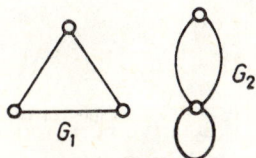

fig. 3.16

Remark. Figure 3.16 shows that Theorem 63 cannot be reversed.

The following questions are obvious. Does any graph have a dual? If not, which are the ones possessing a dual? If a graph does have a dual, is it unique? How can a dual be constructed if the graph is known? The answer to the first question is no. The second question is answered by the following theorem:

66. *A graph possesses a dual if and only if it is planar.*

In connection with this theorem, we present the famous theorem of K Kuratowski [60] yielding the necessary and sufficient condition of a graph being planar. We introduce a concept for the concise formulation of the theorem. The geometrical realisation of graphs has already been mentioned at the beginning of Chapter 1. If, in a geometrical realisation, no distinction is made between the points corresponding to the vertices and the points of the curves representing the edges, then non-isomorphic graphs can have coinciding geometrical realisation. Graphs sharing a geometrical realisation in this sense are called *topologically equivalent*. As an example, consider the common geometrical realisation G_3 of the graphs G_1 and G_2 shown in figure 3.17. Now one can verify that graphs topologically equivalent to a particular graph can be constructed from it by executing in a finite number of times the following two modifications on the realisation of the graph.

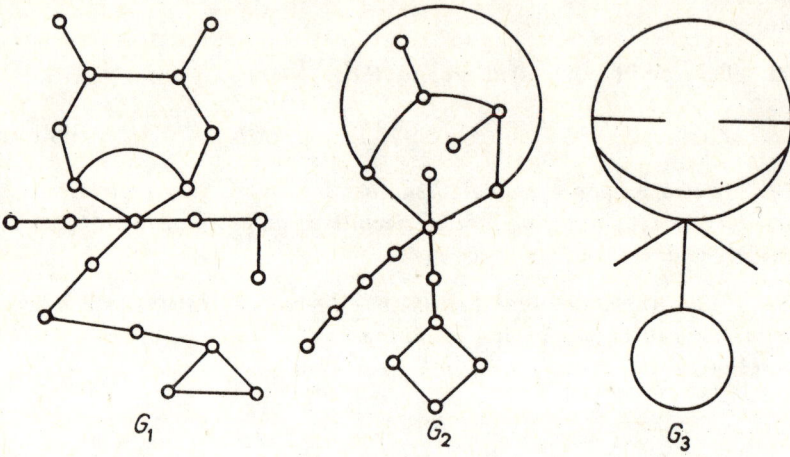

fig. 3.17

I. An arbitrary point p of a curve corresponding to an edge is taken to be the correspondent of a new vertex, i.e. the edge is divided into two edges at p.

II. A point q corresponding to a vertex of degree two is disregarded as the correspondent of a vertex, i.e. the two edges connecting at q are treated as a single edge.

Hereupon, Kuratowski's theorem is as follows:

67. *A graph is planar if and only if it has no subgraph topologically equivalent to one of the two graphs shown in figure 3.18.*

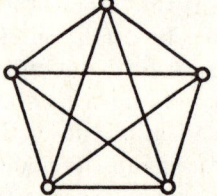 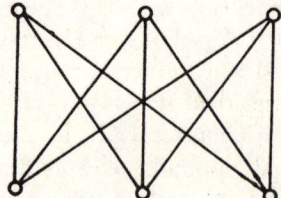

fig. 3.18

(One of these two graphs is the complete graph with five vertices, and the other illustrates the nine paths connecting three houses and three wells already shown in figure 3.13.)

Remarks. 1. A graph is planar if and only if it has a geometrical realisation over the surface of a sphere. This statement can be verified with the aid of stereographic projection. This projection establishes a one-to-one and continuous correspondence between the points of the Euclidean plane S augmented by the point p_∞ at infinity and the points of a spherical surface. Let us place the sphere onto S. Let the point of the sphere diametrically opposite to the tangential point be p_0. The points p_0 and p_∞ correspond to each other and otherwise the correspondent p' of a point p on the sphere is the point of intersection of S with the straight line connecting p_0 and p (figure 3.19).

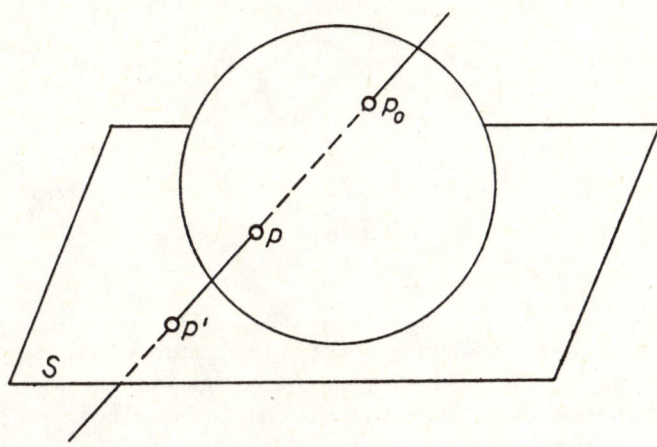

fig. 3.19

2. It is worth mentioning that useful algorithms exist for deciding whether a graph is planar (see, for example, [61] and [62]).

It is now obvious that the dual graph can be constructed in the previously given way since if the geometrical realisations of the graphs are drawn on a sphere when constructing the dual, and this sphere is cut along a circuit of one graph, then the number of components of the other graph is immediately seen to increase on cutting (i.e. deleting) the corresponding edges. Moreover, it increases by no more than one, i.e. this is a (minimal) cutset since if the cutting of a proper subset of it also increased the number of components then the symmetricity of duality would imply that a proper subset of a circuit could constitute a circuit. Therefore, a dual graph can always be constructed on the basis of the rule below:

(40) Let us draw a planar diagram of the graph G_1. G_1 divides the plane into regions. Let us mark a point in the interior of each of these regions: these

are the vertices of the dual G_2 of the graph G_1. Now let a curve intersect each of the edges of G_1 so that the curve joins the two points marked within the two regions bounded by the relevant edge: these are the edges of G_2. If both sides of an edge e_1 of G_1 belong to the same region, then the edge of G_2 intersecting e_1 is a loop of G_2 incident with the vertex marked within the relevant region. A planar diagram of G_2 can thus be constructed.

The *construction of the dual graph* is invariably meant as the construction described in (40). It will be now shown that all the duals of an arbitrary graph cannot in general be constructed. Indeed, consider the graphs G_1, G_2 and G_3 in figure 3.20. It is easy to verify that G_2 and G_3 have been obtained by construction of the dual from G_1 and from G_2, respectively. Then G_1

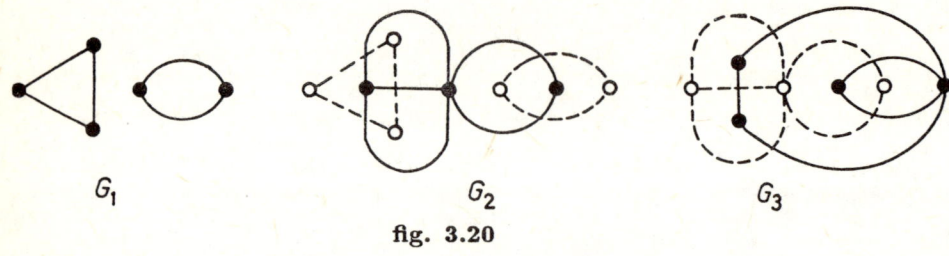

fig. 3.20

is also a dual of G_2 but G_1 can never be obtained from G_2 by construction of the dual, since any planar diagram of G_2 determines four regions in the plane and so, any graph obtained from G_2 by construction of the dual has four vertices whereas G_1 contains five vertices. Observe, however, that if G_3 is separated into its blocks by bisecting its articulation (i.e. if its blocks are drawn separately with no shared vertex) then G_1 is obtained.

Let p be an articulation of a graph G and let T be one of its blocks including p. Let us augment G by a new vertex q. Let us annul the incidences of the edges of T with p and let us make the relevant edges incident to q without changing the incidences of p with the rest of the edges (i.e. with those not in T). This modification of the graph G is called its *separation into blocks*. The separation into blocks can be envisaged as bisecting G at p without modifying T.

However, graphs with isomorphic duals can also be obtained by procedures more complex than the separation into blocks.

The dual G_2 (drawn by dotted lines) of the graph G_1 has been constructed in figure 3.21. The dual G_3 of a graph G_2' isomorphic to G_2 has been constructed in figure 3.22. G_1 and G_3 are immediately seen not to be isomorphic since the vertex q of G_3 is of degree five and G_1 has no vertex of degree five. However, according to the definition of the dual graph, both G_1 and G_3 are duals of G_2. Figure 3.23 illustrates that G_3 has been 'cut' by bisecting the

Graphs and Matrices

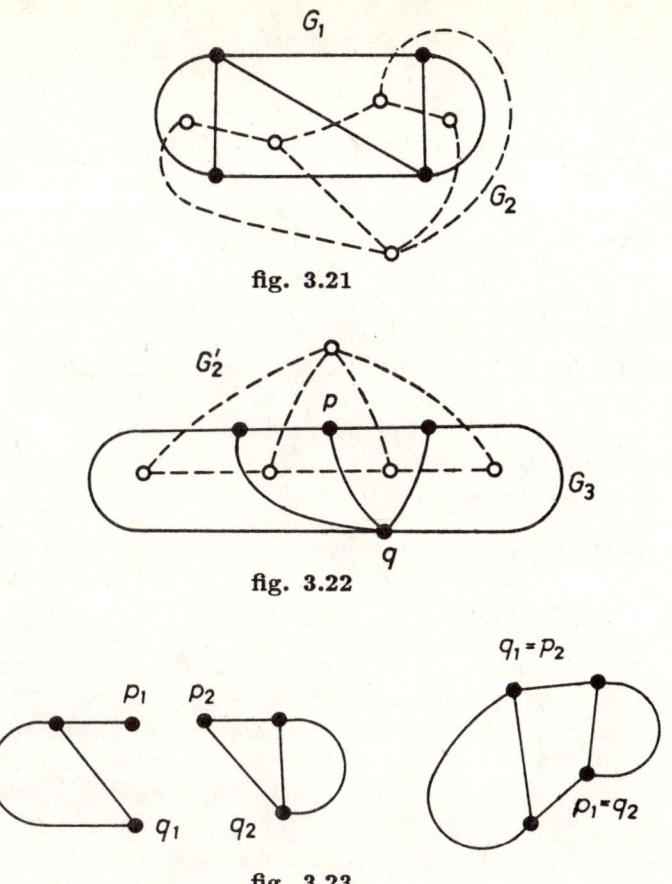

fig. 3.21

fig. 3.22

fig. 3.23

vertex p into p_1 and p_2 and the vertex q into q_1 and q_2 and then, having 'turned' the left part around, the vertices have been 'reunited', namely q_1 with p_2 and p_1 with q_2. This modification will now be precisely described.

Let G_1 and G_2 be two subgraphs of G with the following properties: neither includes isolated vertices, they share no edge, $G_1 \cup G_2$ contains all edges of G, and G_1 and G_2 have exactly two common vertices, p and q. Let us replace p and q by two new vertices each, p by p_1 and p_2 and q by q_1 and q_2. Let us connect the edges in G_1 incident with p or q to p_1 or q_1, respectively, and those in G_2 and incident with p or q to p_2 and q_2, respectively. Let us now unite q_1 with p_2 and p_1 with q_2. This modification of G is called *turning G_1 about p and q*.

The graphs G_1 and G_2 are called *weakly isomorphic* if, allowing a finite number of separations into blocks and turning subgraphs about, we can obtain G_1' from G_1 and G_2' from G_2 so that G_1' and G_2' are isomorphic. Isomorphic graphs are, naturally, weakly isomorphic.

The duals of a graph have been previously seen not to be necessarily isomorphic. The following theorem indicates to what extent the duals of a graph can differ.

68. *If G_1 and G_2 are the duals of the same graph, then G_1 and G_2 are weakly isomorphic.*

Using the forest-like character of the block graph, see Chapter 1, separation into blocks can easily be shown not to affect the circuits of the graph. If some subgraph of a graph is turned about the vertices p and q, only those circuits are effected which include p and q. Even if such a circuit is preserved, only the orientation of one of its pq arcs is changed with respect to the other. Therefore, having executed a weakly isomorphic modification on a graph, a set of edges of the graph determines a circuit of the graph if and only if it has already determined one before the modification. According to the following theorem, this circuit preservation is sufficient to characterise the weak isomorphism.

69. *Two graphs are weakly isomorphic if and only if a one-to-one correspondence can be established between their edges so that the edge-set of any circuit of any one of them corresponds to the edge-set of a circuit in the other.*

The two non-isomorphic duals of the graph G_2 in figure 3.21 could both be constructed. These constructions are shown in figures 3.21 and 3.22. However, the dual G_1 of the graph G_2 in figure 3.20 cannot be constructed on the basis of Instruction (40). This is due to the fact that the graph G_2 of figure 3.20 does not consist of a single block. If the procedure of constructing the dual graph is augmented by allowing the separation into blocks and turning subgraphs about then, in view of Theorem 68, any dual of any graph can be constructed. The following theorem indicates that if the graph is a single block then this augmentation is not needed: if we start from different planar diagrams of the graph then all of its duals can be constructed by means of Instruction (40). The graphs G_2 and G'_2 in figures 3.21 and 3.22 are two different planar diagrams of the same graph and two non-isomorphic duals have indeed been obtained from them.

70. *If the graphs G_1 and G_2 are duals and G_1 is a single block then G_2 is also a single block, and both G_1 and G_2 have a planar diagram constructed in accordance with instruction* (40) *(not necessarily one from the other).*

If, by suitable modification of the instruction, we allowed the introduction of more than one point in a region while constructing the dual graph, then we could construct the dual of any planar graph. However, the limitation for single blocks in Theorem 70 means no essential restriction in investigating electrical networks: the graph corresponding to the network can be investigated by blocks, since the network parts of the different blocks do not affect each other. This is in correspondence with the following theorem:

71. *Separation into blocks does not change the rank or the nullity of the graph.*

The statement of the theorem is obvious, since separation into blocks does not change the number of edges, the number of vertices is increased by one and, in view of the forest-like character of the block graph, so is the number of components.

Now, two special phenomena in electrical networks will be interpreted as duals. Imagine a graph and its dual drawn in the plane with dotted edges and with solid vertices, respectively, in accordance with Instruction (40), and assume these graphs to be the diagrams of electrical networks. If in one of them, say in the solid one, an edge is deleted, i.e. a current is open-circuited, this can also be regarded as a special current-source with zero source-current. If the dotted graph should be modified so that it remains the dual according to Instruction (40) after the deletion then of course the corresponding edge must be deleted in the dotted graph, too, and its endpoints must be united, i.e. short-circuited in the network. This latter can be regarded as a special voltage-source with zero source-voltage. Owing to this interpretation, the open-circuit and the short-circuit are sometimes called dual phenomena in a network. We mention that H Whitney clarified the concept and properties of duality (see [63], [64] and [65]). Before presenting his definition, the concept of the *subtraction of graphs* is introduced: if G' is a subgraph of the graph G, then $G - G'$ is the graph obtained from G by deleting the edges of G' but keeping their endpoints.

Now, let us assume that G_1 and G_2 are dual graphs and that by means of the one-to-one correspondence between their edges, the correspondent of an arbitrary subgraph G'_1 of G_1 is G'_2. How much is lost from the rank of G_2 if the edges of G'_2 are deleted? This loss equals the decrease in the number of the basis cutsets of G_2 due to the deletion. Whitney has shown that this loss is given by the number of basis circuits of G'_1 and this is the definitive property of duality. The graph G_2 is called the *Whitney-dual* of a graph G_1 if a one-to-one correspondence can be established between the edges of G_1 and G_2 so that the following is satisfied for any subgraph G'_1 of G_1 if G'_2 is the

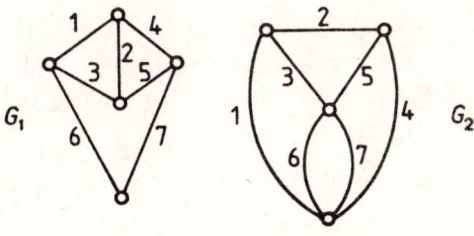

fig. 3.24

subgraph of G_2 with its edges corresponding to the edges of G'_1, then

$$\varrho(G_2 - G'_2) = \varrho(G_2) - \mu(G'_1).$$

As an example, consider the graph G_1 and its dual G_2 in figure 3.24. The correspondence between the edges has been indicated by identical numbers. If, for example, G'_1 is the triangle containing the edges 1, 2 and 3, then $\varrho(G_2 - G'_2) = 2$, $\varrho(G_2) = 3$ and $\mu(G') = 1$.

If duality is introduced using Whitney's abstract definition then Instruction (40) is easily seen to yield an appropriate correspondence and essentially this is the way to prove that a planar graph always has a Whitney-dual. The seemingly complex abstract definition has the advantage of providing a way to prove that a non-planar graph cannot have a dual. However, to prove the latter is rather difficult. In any case, Whitney-duality is a symmetric relationship, too (see Problem 120); moreover, as expected, the two definitions are equivalent:

72. G_1 and G_2 are duals if and only if G_2 is the Whitney-dual of G_1.

Let us now verify that if G_2 is the Whitney-dual of the graph G_1 then the relationships in Theorem 63 hold.

The definitions of the rank and of the nullity immediately imply that for the graph G containing e edges:

$$\varrho(G) + \mu(G) = e. \tag{41}$$

Let us now select the graph G_1 itself as its subgraph G_1. Then the corresponding subgraph G'_2 of G_2 includes all edges of G_2. So, $G_2 - G'_2$ consists of isolated vertices only, and so $\varrho(G_2 - G'_2) = 0$. But, by the definition of the dual graph, $\varrho(G_2 - G'_2) = \varrho(G_2) - \mu(G_1)$ which implies $\varrho(G_2) = \mu(G_1)$. Let e denote the common number of edges in G_1 and G_2. The second relation to be proved is now obtained from the last one and from (41):

$$\varrho(G_1) = e - \mu(G_1) = e - \varrho(G_2) = \mu(G_2).$$

A connection can be established between the features of duality and the properties of the incidence matrix, of the circuit matrix, of the cutset matrix and of the linear vector space constructed by weighting the edges of the graph. The theorem below is from Theorems 35 and 69:

73. $\mathcal{R}_{Vi}$ and $\mathcal{R}_{Ki}$ are the linear vector spaces over the mod 2 field constructed by the linear combinations of the row vectors of a cutset matrix and a circuit matrix, respectively, of a graph G_i $(i = 1, 2)$. If G_1 and G_2 are weakly isomorphic then

$$\mathcal{R}_{V1} = \mathcal{R}_{V2} \quad \text{and} \quad \mathcal{R}_{K1} = \mathcal{R}_{K2}.$$

Conversely: if these equalities hold, then G_1 and G_2 are weakly isomorphic.

This and Theorem 33 imply:

74. B_i *is a reduced incidence matrix of a connected graph* G_i ($i = 1, 2$). G_1 *and* G_2 *are weakly isomorphic if and only if*

$$B_2 = M \cdot B_1$$

where M *is a non-singular matrix.*

Theorem 64 implies the following one:

75. *If* $\mathcal{R}_{Vi}$ *and* $\mathcal{R}_{Ki}$ *are the linear vector spaces over the* mod 2 *field constructed by the linear combinations of the row vectors of a cutset matrix and a circuit matrix, respectively, of a graph* G_i ($i = 1, 2$), *and* G_1 *and* G_2 *are dual graphs then*

$$\mathcal{R}_{V1} = \mathcal{R}_{K2} \quad \text{and} \quad \mathcal{R}_{V2} = \mathcal{R}_{K1}.$$

Finally, as a consequence of this, the following theorem can be stated:

76. *Any reduced incidence matrix of any one of two dual graphs is a reduced circuit matrix of the other.*

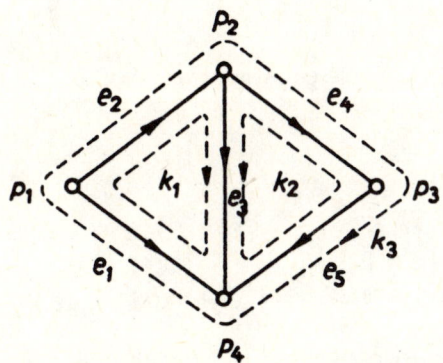

fig. 3.25

Electrical networks have already been treated in Chapter 2 of [2]. The basic investigations are due to G Kirchhoff [52]. The way of associating a directed graph $\vec{G}$ to a network is known. The edges corresponding to the resistive, capacitive, inductive elements, and to the current- and voltage-sources are sometimes also called *branches*. The currents and voltages of the branches satisfy the current and voltage equations written on the basis of Kirchhoff's laws at any instant t. As an example, consider the graph $\vec{G}$ shown in figure 3.25 that represents an electrical network. Let the current and voltage of a network element corresponding to edge e_j be $i_j(t)$ and $v_j(t)$, respectively, at the instant t. The circuits or, with a word more commonly used by engineers, *loops* are denoted by k_1, k_2 and k_3, their orientations are indicated by dotted lines. Kirchhoff's current law for the nodes yield the current equations, and Kirchhoff's voltage law for the loops yield the voltage equations. The equations are written in the order of the subscripts of the vertices and circuits.

The *current equations* are:

$$\begin{aligned} + i_1(t) + i_2(t) &= 0, \\ - i_2(t) + i_3(t) + i_4(t) &= 0, \\ - i_4(t) + i_5(t) &= 0, \\ - i_1(t) \quad - i_3(t) \quad - i_5(t) &= 0. \end{aligned}$$

Here, the matrix of the coefficients is an incidence matrix B of $\vec{G}$. With the notations

$$\mathbf{i}(t) = \begin{bmatrix} i_1(t) \\ i_2(t) \\ i_3(t) \\ i_4(t) \\ i_5(t) \end{bmatrix} \quad \text{and} \quad \mathbf{0} = \begin{bmatrix} 0 \\ 0 \\ 0 \\ 0 \\ 0 \end{bmatrix},$$

the concise form of the current equations is:

$$B \cdot \mathbf{i}(t) = \mathbf{0}.$$

The *voltage equations* are:

$$\begin{aligned} - v_1(t) + v_2(t) + v_3(t) &= 0, \\ + v_3(t) - v_4(t) - v_5(t) &= 0, \\ - v_1(t) + v_2(t) \quad + v_4(t) + v_5(t) &= 0. \end{aligned}$$

Here, the matrix of the coefficients is a circuit matrix K of $\vec{G}$. With the notation

$$\mathbf{v}(t) = \begin{bmatrix} v_1(t) \\ v_2(t) \\ v_3(t) \\ v_4(t) \\ v_5(t) \end{bmatrix},$$

the concise form of the voltage equations is:

$$K \cdot \mathbf{v}(t) = \mathbf{0}.$$

Owing to their physical significance, the vector-scalar functions $\mathbf{i}(t)$ and $\mathbf{v}(t)$ are not arbitrary: for example, obviously all their coordinates must be bounded functions of t in the relevant time interval. However, the boundedness of these functions is by far not sufficient to ensure their realisability. Indeed, consider the function

$$y = f(t) = \sin \frac{1}{t - \frac{1}{\pi}}.$$

This function is obviously bounded since

$$\left| \sin \frac{1}{t - \frac{1}{\pi}} \right| \leq 1.$$

The function oscillates in the time interval $[0, 1/\pi]$ infinitely many times with unit amplitude (see figure 3.26), so the frequency in this interval is not finite,

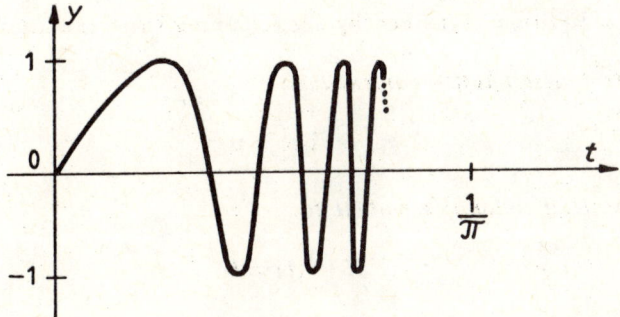

fig. 3.26

consequently there is no real system generating such oscillations. The non-realisability is due to the fact that the formula

$$f\left(\frac{4k-3}{(4k-1)\pi}\right) - f\left(\frac{4k-1}{(4k+1)\pi}\right) = 2$$

with $k = 1, 2, \ldots$ implies that

$$\sum_{k=1}^{n} \left(f\left(\frac{4k-3}{(4k-1)\pi}\right) - f\left(\frac{4k-1}{(4k+1)\pi}\right) \right) = 2n$$

which is greater than any prescribed number if n is sufficiently large, i.e. the function is not of *bounded variation*.

In order to define the concept of electrical network, let us denote the edges of a directed graph $\vec{G}$ by $e_1, e_2, \ldots, e_e$, and let B and K be an incidence matrix and a circuit matrix, respectively, of $\vec{G}$ with their columns listed in the above order. Let us associate to the edge e_j $(j = 1, 2 \ldots, e)$ the real functions $e_j(t)$ and $f_j(t)$ and the functions of bounded variations $i_j(t)$ and $v_j(t)$, all with a real variable t. Let L, R and D be e-order square matrices with D non-singular. Let us introduce the following notations:

$$\mathbf{e}(t) = \begin{bmatrix} e_1(t) \\ e_2(t) \\ \vdots \\ e_e(t) \end{bmatrix}, \quad \mathbf{v}_c(0+) = \begin{bmatrix} \lim_{t \to 0+0} f_1(t) \\ \lim_{t \to 0+0} f_2(t) \\ \vdots \\ \lim_{t \to 0+0} f_e(t) \end{bmatrix}, \quad \mathbf{i}(t) = \begin{bmatrix} i_1(t) \\ i_2(t) \\ \vdots \\ i_e(t) \end{bmatrix},$$

$$\mathbf{v}(t) = \begin{bmatrix} v_1(t) \\ v_2(t) \\ \vdots \\ v_e(t) \end{bmatrix}.$$

An *electrical network* is defined by the following three postulates:

Postulate I. Kirchhoff's current law:

$$B \cdot \mathbf{i}(t) = \mathbf{0}.$$

Postulate II. Kirchhoff's voltage law:

$$K \cdot \mathbf{v}(t) = \mathbf{0}.$$

Postulate III:

$$\mathbf{v}(t) = \mathbf{e}(t) + L \cdot \frac{d\mathbf{i}(t)}{dt} + R \cdot \mathbf{i}(t) + D \cdot \int_0^t \mathbf{i}(x)\, dx + \mathbf{v}_c(0+).$$

This definition of electrical networks is too general. It is, therefore, sometimes restricted to special cases as below:

If the matrices L, R and D are symmetric, then the *network* is called *reciprocal*.

If the elements of the matrices L, R and D are independent of the variables $i_j(t)$ and $v_j(t)$ then the *network* is called *linear*.

If the elements of the matrices L, R and D are functions of t but not of $i_j(t)$ and $v_j(t)$ then the network is *linear, time-variant*.

If the matrices L, R and D contain constants only, then the network is *linear, time-invariant*.

If the matrices L, R and D are positive definite or positive semidefinite[*] and $\mathbf{e}(t) = \mathbf{0}$, then the network is *passive*, otherwise it is *active*.

In what follows, the discussion is restricted to linear, time-invariant, reciprocal networks. The following notations are used: $i_j(t)$ and $v_j(t)$ are the current and voltage, respectively, of the branch, corresponding to the edge e_j at the instant t. $e_j(t)$ is the voltage of the generator forming the jth branch, at the instant t, $f_j(t)$ is the voltage of the capacitor forming the jth edge, while $\lim_{t \to 0+0} f_j(t)$ is the voltage of the latter at $t = 0$. The matrix L includes the mutual inductances, namely the jth element in the ith row of L is the mutual inductance between the inductive elements forming the ith and jth branches. Therefore, it is obvious to restrict the discussion to the case when the matrix L is real and symmetric, and when the matrix obtained from L by deleting its zero rows and columns is positive definite. R and D are presumed to be real, diagonal matrices with non-negative elements. So, the entries in the main diagonal of R are the resistances of the branches and the reciprocal values of the positive numbers in the main diagonal of D are the capacitances of the relevant branches.

Now, some important consequences of the three postulates will be pointed out. For the time being, postulates I and II will only be used. In view of the rank of the matrices B and K, the following theorem can be stated:

77. *If the graph of the electrical network contains c vertices, e edges and it consists of k components then the maximal number of Kirchhoff's independent current and voltage laws are $c - k$ and $e - c + k$, respectively.*

Let the graph $\overrightarrow{G}$ of the network considered be connected, let B_0 be a reduced incidence matrix obtained from B and V_0 be a reduced cutset matrix with the edges in the same order as in B. Then the equation in Postulate I is equivalent to

$$B_0 \cdot \mathbf{i}(t) = \mathbf{0}. \tag{42}$$

According to §3.4, B_0 and V_0 can be obtained from one another by means of a multiplication by a non-singular square matrix. Consequently, equation (42) is equivalent to the one below:

$$V_0 \cdot \mathbf{i}(t) = \mathbf{0}. \tag{43}$$

Therefore, it is customary in network theory to use the cutset matrix instead of the incidence matrix.

---

[*] The n-order, symmetric matrix M is *positive definite* or *positive semidefinite* if for any n-dimensional column vector $\mathbf{r} \neq \mathbf{0}$:

$$\mathbf{r}^* \cdot M\mathbf{r} > 0 \quad \text{or} \quad \mathbf{r}^* \cdot M\mathbf{r} \geq 0.$$

According to Theorem 38, the linear vector spaces $\mathcal{R}_V^{c-1}$, formed by the linear combinations of the row vectors of B_0 or V_0, and $\mathcal{R}_K^{e-c+1}$, formed by the linear combinations of the row vectors of K_0, are orthogonal components of the linear vector space $\mathcal{R}_{\vec{G}}^e$, and their direct sum is $\mathcal{R}_{\vec{G}}^e$. Obviously, at any fixed value of t:

$$\mathbf{i}(t) \in \mathcal{R}_{\vec{G}}^e.$$

According to (43), $\mathbf{i}(t)$ is orthogonal to any row vector of V_0. Therefore, $\mathbf{i}(t)$ is also orthogonal to any linear combination of the row vectors of V_0. Hence

$$\mathbf{i}(t) \in \mathcal{R}_K^{e-c+1}.$$

So $\mathbf{i}(t)$ can be written as a linear combination of any base of $\mathcal{R}_K^{e-c+1}$. The elements of K being constants, and $\mathbf{i}(t)$ being a function of t, the coefficients in such a linear combination must be functions of t. The row vectors of K_0 are known to form a base of $\mathcal{R}_K^{e-c+1}$. Let us denote the row vectors of K_0 as

$$\mathbf{k}_1^*, \mathbf{k}_2^*, \ldots, \mathbf{k}_\mu^* \quad \text{where} \quad \mu = e - c + 1.$$

Hence the functions

$$i_{m1}(t), i_{m2}(t), \ldots, i_{m\mu}(t)$$

exist, so that

$$\mathbf{i}^*(t) = i_{m1}(t)\mathbf{k}_1^* + i_{m2}(t)\mathbf{k}_2^* + \ldots + i_{m\mu}(t)\mathbf{k}_\mu^*. \tag{44}$$

Let us introduce the following notation:

$$\mathbf{i}_m(t) = \begin{bmatrix} i_{m1}(t) \\ i_{m2}(t) \\ \vdots \\ i_{m\mu}(t) \end{bmatrix}.$$

Then (44) can be written as

$$\mathbf{i}^*(t) = \begin{bmatrix} i_{m1}(t) & i_{m2}(t) & \ldots & i_{m\mu}(t) \end{bmatrix} \cdot \begin{bmatrix} \mathbf{k}_1^* \\ \mathbf{k}_2^* \\ \vdots \\ \mathbf{k}_\mu^* \end{bmatrix} = \mathbf{i}_m^*(t) \cdot K_0.$$

Transposing this we obtain the following equation which is usually called the *loop transformation*:

$$\mathbf{i}(t) = K_0^* \cdot \mathbf{i}_m(t). \tag{45}$$

This formula has been deduced from Postulate I employed, so (45) is a necessary condition of Postulate I. It will be shown to be sufficient as well. Using (45), the relation

$$B_0 \cdot K_0^* = N \tag{46}$$

due to Theorem 19 (N is zero matrix), and the fact that (42) is equivalent to Postulate I, the following can be written:

$$B_0 \cdot \mathbf{i}(t) = B_0 \left(K_0^* \cdot \mathbf{i}_m(t) \right) = (B_0 \cdot K_0^*) \cdot \mathbf{i}_m(t) = N \cdot \mathbf{i}_m(t) = 0,$$

which proves our assertion. So, the following theorem can be stated:

78. *If the current function of an electrical network is $\mathbf{i}(t)$, the graph $\vec{G}$ of the network is connected, $\mu(\vec{G}) = \mu$ and B_0 and K_0 — columns of which are in the same order of $\mathbf{i}(t)$ — are a reduced incidence matrix and a reduced circuit matrix, respectively, of $\vec{G}$ then the necessary and sufficient condition for the validity of Kirchhoff's current law*

$$B_0 \cdot \mathbf{i}(t) = 0$$

is the existence of a μ-dimensional vector-scalar function $\mathbf{i}_m(t)$ fulfilling the loop transformation

$$\mathbf{i}(t) = K_0^* \cdot \mathbf{i}_m(t).$$

In a similar way one can prove

$$\mathbf{v}(t) \in \mathcal{R}_V^{c-1},$$

and the following theorem.

79. *If the voltage function of an electrical network is $\mathbf{v}(t)$, the graph $\vec{G}$ of the network is connected, $\varrho(\vec{G}) = \varrho$ and B_0 and K_0 — columns of which are in the same order of $\mathbf{i}(t)$ — are a reduced incidence matrix and a reduced circuit matrix, respectively, of $\vec{G}$ then the necessary and sufficient condition for the validity of Kirchhoff's voltage law*

$$K_0 \cdot \mathbf{v}(t) = 0$$

is the existence of a ϱ-dimensional vector-scalar function $\mathbf{v}_n(t)$ fulfilling the node transformation

$$\mathbf{v}(t) = B_0^* \cdot \mathbf{v}_n(t).$$

The coordinates of the functions

$$\mathbf{i}_m(t) = \begin{bmatrix} i_{m1}(t) \\ i_{m2}(t) \\ \vdots \\ i_{m\mu}(t) \end{bmatrix} \quad \text{and} \quad \mathbf{v}_n(t) = \begin{bmatrix} v_{n1}(t) \\ v_{n2}(t) \\ \vdots \\ v_{n\varrho}(t) \end{bmatrix}$$

are known as *loop-currents* and *node-voltages* in network theory. Using the loop and node transformations, $\mathbf{i}(t)$ and $\mathbf{i}_m(t)$ or $\mathbf{v}(t)$ and $\mathbf{v}_n(t)$ can be calculated from one another. The functions $i_{mj}(t)$ or $v_{nj}(t)$ can be uniquely determined from the transformation equations since the rank of the coefficient matrix is in both cases equal to the number of unknowns. If a reduced cutset matrix of $\vec{G}$ is V_0 then B_0 can obviously be substituted in Theorem 79 by V_0. Then the transformation is

$$\mathbf{v}(t) = V_0^* \cdot \mathbf{v}_p(t).$$

It can be verified that if a cutset matrix V_f is applied then $\mathbf{v}_p(t)$ can be associated with cutset-voltages, so the transformation

$$\mathbf{v}(t) = V_f^* \cdot \mathbf{v}_p(t)$$

is called *cutset-transformation*.

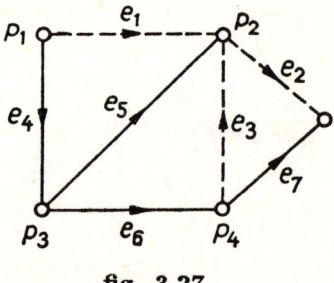

fig. 3.27

To illustrate the use of the loop and node transformation, consider the graph of an electrical network shown in figure 3.27. The edges drawn by dotted lines constitute a system of chords. The reduced circuit matrix is given by the f-circuits determined by this system of chords. The rows of the matrix B_0 are ordered in the sequence of the subscripts of the nodes. So

$$B_0 = \begin{bmatrix} 1 & 0 & 0 & 1 & 0 & 0 & 0 \\ -1 & 1 & -1 & 0 & -1 & 0 & 0 \\ 0 & 0 & 0 & -1 & 1 & 1 & 0 \\ 0 & 0 & 1 & 0 & 0 & -1 & 1 \end{bmatrix}$$

and

$$K_f = \begin{bmatrix} 1 & 0 & 0 & -1 & -1 & 0 & 0 \\ 0 & 1 & 0 & 0 & 1 & -1 & -1 \\ 0 & 0 & 1 & 0 & -1 & 1 & 0 \end{bmatrix}.$$

The detailed form of the loop transformation

$$\mathbf{i}(t) = K_f^* \cdot \mathbf{i}_m(t)$$

is

$$\begin{bmatrix} i_1(t) \\ i_2(t) \\ i_3(t) \\ i_4(t) \\ i_5(t) \\ i_6(t) \\ i_7(t) \end{bmatrix} = \begin{bmatrix} 1 & 0 & 0 \\ 0 & 1 & 0 \\ 0 & 0 & 1 \\ -1 & 0 & 0 \\ -1 & 1 & -1 \\ 0 & -1 & 1 \\ 0 & -1 & 0 \end{bmatrix} \cdot \begin{bmatrix} i_{m1}(t) \\ i_{m2}(t) \\ i_{m3}(t) \end{bmatrix},$$

and so:

$$\begin{aligned}
i_1(t) &= i_{m1}(t), \\
i_2(t) &= i_{m2}(t), \\
i_3(t) &= i_{m3}(t), \\
i_4(t) &= -i_{m1}(t), \\
i_5(t) &= -i_{m1}(t) + i_{m2}(t) - i_{m3}(t), \\
i_6(t) &= -i_{m2}(t) + i_{m3}(t), \\
i_7(t) &= -i_{m2}(t).
\end{aligned} \qquad (47)$$

The detailed form of the node transformation

$$\mathbf{v}(t) = B_0^* \cdot \mathbf{v}_n(t)$$

is

$$\begin{bmatrix} v_1(t) \\ v_2(t) \\ v_3(t) \\ v_4(t) \\ v_5(t) \\ v_6(t) \\ v_7(t) \end{bmatrix} = \begin{bmatrix} 1 & -1 & 0 & 0 \\ 0 & 1 & 0 & 0 \\ 0 & -1 & 0 & 1 \\ 1 & 0 & -1 & 0 \\ 0 & -1 & 1 & 0 \\ 0 & 0 & 1 & -1 \\ 0 & 0 & 0 & 1 \end{bmatrix} \cdot \begin{bmatrix} v_{n1}(t) \\ v_{n2}(t) \\ v_{n3}(t) \\ v_{n4}(t) \end{bmatrix},$$

and so:

$$\begin{aligned}
v_1(t) &= v_{n1}(t) - v_{n2}(t), \\
v_2(t) &= v_{n2}(t), \\
v_3(t) &= -v_{n2}(t) + v_{n4}(t), \\
v_4(t) &= v_{n1}(t) - v_{n3}(t), \\
v_5(t) &= -v_{n2}(t) + v_{n3}(t),
\end{aligned} \qquad (48)$$

$$\begin{aligned}
v_6(t) &= v_{n3}(t) - v_{n4}(t), \\
v_7(t) &= v_{n4}(t).
\end{aligned} \qquad (48)$$

Let us return to the result that

$$\mathbf{i}(t) \in \mathcal{R}_K^{e-c+1}.$$

This and Theorem 21 imply the following theorem:

80. *If the graph of an electrical network is connected and T is a spanning tree of the graph, then the current function corresponding to any edge of T can be expressed as a linear combination of the current functions corresponding to the chords belonging to T.*

Applying this theorem for the above example, (47) yields

$$i_4(t) = -i_1(t),$$
$$i_5(t) = -i_1(t) + i_2(t) - i_3(t),$$
$$i_6(t) = -i_2(t) + i_3(t),$$
$$i_7(t) = -i_2(t).$$

Similarly, Theorem 14 and

$$\mathbf{v}(t) \in \mathcal{R}_V^{c-1}$$

imply the following theorem:

81. *If the graph of an electrical network is connected and T is a spanning tree of the graph then the voltage function corresponding to any chord belonging to T can be expressed as a linear combination of the voltage functions corresponding to the edges of T.*

Applying this theorem for the above example, (48) yields

$$v_1(t) = v_4(t) + v_5(t),$$
$$v_2(t) = -v_5(t) + v_6(t) + v_7(t),$$
$$v_3(t) = v_5(t) - v_6(t).$$

Now, Postulates I and II will be shown to imply the principle of the conservation of energy. Indeed, consider the following power functions:

$$p_j(t) = v_j(t) \cdot i_j(t) \qquad (j = 1, 2, \ldots, e). \tag{49}$$

If the energy accumulated in the jth branch is denoted by $w_j(t)$, then

$$\frac{dw_j(t)}{dt} = p_j(t). \tag{50}$$

We use the following theorem:

82. *The fact that an electrical network satisfies Kirchhoff's current and voltage laws implies that*

$$\sum_{j=1}^{e} p_j(t) = 0.$$

Proof. Employing (49), Theorems 78 and 79 as well as (46), the following can be written:

$$\sum_{j=1}^{e} p_j(t) = \sum_{j=1}^{e} v_j(t) \cdot i_j(t) = \mathbf{i}^*(t) \cdot \mathbf{v}(t) = (K_0^* \mathbf{i}_m(t))^* \cdot (B_0 \cdot \mathbf{v}_n(t))$$
$$= (\mathbf{i}_m^*(t) \cdot K_0) \cdot (B_0^* \cdot \mathbf{v}_n(t)) = \mathbf{i}_m^*(t) \cdot (K_0 \cdot B_0^*) \cdot \mathbf{v}_n(t) = 0.$$

Now, as a consequence of this theorem, the integration of (50) yields the theorem below, that expresses the law of conservation of energy:

83. *The fact that an electrical network satisfies Kirchhoff's current and voltage laws and that the functions $p_j(t)$ ($j = 1, 2, \ldots, e$) are continuous at any relevant t implies that*

$$\sum_{j=1}^{e} w_j(t) = 0.$$

The interpretation of Postulate III is presented in an example. The connection diagram and the graph of an electrical network are shown in figure 3.28. The association of the nodes of the network with the vertices of the graph is

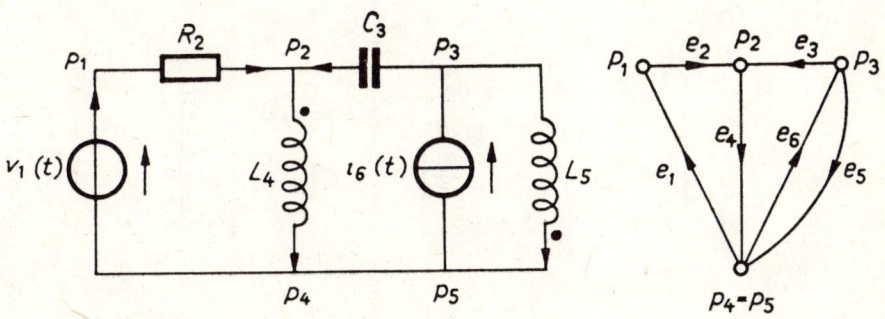

fig. 3.28

indicated by identical subscripts of the letters p. The subscript of any letter denoting a network element (or one of its parameters) coincides with the subscript of the letter denoting the edge corresponding to the network element. Between the two points p_4 and p_5 of the network there is wire of negligible resistance, it can be substituted by a shorter one or the two points can even be drawn into one point without changing anything. By Postulate III:

$$\begin{bmatrix} v_1(t) \\ v_2(t) \\ v_3(t) \\ v_4(t) \\ v_5(t) \\ v_6(t) \end{bmatrix} = \begin{bmatrix} v_1(t) \\ 0 \\ 0 \\ 0 \\ 0 \\ v_6(t) \end{bmatrix} + \begin{bmatrix} 0 & 0 & 0 & 0 & 0 & 0 \\ 0 & 0 & 0 & 0 & 0 & 0 \\ 0 & 0 & 0 & 0 & 0 & 0 \\ 0 & 0 & 0 & L_{44} & L_{45} & 0 \\ 0 & 0 & 0 & L_{45} & L_{55} & 0 \\ 0 & 0 & 0 & 0 & 0 & 0 \end{bmatrix} \cdot \begin{bmatrix} \frac{d}{dt} i_1(t) \\ \frac{d}{dt} i_2(t) \\ \frac{d}{dt} i_3(t) \\ \frac{d}{dt} i_4(t) \\ \frac{d}{dt} i_5(t) \\ \frac{d}{dt} i_6(t) \end{bmatrix}$$

$$+ \begin{bmatrix} 0 & 0 & 0 & 0 & 0 & 0 \\ 0 & R_2 & 0 & 0 & 0 & 0 \\ 0 & 0 & 0 & 0 & 0 & 0 \\ 0 & 0 & 0 & 0 & 0 & 0 \\ 0 & 0 & 0 & 0 & 0 & 0 \\ 0 & 0 & 0 & 0 & 0 & 0 \end{bmatrix} \cdot \begin{bmatrix} i_1(t) \\ i_2(t) \\ i_3(t) \\ i_4(t) \\ i_5(t) \\ i_6(t) \end{bmatrix}$$

$$+ \begin{bmatrix} 0 & 0 & 0 & 0 & 0 & 0 \\ 0 & 0 & 0 & 0 & 0 & 0 \\ 0 & 0 & D_3 & 0 & 0 & 0 \\ 0 & 0 & 0 & 0 & 0 & 0 \\ 0 & 0 & 0 & 0 & 0 & 0 \\ 0 & 0 & 0 & 0 & 0 & 0 \end{bmatrix} \cdot \begin{bmatrix} \int_0^t i_1(x)\, dx \\ \int_0^t i_2(x)\, dx \\ \int_0^t i_3(x)\, dx \\ \int_0^t i_4(x)\, dx \\ \int_0^t i_5(x)\, dx \\ \int_0^t i_6(x)\, dx \end{bmatrix} + \begin{bmatrix} 0 \\ 0 \\ v_3(0+) \\ 0 \\ 0 \\ 0 \end{bmatrix} \cdot$$

According to the notations, the mutual inductance L_{45} is negative. $v_3(0+)$ is the (initial) voltage of the capacitor at $t = 0$. With Laplace transformation in mind (see [66]), we introduce the following notations:

$$s i_j(t) = \frac{d i_j(t)}{dt} \qquad \text{and} \qquad \frac{1}{s} i_j(t) = \int_0^t i_j(x)\, dx.$$

Then, our matrix equation can be written in the following concise form as well:

$$\begin{bmatrix} v_1(t) \\ v_2(t) \\ v_3(t) \\ v_4(t) \\ v_5(t) \\ v_6(t) \end{bmatrix} = \begin{bmatrix} v_1(t) \\ 0 \\ 0 \\ 0 \\ 0 \\ v_6(t) \end{bmatrix} + \begin{bmatrix} 0 & 0 & 0 & 0 & 0 & 0 \\ 0 & R_2 & 0 & 0 & 0 & 0 \\ 0 & 0 & \frac{1}{s}D_3 & 0 & 0 & 0 \\ 0 & 0 & 0 & sL_{44} & sL_{45} & 0 \\ 0 & 0 & 0 & sL_{45} & sL_{55} & 0 \\ 0 & 0 & 0 & 0 & 0 & 0 \end{bmatrix} \cdot \begin{bmatrix} i_1(t) \\ i_2(t) \\ i_3(t) \\ i_4(t) \\ i_5(t) \\ i_6(t) \end{bmatrix}$$

$$+ \begin{bmatrix} 0 \\ 0 \\ v_3(0+) \\ 0 \\ 0 \\ 0 \end{bmatrix}.$$

Our example indicates that the inductive, capacitive and resistive parameters can generally be concentrated into a coefficient matrix of $\mathbf{i}(t)$. Laplace transformation is usually used for the solution since this renders the system a set of linear algebraic equations with the complex variable s instead of the real variable t. If we denote the Laplace transforms by the corresponding capital letters, the transform of the equation in III is:

$$\mathbf{V}(s) = \mathbf{E}(s) + Z(s) \cdot \mathbf{I}(s) - L \cdot \lim_{t \to 0+0} \mathbf{i}(t) + \frac{1}{s}\mathbf{v}_c(0+).$$

Here the square matrix $Z(s)$ is the *impedance matrix*. If $Z(s)$ is non-singular then the *admittance matrix*

$$Y(s) = Z^{-1}(s)$$

exists, too.

Let us now restrict the discussion to the case if the following three requirements hold:

1° The initial conditions are zero, i.e.

$$\lim_{t \to 0+0} \mathbf{i}(t) = \mathbf{0} \quad \text{and} \quad \mathbf{v}_c(0+) = \mathbf{0}.$$

This is equivalent to the statement that the network is free of energy at $t = 0$.

2° All the sources in the network are voltage-sources.

3° All voltage-sources appear with an internal impedance. In this case $Z(s)$ can be shown to be invertible.

Consider the transforms of the equations appearing in Postulates I, II and III in the case of the three above requirements being satisfied. The relevant reduced matrices are written in I and II, and, for brevity, the dependence upon the variable s is not expressly indicated. So

$$B_0 \mathbf{I} = \mathbf{0}, \qquad (51)$$

$$K_0 \mathbf{V} = \mathbf{0}, \qquad (52)$$

$$\mathbf{V} = \mathbf{E} + Z\mathbf{I}. \qquad (53)$$

The Laplace transform of (45) is:

$$\mathbf{I} = K_0^* \mathbf{I}_m. \qquad (54)$$

Let us substitute this into (53) and then into (52). This yields

$$K_0 Z K_0^* \mathbf{I}_m = -K_0 \mathbf{E}.$$

This equation is called the *loop equation*. $\mathbf{I}_m$ can hence be calculated and then $\mathbf{I}$ can be obtained from (54) and so $\mathbf{V}$ from (53). Consequently, $\mathbf{i}(t)$ and $\mathbf{v}(t)$ can be obtained by inverse Laplace transformation.

An analogous method of solution is the following: observe that Z is invertible in view of 3°. So (53) yields

$$\mathbf{I} = Y\mathbf{V} - Y\mathbf{E}. \qquad (55)$$

The Laplace transform of the node transformation in Theorem 79 is:

$$\mathbf{V} = B_0^* \mathbf{V}_n. \qquad (56)$$

Let us substitute this into (55) and then into (51). This yields

$$B_0 Y B_0^* \mathbf{V}_n = B_0 Y \mathbf{E}.$$

This equation is called the *node equation*. $\mathbf{V}_n$ can hence be calculated and then $\mathbf{V}$ can be obtained from (56) and so $\mathbf{I}$ from (55).

In case of more complex networks the calculations involved by the above procedure are rather lengthy, but appropriate codes for computers exist.

Naturally, Postulate III does not describe the most general case since, for example, current-sources could be allowed as constituents of an electrical network. However, accepting assumption 3° above renders this restriction insignificant since, in view of the well-known theorem of Thévenin and Norton equivalents, a current-source with a parallel internal conductance is equivalent to a voltage-source (of suitable parameter) with a series internal resistance.

If voltage-sources with zero internal resistance or current-sources with zero internal conductance are also allowed, the procedure becomes more complicated. The papers [67] and [68] and, first of all, the book of R A Rohrer [69] are recommended for the interested reader. A different approach can also be learned from this book. Namely, if only the currents of the coils and the voltages of the capacitors are treated as unknowns, then differential equations only, rather than integro-differential equations of the form III, are required.

This so-called state variable approach has numerous further advantages and its significance is beyond the theory of electrical networks.

About matroids. The voltage on the speaker of a radio or a tape recorder is not simply a function of time (as, for example, the supply voltage) but it also depends upon the programme broadcasted on the device. The above classical model is usually not applicable to networks containing constituents modelled by such so-called controlled sources: the existence of an equivalent graph model is not ensured. The appropriate combinatorial structure is defined below.

Let S be a finite set and $\mathcal{M}$ a set of the subsets of S. The pair $(S, \mathcal{M})$ is called a *matroid* if the following three axioms are satisfied:

(1) $\emptyset \in \mathcal{M}$.

(2) If $X \in \mathcal{M}$ and $Y \subset X$, then $Y \in \mathcal{M}$.

(3) If $X \in \mathcal{M}$, $Y \in \mathcal{M}$ and $|X| > |Y|$, then there is an $x \in X - Y$ with $Y \cup \{x\} \in \mathcal{M}$.

Among the subsets of S, those in $\mathcal{M}$ are also called *independent*. If the set S is the set of edges of a graph G, and the elements of $\mathcal{M}$ are selected as the sets of edges in the subgraphs of G with no circuit, then a matroid is obtained, namely the so-called circuit matroid of G. Indeed, according to axiom (1), the empty set is independent, i.e. the subgraph without edges contains no circuits. According to axiom (2), a subset of an independent set is also independent, i.e. if a subgraph G' of G contains no circuit then neither does any graph obtained from G' by deleting edges. Axiom (3) expresses the fact that if one of two independent sets has more elements then the other can be augmented by some of its elements with the independence of the latter retained. This also means that the independent sets of a maximal number of elements, the so-called *bases*, are all equicardinal. These requirements are also fulfilled by the subgraphs of G with no circuit.

As already mentioned, matroids, which constitute a concept more general than that of graphs, can be used to treat networks containing controlled sources as well. A further advantage of the matroid model is that two previously discussed phenomena can be described in a more uniform way. One of these is that the same matroid corresponds to weakly isomorphic graphs. The other is that, among matroids, the dual always exists. If some graph has no dual, the consequence is merely that the dual of the corresponding matroid cannot be obtained as the matroid of a graph.

The introductory papers [70], and [71] are recommended for a study of matroids and their applications.

3.8 Further matrices associated with graphs

Matrices have been previously seen to be utilisable in investigating graphs in several ways. The discussed cases, however, do not exhaust all possibilities. Without straining for completeness, some further matrices associated with graphs are mentioned along with a possible application or connection with the previous considerations of each.

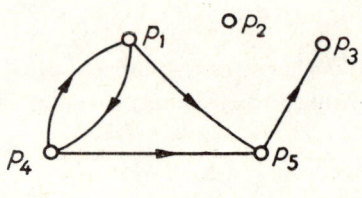

fig. 3.29

If the vertices of a graph G or $\vec{G}$ are $p_1, p_2, \ldots, p_c$ and the distance of p_i from p_j is d_{ij} then the c-order square matrix $D = [d_{ij}]$ is called the *distance matrix* of the graph G or $\vec{G}$. If the word 'shortest' is replaced by the word 'longest' in the definition of distance and the 'greatest distance' of p_i from p_j thus obtained is f_{ij} provided $i \neq j$ and $f_{ii} = 0$, then the c-order square matrix $F = [f_{ij}]$ is called the *detour matrix* of the graph G or $\vec{G}$. Let $h_{ij} = 1$ if there is a $p_i p_j$-path in the graph or a $\overrightarrow{p_i p_j}$-path in $\vec{G}$, and otherwise let $h_{ij} = 0$. The c-order square matrix $H = [h_{ij}]$ is called the *attainability matrix* of the graph G or $\vec{G}$. For the graph G or $\vec{G}$ shown in figure 3.29, these matrices are the following:

$$G: \quad D = \begin{bmatrix} 0 & \infty & 2 & 1 & 1 \\ \infty & 0 & \infty & \infty & \infty \\ 2 & \infty & 0 & 2 & 1 \\ 1 & \infty & 2 & 0 & 1 \\ 1 & \infty & 1 & 1 & 0 \end{bmatrix} \quad F = \begin{bmatrix} 0 & \infty & 3 & 2 & 2 \\ \infty & 0 & \infty & \infty & \infty \\ 3 & \infty & 0 & 3 & 1 \\ 2 & \infty & 3 & 0 & 2 \\ 2 & \infty & 1 & 2 & 0 \end{bmatrix} \quad H = \begin{bmatrix} 1 & 0 & 1 & 1 & 1 \\ 0 & 1 & 0 & 0 & 0 \\ 1 & 0 & 1 & 1 & 1 \\ 1 & 0 & 1 & 1 & 1 \\ 1 & 0 & 1 & 1 & 1 \end{bmatrix}$$

$$\vec{G}: \quad \begin{bmatrix} 0 & \infty & 2 & 1 & 1 \\ \infty & 0 & \infty & \infty & \infty \\ \infty & \infty & 0 & \infty & \infty \\ 1 & \infty & 2 & 0 & 1 \\ \infty & \infty & 1 & \infty & 0 \end{bmatrix} \quad \begin{bmatrix} 0 & \infty & 3 & 1 & 2 \\ \infty & 0 & \infty & \infty & \infty \\ \infty & \infty & 0 & \infty & \infty \\ 1 & \infty & 3 & 0 & 2 \\ \infty & \infty & 1 & \infty & 0 \end{bmatrix} \quad \begin{bmatrix} 1 & 0 & 1 & 1 & 1 \\ 0 & 1 & 0 & 0 & 0 \\ 0 & 0 & 1 & 0 & 0 \\ 1 & 0 & 1 & 1 & 1 \\ 0 & 0 & 1 & 0 & 1 \end{bmatrix}$$

Theorem 6 establishes a connection between the distance matrix and the adjacency matrix. It is also obvious that $h_{ij} = 1$ if and only if there is an n

with $a_{ij}^{(n)} > 0$. The construction of the detour matrix involves the solution of the problem of seeking Hamiltonian paths or directed Hamiltonian paths. The attainability matrix with respect to G indicates the components of G and the one relating to $\vec{G}$ permits discovering the strongly connected components of $\vec{G}$ as described below. The matrix $H^+ = [h_{ij}^+]$ is obtained from the matrix $H = [h_{ij}]$ relating to $\vec{G}$ as follows:

$$h_{ij}^+ = h_{ij} \cdot h_{ji}.$$

Now, p_j is included in a strongly connected component of $\vec{G}$ containing p_i if and only if $h_{ij}^+ = 1$. With respect to the graph shown in figure 3.29:

$$H^+ = \begin{bmatrix} 1 & 0 & 0 & 1 & 0 \\ 0 & 1 & 0 & 0 & 0 \\ 0 & 0 & 1 & 0 & 0 \\ 1 & 0 & 0 & 1 & 0 \\ 0 & 0 & 0 & 0 & 1 \end{bmatrix}.$$

Let us select two vertices x and y in a connected graph G. If the edges of G are $e_1, e_2, \ldots, e_e$ and the xy-paths in G are $l_1, l_2, \ldots, l_u$, then the *xy-path matrix* $L = [l_{ij}]$ of the graph G is the $u \times e$ dimensional matrix with

$$l_{ij} = \begin{cases} 1 & \text{if } e_j \text{ is included in the path } l_i \text{ and} \\ 0 & \text{if } e_j \text{ is not included in the path } l_i. \end{cases}$$

Observe that among the edges incident to the vertices x and y, one edge is included in any xy-path, and among the edges incident to any other vertex, 0 or 2 are included in any xy-path. Consequently, the following theorem is true [72]:

84. *Arrange the columns of the incidence matrix B and those of the xy-path matrix L of a connected graph G in the same order of the edges and consider the mod 2 product matrix BL^*. Then all entries are one in the rows corresponding to the vertices x and y and the entries in any other row are all 0.*

We have seen in the initial part of this chapter that the kth power of the adjacency matrix yields the number of edge-sequences of length k in a graph or directed graph. By modifying the procedure presented there, these edge-sequences can actually be found, too. Moreover, the paths can be 'filtered out' from among the edge sequences connecting pairs of vertices. This will now be presented in a directed graph $\vec{G}$ whose vertices are $p_1, p_2, \ldots, p_c$.

Consider the adjacency matrix $A = [a_{ij}]$ of a graph $\vec{G}$, and replace each entry $a_{ij} \neq 0$ by a_{ij} times the symbol (p_i, p_j) indicating the relevant edges or simply the pair i, j (if no ambiguity can occur, the comma is omitted). This yields the so-called *latin matrix* M of the graph $\vec{G}$. Let us now omit the first term of each pair in M, thus obtaining the matrix $\hat{M}$. For example, for the directed graph shown in figure 3.30:

$$M = \begin{bmatrix} 0 & 12 & 0 & 0 & 15 \\ 0 & 0 & \begin{bmatrix} 23 \\ 23 \end{bmatrix} & 0 & 0 \\ 0 & 0 & 0 & 34 & 0 \\ 41 & 0 & 0 & 0 & 45 \\ 51 & 52 & 53 & 0 & 0 \end{bmatrix}, \quad \hat{M} = \begin{bmatrix} 0 & 2 & 0 & 0 & 5 \\ 0 & 0 & \begin{bmatrix} 3 \\ 3 \end{bmatrix} & 0 & 0 \\ 0 & 0 & 0 & 4 & 0 \\ 1 & 0 & 0 & 0 & 5 \\ 1 & 2 & 3 & 0 & 0 \end{bmatrix}.$$

The matrices M and $\hat{M}$ yield no more information about $\vec{G}$ than A, i.e. both indicate the walkways of length 1, i.e. the edges. Let us now start building the product $M \cdot \hat{M}$ with the usual row–column compositions, but each multiplication by an element of a column is replaced by simply writing the elements of the product beside each other and each addition is replaced

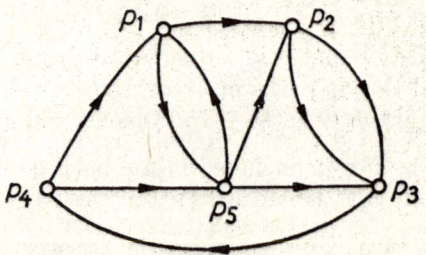

fig. 3.30

by writing the addends one beneath the other. Sequences with no recurrence are written only. This yields the matrix $M * \hat{M}$. In our example:

$$M * \hat{M} = \begin{bmatrix} 0 & 152 & \begin{bmatrix} 123 \\ 123 \\ 153 \end{bmatrix} & 0 & 0 \\ 0 & 0 & 0 & \begin{bmatrix} 234 \\ 234 \end{bmatrix} & 0 \\ 341 & 0 & 0 & 0 & 345 \\ 451 & \begin{bmatrix} 412 \\ 452 \end{bmatrix} & 453 & 0 & 415 \\ 0 & 512 & \begin{bmatrix} 523 \\ 523 \end{bmatrix} & 534 & 0 \end{bmatrix}.$$

It is easy to verify that the elements of $M * \hat{M}$ indicate the vertices of the directed paths of length 2 in $\vec{G}$ in the order following the direction. A similar reasoning proves that the elements of the 'product' $(M * \hat{M}) * \hat{M} = M * \hat{M}^2$

Graphs and Matrices

indicate the directed paths of length 3 in $\vec{G}$, the elements of $M * \hat{M}^3$ indicate the directed paths of length 4, etc. In our example:

$$M * \hat{M}^2 = \begin{bmatrix} 0 & 0 & \begin{bmatrix}1523\\1523\end{bmatrix} & \begin{bmatrix}1234\\1234\\1534\end{bmatrix} & 0 \\ \begin{bmatrix}2341\\2341\end{bmatrix} & 0 & 0 & 0 & \begin{bmatrix}2345\\2345\end{bmatrix} \\ 3451 & \begin{bmatrix}3412\\3452\end{bmatrix} & 0 & 0 & 3415 \\ 0 & \begin{bmatrix}4512\\4152\end{bmatrix} & \begin{bmatrix}4123\\4123\\4523\\4523\\4153\end{bmatrix} & 0 & 0 \\ 5341 & 0 & \begin{bmatrix}5123\\5123\end{bmatrix} & \begin{bmatrix}5234\\5234\end{bmatrix} & 0 \end{bmatrix},$$

$$M * \hat{M}^3 = \begin{bmatrix} 0 & 0 & 0 & \begin{bmatrix}15234\\15234\end{bmatrix} & \begin{bmatrix}12345\\12345\end{bmatrix} \\ \begin{bmatrix}23451\\23451\end{bmatrix} & 0 & 0 & 0 & \begin{bmatrix}23415\\23415\end{bmatrix} \\ 0 & \begin{bmatrix}34512\\34152\end{bmatrix} & 0 & 0 & 0 \\ 0 & 0 & \begin{bmatrix}45123\\45123\\41523\\41523\end{bmatrix} & 0 & 0 \\ \begin{bmatrix}52341\\52341\end{bmatrix} & 53412 & 0 & \begin{bmatrix}51234\\51234\end{bmatrix} & 0 \end{bmatrix}.$$

Obviously, the elements of the matrix $M * \hat{M}^{c-2}$ indicate the directed Hamiltonian paths in $\vec{G}$ and all elements of the matrix

$$M * \hat{M}^{c-1}$$

are zero since there is no path of length c in $\vec{G}$ containing c vertices. However, if in the course of constructing the matrix

$$M_c = M * \hat{M}^{c-1},$$

the last term in a sequence forming an element is allowed to coincide with the first one, then each element in the main diagonal of these matrices includes as many sequences as the number of directed Hamiltonian circuits in $\vec{G}$ and the order of the numbers in these sequences indicates the traversal of the Hamiltonian circuits. The elements off the main diagonal of M_c are obviously all

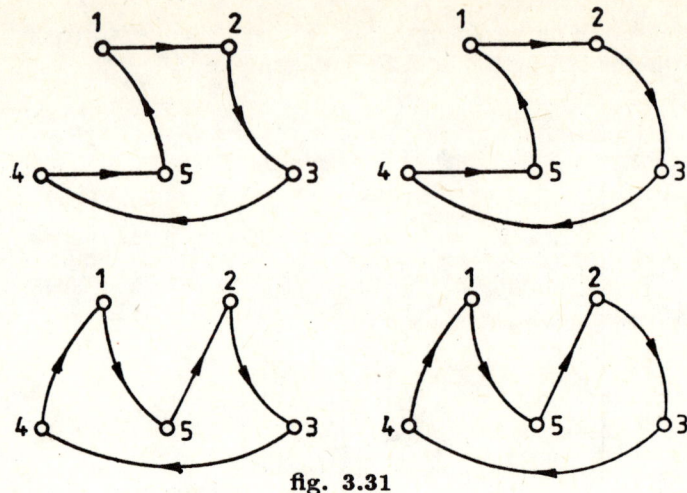

fig. 3.31

zero. The last matrix constructed in connection with our example indicates the directed Hamiltonian paths of the graph. All elements in the main diagonal of M_5 would be formed by sequences of four terms indicating the four Hamiltonian circuits shown in figure 3.31.

3.9 Problems

85. Let $A = [a_{ij}]$ be the adjacency matrix of a graph G or $\vec{G}$ without loops, $A^k = [a_{ij}^{(k)}]$ and let the vertices of G be $p_1, p_2, \ldots, p_c$. Prove that in the case $i \neq j$, $a_{ij}^{(2)}$ yields the number of $p_i p_j$-paths or $\overrightarrow{p_i p_j}$-paths of length 2 in the graph. For what kind of $\vec{G}$ does $a_{ij}^{(k)}$ yield the number of directed $\overrightarrow{p_i p_j}$-paths of length k?

86. Prove that if A is the adjacency matrix of a graph G without loops then the trace of A^3 (i.e. the sum of the elements in the main diagonal of A^3) is six times the number of triangles in G. (For the generalisation of the problem, see Harary–Manvel [73] and [74], page 158, Problem 13.3.)

87. A binary relationship $\mathcal{R}$ defined over a finite set P can be represented by a directed graph $\vec{G}$: the vertices of $\vec{G}$ correspond to the elements of P and (p, q) is an edge of $\vec{G}$ if and only if the correspondents in P satisfy

$$p \mathcal{R} q.$$

Use the adjacency matrix of $\overrightarrow{G}$ and the square of that matrix to express the facts that $\mathcal{R}$ is reflexive, irreflexive, symmetric, asymmetric or transitive.

88. For the non-empty subsets P_i of the set of vertices P of a simple graph G:

$$P = \bigcup_{i=1}^{c} P_i \quad (c \geq 2) \quad \text{and} \quad P_i \cap P_j = 0 \quad (i \neq j),$$

and two vertices of G are adjacent if and only if they belong to distinct sets P_i. Calculate the determinant of the adjacency matrix of G if G is the complete graph (i.e. any P_i contains a single element), and also if G is not the complete graph.

89. The matrix A is the adjacency matrix of a graph G containing c vertices, and E is a c-order unit matrix. Prove that G is connected if and only if no entry of the matrix

$$(A + E)^{c-1}$$

is zero.

90. Show that all maximal subdeterminants of the matrix $D - A$ in Theorem 42 (i.e. the determinants of the matrices obtained from $D - A$ by deleting one row and one column) are zero if the graph G is disconnected.

91. Consider the adjacency matrix of a simple graph; prove that the scalar product of any two distinct rows is less than or equal to 1 if and only if the graph contains no circuit of length 4.

92. Prove that a graph G with adjacency matrix A is bipartite if and only if all entries in the main diagonal of A^n are zero for all odd n.

93. The graph $\overrightarrow{T}$ with adjacency matrix A has been obtained by directing the complete graph T with n vertices. Let s_i denote the sum of the elements in the ith row of A ($i = 1, 2, \ldots, n$). Show that the number of directed circuits of length 3 (i.e. of directed triangles) in $\overrightarrow{T}$ is

$$\binom{n}{3} + \frac{1}{2}\binom{n}{2} - \frac{1}{2}\sum_{i=1}^{n} s_i^2.$$

Find the realisable maximum of this number.

94. Solve the following modification of the missionary–cannibal problem presented in §3.1: three missionaries and three cannibals are to cross from the right bank to the left bank of a river. They have a single boat at their disposal which carries no more than two people. The missionaries are able to row, but only one of the cannibals is. Naturally, the cannibals may not outnumber the missionaries at any bank. Is the crossing possible, and if so, how?

95. Can the previous problem be solved by fewer crossovers if every cannibal is able to row?

96. Consider the course of solving the missionary–cannibal problem in §3.1. Find a procedure serving merely to decide whether the problem is solvable which requires the matrices A and AA^* only. Simultaneously decide whether the problem with four missionaries and four cannibals with each cannibal able to row, is solvable.

97. Does there exist a connected graph with identical adjacency and incidence matrices if the vertices are numbered in the same way?

98. Prove that the rank of the incidence matrix B of a tree F with c vertices is equal to the rank of F in the vector space over the field of the reals, i.e.

$$r(B) = \varrho(F) = c - 1.$$

99. Prove that the rank of the incidence matrix B of a loopless connected graph G with c vertices is

$$r(B) \geq \varrho(G)$$

in the vector space over the field of the reals, and equality holds here if and only if G is a bipartite graph.

100. Show that a loopless graph is bipartite if and only if any non-singular square minor M of its incidence matrix satisfies

$$|\det M| = 1$$

in the vector space over the field of the reals (see [75], page 17).

101. The vertices of a tree T are $p_1, p_2, \ldots, p_c$ and its edges are $e_1, e_2, \ldots, e_e$. The graph $\vec{T}$ is obtained by arbitrarily directing T. The reduced incidence matrix of $\vec{T}$ is B, excluding the row corresponding to p_c. Let b_{ij} be the entry of the matrix B^{-1} in the position (i, j). Prove that

$$|b_{ij}| = \begin{cases} 1 & \text{if } e_i \text{ is an edge of the } p_j p_c\text{-path in } T, \\ 0 & \text{otherwise} \end{cases}$$

(see F H Branin Jr [76]).

102. Prove that a set of edges of a graph G constitutes the edges of a circuit of G if and only if it includes at least one edge from the set of chords with respect to any spanning forest of G and is minimal with respect to this property.

103. Let L be a spanning forest of the graph G and V be the f-cutset of G defined by the edge e in L. Prove that a chord of G with respect to L belongs to V if and only if the corresponding f-circuit contains e.

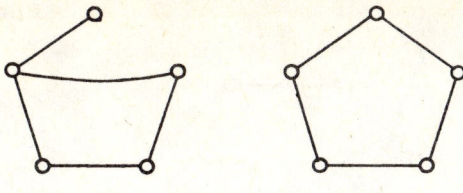

fig. 3.32

104. Calculate the determinants of the adjacency matrices of the graphs shown in figure 3.32 and, knowing these, write the characteristic polynomials of these graphs.

105. Do two (non-isomorphic) simple graphs with four vertices exist with identical characteristic polynomials?

106. At most how many simple graphs with five vertices can exist with identical characteristic polynomials?

107. Determine the spectrum of the star F with n edges. The vertices of F are $p_1, p_2, \ldots, p_{n+1}$ and its edges are $\{p_1, p_i\}$ in the case $i = 2, 3, \ldots, n+1$.

108. Determine the spectrum of the complete graph with n vertices.

109. Determine the spectrum of the so-called complete bipartite graph G. This is a simple graph $G = G(P, Q)$ where all the PQ-edges exist.

110. Write the spectrum of the edge graph of the complete graph with n vertices.

111. Determine the spectrum of the complement of the edge graph of the complete graph with five vertices.

112. Determine the spectrum of the circuit with five vertices.

113. $e = \{p, q\}$ is an arbitrary edge of a forest L. L_1 is obtained from L by deleting the edge e and L_2 by deleting the vertices p and q. Prove the following relationship between the characteristic polynomials of these graphs:

$$p(L, \lambda) = p(L_1, \lambda) - p(L_2, \lambda).$$

(A Mowshowitz [77] and L Lovász and J Pelikán [48].)

114. Prove that if zero is not an eigenvalue of the bipartite graph G then G possesses a one-factor.

115. The set of the vertices of a three-regular graph G can be partitioned into the pairwise disjoint sets $P_1, P_2, \ldots, P_m$ so that for any i, $|P_i| = 6$ and P_i constitutes the vertices of a subgraph of G shown in figure 3.33. Prove that zero is an eigenvalue of G. (H J Finck and H Sachs [78].)

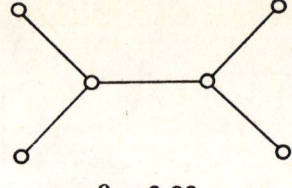

fig. 3.33

116. Prove that the edge graph $L(G)$ of the simple graph $G = (P, E, \mathcal{G})$ has no eigenvalue less than (-2), and if $|E| > |P|$, then the least eigenvalue of $L(G)$ is (-2). (A J Hoffmann [79].)

117. Calculate the complexity of the graphs formed by the network of edges of an octahedron and of a cube.

118. In case of n even, $n/2$ independent edges are deleted from the complete graph with n vertices. Calculate the complexity of the graph obtained.

119. Prove that dual graphs have the same complexity.

120. Prove that Whitney-duality is a symmetric relation, without using Theorem 72.

121. The graphs G_1 and G_2 are Whitney-duals with a one-to-one correspondence M between their edges. $\{a_1, b_1\}$ and $\{a_2, b_2\}$ are two corresponding edges of G_1 and G_2. The graph G_1' is obtained from G_1 by deleting the edge $\{a_1, b_1\}$. Delete the edge $\{a_2, b_2\}$ from G_2 and unite the vertices a_2 and b_2; this yields the graph G_2' from G_2. Prove that the graphs G_1' and G_2' are Whitney-duals with the correspondence M between their edges.

122. Prove that the complete graph with five vertices is not planar.

123. Is the graph shown in figure 4.20 planar?

124. The DC voltage-source v_1 is connected to the network of figure 3.34 at $t = 0$. Determine the branch currents and branch voltages as functions of time.

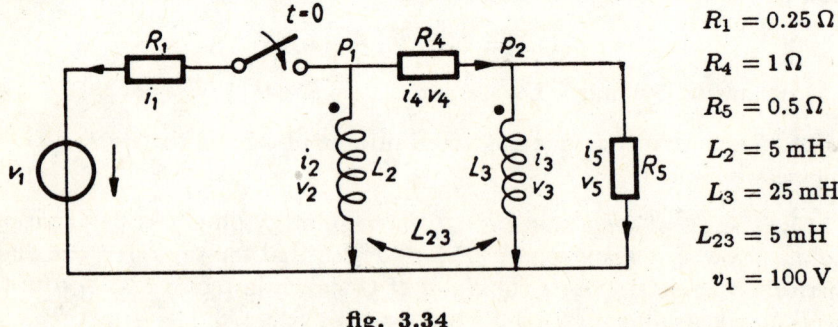

$R_1 = 0.25\,\Omega$
$R_4 = 1\,\Omega$
$R_5 = 0.5\,\Omega$
$L_2 = 5\,\text{mH}$
$L_3 = 25\,\text{mH}$
$L_{23} = 5\,\text{mH}$
$v_1 = 100\,\text{V}$

fig. 3.34

125. Two DC voltage-sources are connected to the network of figure 3.35 at $t = 0$. Determine the branch currents and branch voltages as functions of time.

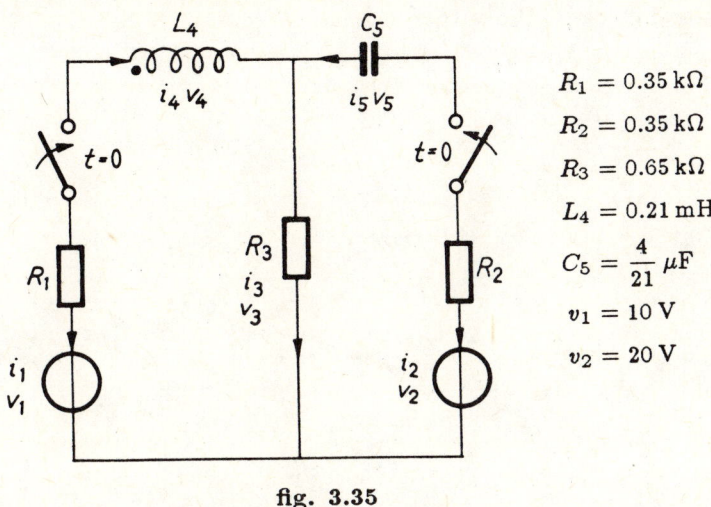

$R_1 = 0.35 \text{ k}\Omega$
$R_2 = 0.35 \text{ k}\Omega$
$R_3 = 0.65 \text{ k}\Omega$
$L_4 = 0.21 \text{ mH}$
$C_5 = \dfrac{4}{21} \mu\text{F}$
$v_1 = 10 \text{ V}$
$v_2 = 20 \text{ V}$

fig. 3.35

126. Find the directed Hamiltonian paths of minimum and maximum value in the graph shown in figure 3.36. The value of a subgraph is the sum of the values of its edges.

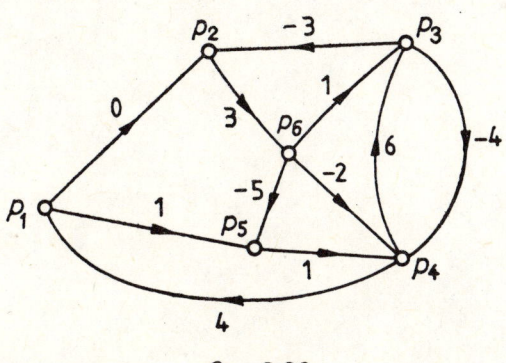

fig. 3.36

127. Prove that a c-order square matrix $D = [d_{ij}]$ is the distance matrix of a connected graph with c vertices if and only if it has the following properties:
(1) d_{ij} is a non-negative integer for any pair i, j.
(2) $d_{ij} = 0$ if and only if $i = j$.
(3) D is symmetric.

(5) If $d_{ij} > 1$ then there is a number k different from i and j so that $d_{ij} = d_{ik} + d_{kj}$.

128. Let L be the xy-path matrix of a connected graph G containing c vertices and e edges. Prove that

$$r(L) \leq e - c + 2 \qquad (\mod 2).$$

4

Solution of Problems

Chapter 1

24. It suffices to prove the statement for complete graphs with n vertices. We use induction on n. The statement is obviously true for $n = 1$. Let us assume the statement to be true for the complete graph with n vertices at some fixed n, and consider a corresponding diagram in space of the complete graph with n vertices. Now, a further point p can be selected so that it is not incident with any of the $\binom{n}{3}$ planes determined by the n vertices of the complete graph. (All these planes are unique and distinct.) It is easy to see that the straight lines connecting the point p with the n vertices of the complete graph cannot intersect any edge of the complete graph with n vertices.

25. The relation $\mathcal{P}$ is obviously reflexive and symmetric. It remains to be proved that it is also transitive. Let p, q and r be three distinct vertices of G and assume that $p\mathcal{P}q$ and $q\mathcal{P}r$. Indirectly, assume that the path F in G is a pr-path of odd length. The path F cannot include the vertex q since otherwise either the pq-path part or the qr-part of F would be of odd length, contrary to our assumption. Let F_1 be a qr-path. If F_1 had no common inner vertex with F then F_1 and F together would constitute a pq-path of odd length, a contradiction. Let F_2 denote the maximal qs-path part of F_1 with no inner vertex in F. The vertex s is in F, and either the ps part or the sr part of F is of even length and the other is of odd length. Now, regardless of F_2 being of odd or even length, it constitutes a pq-path or qr-path of odd length together with the ps part or the sr part of F. Both cases contradict our assumption. Therefore, $\mathcal{P}$ is in fact an equivalence relation.

It is easy to verify that if the connected graph is a bipartite graph $G(A, B)$ then the partition of P induced by $\mathcal{P}$ is $\{A, B\}$.

It is noted that if G is disconnected but it does contain at least one edge then $\mathcal{P}$ is not transitive, since both endpoints of an edge $\{a, b\}$ are in the relation $\mathcal{P}$ with all the vertices of any component not containing a and b, but $a\mathcal{P}b$ fails to hold. In the case of a graph with no edges, however, $\mathcal{P}$ is obviously an equivalence relation.

26. The statement of the problem is evident since no circuits at all are included in a tree.

27. It suffices to show that $p\mathcal{R}_2 q$ for any vertices p and q of the graph. Since the graph is connected, it includes a pq-path. Let F be such a path and let $\{r, s\}$ be an arbitrary edge in F. According to the condition in the statement of the problem, the edge $\{r, s\}$ is included in some circuit K of the graph. Taking the two rs-arcs of K into account, $r\mathcal{R}_2 s$ is implied. Now, using the transitive property of the relation $\mathcal{R}_2$, we obtain our above assertion and hence the solution of the problem as well.

28. It will be shown that two vertices of any graph are related by the relation $\mathcal{V}_1$ if and only if they are related by the relation $\mathcal{R}_1$. This immediately implies that $\mathcal{V}_1$ is an equivalence relation and that the subgraphs in question are the components of G.

It is obvious that if two vertices of a graph G are related by the relation $\mathcal{R}_1$ then they are also related by the relation $\mathcal{V}_1$. Let us now assume that two elements p and q of P are not related by $\mathcal{R}_1$. Let $\{P_1, P_2\}$ be the partition of P with P_1 formed by the elements of P related to p by the relation $\mathcal{R}_1$. The transitivity of the relation $\mathcal{R}_1$ implies that G cannot contain any $P_1 P_2$-edge. Since $q \in P_2$, p and q are not related by $\mathcal{V}_1$ either.

29. In view of Theorem 9, it suffices to verify that $p\mathcal{L}q$ if and only if $p\mathcal{R}_2 q$, i.e. $p\mathcal{V}_2 q$. Assume first that $p\mathcal{L}q$ and let F be a pq-path with any of its edges included in some circuit of G. At any partition $\{P_p, P_q\}$ of the set P, F includes a $P_p P_q$-edge, and hence, since any edge of F is contained in some circuit of G, G includes at least two $P_p P_q$-edges. Therefore $p\mathcal{V}_2 q$.

Assume now that $p\mathcal{R}_2 q$. Then there are two edge-disjoint pq-paths in G, say F_1 and F_2. These two paths together constitute a subgraph G' of G where the degree of any vertex is even. According to the assertion of [2], Problem 3.5, any edge of G', and hence of F_1 and F_2, is included in some circuit of G'. Consequently, $p\mathcal{L}q$.

30. Since a bridge in G is incident to p and is not included in any circuit of G, the statement of the problem immediately follows in view of $\varphi(p) \geq 2$.

31. The problem is easy to solve; Problem 1.36 of [2] can also be applied.

32. In view of the relation $\mathcal{T}$, all blocks of G are circuits.

33. According to [2], a vertex p of a connected graph is its articulation i.e. cut-vertex if the removal of p results in a disconnected graph or if a loop is incident with p and the graph contains at least two edges.

Let p be an articulation of a connected graph G according to [2]. Assume that the disconnected graph G' is obtained from G by the deletion of p. Let q and r be two vertices of the graph G and let us assume that they do not belong to the same component of G'. G being connected, it contains a qr-path, say F. The vertex p is evidently an inner vertex of F. Let the two edges of F incident with p be $\{q_1, p\}$ and $\{p, r_1\}$. Now, no circuit K of G can include these two edges since otherwise the q_1r_1-arc of K not containing p would be included in G' and so, using the transitive property of the relationship $\mathcal{R}_1$, the presence of a qr-path in G' would follow. This, however, contradicts the fact that q and r are in distinct components of G'.

If a loop e in G is incident with p and G has at least two edges, then there is an edge f different from e incident to p. However, e and f cannot belong to the same circuit of G.

Thus, p has been shown to be an articulation of G in the sense of the definition given in Chapter 1. The other part of the problem, — i.e. showing that if a vertex is an articulation in the sense of the definition given in Chapter 1 then it is an articulation in the sense of the definition given in [2] — can be proved by means of a similar reasoning.

34. The problem can easily be solved using Theorems 10 and 14.

35. Assume T to have more than one edge. Then T contains at least two vertices. According to Theorem 11, any two vertices of T are included in some circuit of T. Consequently, T contains a circuit K of odd length. In order to solve the problem, in view of Theorem 10 it suffices to verify that no vertex of K is incident to any edge of T not in K. This statement is proved indirectly.

Assume that the edge $e = \{p, q\}$ in T is not an edge of K but the vertex p is in K. If q is also in K then one pq-arc of K is of odd length and this forms a circuit of even length together with e, contrary to the fact that the graph cannot contain any circuit of even length. So, q can be assumed not to be a vertex of K. According to Theorem 11, the vertex q is connected to the vertex r of K different from p by at least two edge-disjoint paths in T. p is not contained in one of these, say in F. Let F_1 denote the qs part of F of minimal length with its endpoint s in K. Let us attach the edge e to F_1. This results in the ps-path denoted by F_2 (this is a chord of K). One of the ps-arcs of K is of even length and the other is of odd length, so one of them constitutes a circuit of even length together with F_2 which is, however, impossible.

36. The block T has at least two edges. Consequently, according to Theorem 11, there are at least two edge-disjoint pr-paths in T, say F_1 and F_2. Both of these two paths cannot contain the vertex q. If one of them contains q, then one can easily find a pq-path including r. So, q can be assumed not to be a vertex of either F_1 or F_2. According to Theorem 10, there is a qr-path in T, say F_3. Let F_3' denote the sq-path part of F_3 of maximal length with no inner vertex in either F_1 or F_2. The vertex s is a vertex of either F_1 or F_2, it can be assumed to be a vertex of F_2 without loss of generality. Now, F_1, the rs part of F_2 and F_3' together constitute a pq-path containing r.

As for the other part of the problem, three suitable vertices can be found in the graph shown in figure 1.9. Namely, consider one internal vertex of each of the three vertex-disjoint a_2a_3-paths of the block formed by the edges marked by 4.

37. A suitable example is the strongly connected component containing the vertices marked by 9 in the graph shown in figure 1.13. Let p and q be the two vertices with both their indegree and outdegree equal to 1 in this component.

38. It suffices to show that two vertices of $\vec{G}$ are related by the relation $\mathcal{V}'$ if and only if they are related by the relation $\mathcal{E}$. Indeed, it will imply that $\mathcal{V}'$ is an equivalence relation and that the relevant subgraphs are the leaves of $\vec{G}$.

It is obvious that if two vertices of the graph $\vec{G}$ are related by the relation $\mathcal{E}$ then they are also related by $\mathcal{V}'$. Now, assume that two elements p and q of P are not related by $\mathcal{E}$. Then p and q do not belong to the same leaf of $\vec{G}$, let p belong to the leaf $\vec{L}_1$ and q to the leaf $\vec{L}_2$. Consider the structure of the directed leaf graph of $\vec{G}$ (it is an acyclic graph) as described in Theorem 23. The subscript of the set T_i containing the vertex corresponding to $\vec{L}_1$ can be assumed not to be higher than that of the set containing the vertex corresponding to $\vec{L}_2$, since otherwise the reasoning is entirely similar. Now, let $\{P_1, P_2\}$ be the partition of the set of vertices of the graph G with P_1 formed by the vertices of $\vec{L}_1$ and by the vertices of those leaves whose corresponding subscripts in the directed leaf graph of $\vec{G}$ are less than i. Obviously, $p \in P_1$ and $q \in P_2$. Theorem 23 implies that $\vec{G}$ contains no $\overrightarrow{P_2P_1}$-edge. Consequently, p and q are not related by the $\mathcal{V}'$ relation either.

Chapter 2

48. Since the vertex cut in question is minimal, there is an xy-path in G, say L. Let x_1 and y_1 be the two neighbours of z in L and let x_1 be contained in the xz part of L. We only have to verify that there is no circuit in G that contains the edges $\{x_1, z\}$ and $\{z, y_1\}$ of L. Assume indirectly that the circuit K in G contains both of these edges. Consider the x_1y_1-path formed by edges in K only but not containing the vertex z, as well as the xx_1-path and the y_1y-path formed by edges of L only (each of the two latter may degenerate into a single vertex). An xy-path in G not containing the vertex z can be formed from these. This is, however, contrary to the fact that z constitutes a vertex xy-cut.

49. Figure 4.1 illustrates that in general at least two vertices are necessary to 'cover' all paths in H.

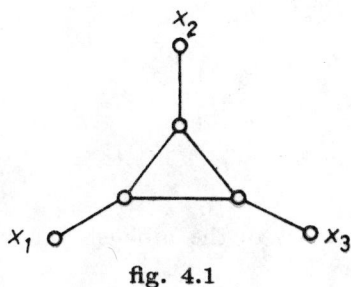

fig. 4.1

Let us augment the graph as follows: introduce a new vertex y as well as the edges $\{x_2, y\}$ and $\{x_3, y\}$. This yields the graph G_y. It is not hard to see that, with the x_1y-paths regarded, the conditions of Theorem 1 hold for G_y. Consequently, G_y has a vertex p constituting an inner vertex of each x_1y-path. Clearly, p is also an inner vertex of any x_1x_2-path or x_1x_3-path in H. Let us now augment G to obtain the graph G_z as follows: introduce a new vertex z as well as the edges $\{x_3, z\}$ and $\{x_1, z\}$. Let us repeat the above reasoning with the x_2z-paths in G_z. The vertex obtained, q, and the above vertex, p, satisfy the requirements in the problem.

50. Evidently, for any i there is a j so that H includes an x_ix_j-path. Let L be an x_1x_2-path. Let us start from x_3 along an arbitrary x_3x_j-path. Let p be the vertex in L first encountered in the course of this route. Obviously, p exists and it is an inner vertex of L. Let L_1 and L_2 denote the x_1p part and the x_2p part of L, respectively, and let L_3 be the x_3p-path traversed. Let us start from x_4 along an arbitrary x_4x_j-path and let us stop as soon as a vertex in L_i ($i = 1, 2, 3$) is encountered. Such a vertex does exist, and can only be p, otherwise there would be two vertex-disjoint paths in H. A similar procedure is followed when starting from every further x_i ($i > 4$). The result is that the vertex-disjoint x_ip-paths ($i = 1, 2 \ldots, k$) exist in G. Now, the assertion is that p is an inner vertex of every path in H. Indeed, assume that p is not included in a path L_0 in H. Then, there is a partial path L_0' of L_0 whose endpoints are included in two distinct paths among the x_ip-paths enumerated, so that no inner vertex of L_0' is a vertex of any of these x_ip-paths. This, however, implies the existence of two vertex-disjoint paths in H, and this is impossible.

51. Indirectly, assume that $p \in P_x$ and $p \in Q_1$. In any case, the existence of a bypass xp-path relative to H is assured. This and each of the k xy-paths in H contain a P_1Q_1-edge, but this is a contradiction.

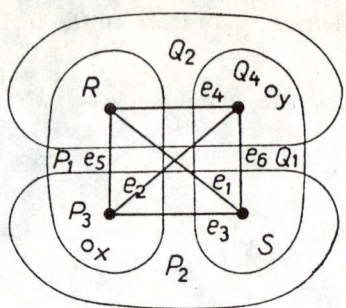

fig. 4.2

It is noted that the statement of the problem can be reformulated for directed graphs as well.

52. Let the minimal number of xy-cut edges be k. A so-called Venn diagram is shown in figure 4.2. $R = P_1 \cap Q_2$, $S = Q_1 \cap P_2$, and $e_1, e_2, \ldots, e_6$ denote the number of RS-, P_3Q_4-, P_3S-, RQ_4-, P_3R- and SQ_4-edges, respectively. With our new notations: $Q_3 = R \cup S \cup Q_4$ and $P_4 = R \cup S \cup P_3$. Since W_1 and W_2 are minimal edge xy-cuts,

$$e_1 + e_2 + e_3 + e_4 = k,$$
$$e_1 + e_2 + e_5 + e_6 = k.$$

The sets of P_3Q_3-edges and P_4Q_4-edges are both edge xy-cuts, so

$$e_2 + e_3 + e_5 \geq k,$$
$$e_2 + e_4 + e_6 \geq k.$$

The sum of the two former equalities and that of the latter two inequalities are

$$2e_1 + 2e_2 + e_3 + e_4 + e_5 + e_6 = 2k$$

and

$$2e_2 + e_3 + e_4 + e_5 + e_6 \geq 2k.$$

Hence $e_1 = 0$. Thus both of the previous inequalities were met by equality, as requested.

It is noted that the statement of this problem can also be reformulated for directed graphs.

53. Clearly, any n-connected graph is simultaneously $(n-1)$-connected. Let G' denote the graph obtained from G by deleting the edges $\{p, q\}$ and, indirectly assume that G' is not $(n-1)$-connected. G' is obviously connected. Therefore, with some $k \leq n-2$, k vertices can be deleted from G' so that the remaining graph G'' is disconnected. Since G is n-connected, G'' consists of exactly two components, one containing the vertex p and the other the vertex

q. Consequently, in at least one of these two components, say in the one containing p, there are at least two vertices. Now, if the relevant k vertices as well as p are deleted from G, a disconnected graph is obtained, contrary to the n-connectedness of G.

54. The graph G is n-connected, so if V is a vertex pq-cut then $|V| \geq n$. If p and q are non-adjacent in G then the solution of the problem is immediately obtained from Theorem 8. If p and q are adjacent then let us delete the edges $\{p, q\}$ from G. So, the statement of Problem 53 and Theorem 8 imply that there are $n-1$ vertex-disjoint pq-paths in G. If a pq-path consisting of a single edge only is attached to these, then n vertex-disjoint pq-paths are obtained in G.

55. Let, in the graph G, L be an xy-path containing a maximum number of the vertices z_i and, indirectly assume that L does not include the vertex z_1 (the subscripts can be subsequently so altered). Let us mark the vertices x, y and z_i in L. Let us decompose L into partial paths with their endpoints marked but their inner vertices not marked. Let the order of these partial paths following the traversal of L starting from x be $L_1, L_2, \ldots, L_k$. In any case $k \leq n - 1$.

First, let us assume the existence of a marked vertex p in L with the following property: no more than one element of the set H, formed by the n vertex-disjoint $z_1 p$-paths existing in view of Problem 54, includes a vertex of L as inner vertex. Let us select the partial paths of each element of H which have z_1 as one of their endpoints (the other endpoint of each is in L) and which contain no vertex in L as inner vertex. Let these partial paths be denoted as $F_1, F_2, \ldots, F_m$ ($m \geq n$), and their endpoints different from z_1 as $q_1, q_2, \ldots, q_m$ ($q_i = p$ may hold for some i). According to the pigeonhole principle there is a path L_h containing at least two vertices q_i (either as an inner vertex or as an endpoint), let q_i and q_j be included in such a path. Let us delete from L the inner vertices of its $q_i q_j$-subpath and let us augment L by the paths F_i and F_j. This yields an xy-path in G which contradicts the maximality of L.

It can be now assumed that there are two vertex-disjoint $z_1 x$-paths in G with no one of them containing any vertex of L as inner vertex. Let the endpoint of L_1 different from x be q. In any case, there is a $z_1 q$-path in G containing no vertex of L as inner vertex. Let the two former paths be M_1 and M_2, and the latter one M_3. Let M_3' denote the qr part of M_3 which contains no vertex in M_1 or M_2 as inner vertex, but its endpoint r is either in M_1 or M_2, say in M_2. Consider the path formed by the following partial paths: M_1, the $z_1 r$-path part of M_2 (if $r = z_1$ it reduces to z_1) and M_3'. Substituting the part L_1 of L by this path yields a contradiction to the maximality of L.

56. Let Y denote the set of those vertices q in $A \cup B$ for which there exists a qy-path in G containing no vertex of $A \cup B$ as inner vertex. It will be shown that $|X| = |Y| = n$. Since both X and Y are vertex xy-cuts, we have $|X| \geq n$,

$|Y| \geq n$. If $X \cap Y \neq \emptyset$, then let $r \in X \cap Y$. Hence an xr-path L_1 and an ry-path L_2 containing no vertex of $A \cup B$ as inner vertex exist. L_1 and L_2 can have no common vertex beside r since that could not be included in $A \cup B$ and so an xy-path could be found in G containing no vertex of $A \cup B$, contrary to the fact that both A and B are vertex xy-cuts in G. Therefore, L_1 and L_2 together constitute an xy-path in G, and so $r \in A \cap B$. Consequently, $X \cap Y \subset A \cap B$, and so $|X \cap Y| \leq |A \cap B|$. But $X \cup Y \subset A \cup B$ holds, and so $|X \cup Y| \leq |A \cup B|$. Hence:

$$2n \leq |X| + |Y| = |X \cup Y| + |X \cap Y| \leq |A \cup B| + |A \cap B| = |A| + |B| = 2n.$$

This implies the solution of the problem.

57. In the graph G of Theorem 8 let us divide each edge incident with x or y into two by a new vertex along the edge. Let A and B denote the sets of the vertices introduced along the edges incident to x and to y, respectively. Now let us delete from the obtained graph the vertices x and y along with the 'half edges in G' incident to them. Let G^* denote the remaining graph. Clearly, a one-to-one correspondence can be established between the systems of vertex-disjoint xy-paths in G and the systems of independent AB-paths in G^*. Let k denote the maximal number of vertex-disjoint xy-paths. Let R^* be a set of vertices covering the AB-paths in G^* with k elements. This exists in view of Theorem 10. Observe that there is no AB-edge in G^* and that $\varphi(p) = 1$ for any vertex $p \in A \cup B$ in G^*. Now, if the vertex $r \in R^*$ is in $A \cup B$, let us substitute it by its neighbour. The set of vertices R obtained from R^* through the modifications constitutes a vertex xy-cut in G, and obviously $|R| \leq k$. Consequently, in view of Theorem 7, the statement of Theorem 8 has been obtained.

58. For proving the necessity, assume that the graph G with at least $2n$ vertices is n-connected. Let A and B denote two arbitrary disjoint subsets with n elements each from the set of vertices of G. Let us augment G as follows: let us introduce the new vertices x and y and let us join x to all vertices of A and y to all vertices in B, each by a new edge. The new graph G^* is clearly n-connected, too. So, the statement of Problem 54 guarantees the existence of n vertex-disjoint xy-paths in G^*. The subpaths of such a system of paths within G yield n independent AB-paths.

To prove sufficiency let us indirectly assume the following: the graph $G = (P, E, \mathcal{G})$ with at least $2n$ vertices fulfils the condition and the set of vertices of G has a subset S with $n-1$ vertices so that its deletion from G results in a disconnected graph G'. Let G_1 denote an arbitrary component of G' and let G_2 be the graph obtained from G' by deleting G_1. Let, further, P_i' denote the set of vertices of G_i $(i = 1, 2)$. Since $|S| = n - 1$, $|P_1'| + |P_2'| + |S| = |P| \geq 2n$ and $|P_i'| \geq 1$ $(i = 1, 2)$, the existence of a partition $\{S_1, S_2\}$ of S is ensured so that $|P_i' \cup S_i| \geq n$ $(i = 1, 2)$ (one S_i may also be the empty set). Let P_i denote a subset of $P_i' \cup S_i$ with n elements $(i = 1, 2)$. According to the condition, there exist n independent P_1P_2-paths in G. Any P_1P_2-path necessarily contains a vertex of S since the endpoints of any P_1P_2-path are in distinct components of G'. Hence $|S| \geq n$, and this contradicts our indirect assumption.

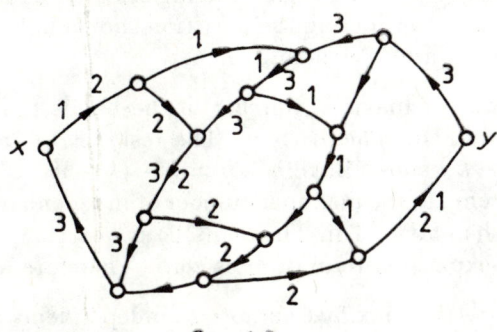

fig. 4.3.

59. The negative answer to the question can be seen from figure 4.3: it is easy to see that any two $\overleftrightarrow{xy}$-paths in this graph share at least one common edge. Three $\overleftrightarrow{xy}$-paths have been marked in the diagram by numbering the edges (edges marked by the same number define an $\overleftrightarrow{xy}$-path). The graph has no edge included in all of these three $\overleftrightarrow{xy}$-paths.

60. It suffices to verify the following statement: there exists a subset H of the set of edges of G so that every vertex of G is incident to exactly one edge of H. Indeed, in this case the edges of H are independent in G and each vertex of G is incident with an edge of H. Let us thereupon delete the edges of H from G. In the remaining bipartite graph, the degree of every vertex is $k-1$ and so, the repeated application of our statement leads to the solution of the problem.

Our assertion is proved as follows: clearly, both the number of edges incident to the vertices in A and that of the edges incident to the vertices in B equal the number of edges of G, i.e. e. Therefore, $e = k|A| = k|B|$ and so, $|A| = |B|$. Let R be a minimal set of covering vertices of G. The number of edges incident to the vertices in R is no more than $k|R|$. R being a covering set, $e = k|B| \leq k|R|$, i.e. $|B| \leq |R|$. But B is also a set of covering vertices of G and, R being minimal, $|B| \geq |R|$. Consequently, $|R| = |B| = |A|$. Hence, Theorem 14 ensures the existence of the appropriate set H.

Remark. Theorem 31 has been obtained as a consequence of Theorem 14. The solution of the problem can be easily obtained by means of Theorem 31 as well (see [2], page 125).

61. Let us associate a bipartite graph $G = G(A, B)$ to the problem as follows: the vertices of the set A correspond to the persons, the vertices of the set B to the jobs, and an edge indicates that the person corresponding to one of its endpoints takes part in the job corresponding to its other endpoint.

Now, it is evident that a minimal set of covering vertices of the graph is to be found and a questionnaire can be associated with each of its elements. Let us employ Theorem 14 and the method of alternating paths to select a maximal number of independent edges. Having selected a maximal number of independent edges and considering the partition shown in figure 2.10, $B_2 \cup A_3$ is a minimal set of covering vertices in G.

62. Let k denote the maximal number of the bands that can be selected as stated in the problem. Then $k > n$. The rest, let us call them 'x's, can be covered by $2n - k$ bands. But $2n - k = n - (k - n) < n$. Consequently, according to Theorem 16, the maximal number of independent entries x (those corresponding to the entries 1 in Theorem 16) is less than n and so at least one factor in each expansion term of M is zero. Therefore, $\det M = 0$.

63. Let m denote the maximal number of independent 'x's (non-zero elements) in M. Then $m < n$. According to Theorem 16, the 'x's can be covered by m bands. Therefore, there are at least $2n - m$ bands so that all elements included in at least two of them are zero. But $2n - m = n + (n - m) > n$.

	3		1		1	3
95	**98**	85	0	89	0	98
95	97	0	0	94	**96**	0
95	0	94	96	**95**	82	98
94	**97**	91	0	93	0	0
94	0	88	**95**	0	95	0

fig. 4.4

64. Instead of the percentage s of faulty products, the number $100 - s$ can be used for efficiency. So, we have an optimal assignment problem and apply Algorithm 18. The solution has been given in figure 4.4 with the optimal assignments marked by heavy frames. The maximal total efficiency is 481, so the minimum of the sum of the percentages of faulty products is 19. To check the solution, a covering set of weights with a total weight of 481 has also been indicated: the numbers in front of the rows and over the columns are the partial weights.

65. Since $\kappa(P_1, Q_1) = \kappa(P_2, Q_2)$ and, due to the minimal property of W_1 and W_2, $\kappa(P_3, Q_3) \geq \kappa(P_1, Q_1)$ and $\kappa(P_4, Q_4) \geq \kappa(P_2, Q_2)$, the statement immediately follows from the trivial inequality

$$\kappa(P_1, Q_1) + \kappa(P_2, Q_2) \geq \kappa(P_3, Q_3) + \kappa(P_4, Q_4).$$

66. Let us assume indirectly $p \in X$ and $p \notin X_1$ for a vertex $p \in P$. Let $Y = P - X$, $X_2 = X \cap X_1$ and $Y_2 = P - X_2$. The definition of the set X implies that

$$f(X, Y) = \kappa(X, Y) \quad \text{and} \quad f(Y, X) = 0.$$

Therefore, according to Theorem 28, the set of $\overrightarrow{XY}$-edges constitutes an edge $\overrightarrow{xy}$-cut of minimal capacity. So, in view of the statement of Problem 65, the set of $\overrightarrow{X_2Y_2}$-edges is also an edge $\overrightarrow{xy}$-cut of minimal capacity. Since $p \neq x$, the recursive definition of the set X implies that there is a pair of vertices $p_1, p_2 \in X$ with $p_1 \in X_2$, $p_2 \in Y_2$, and further, either $(p_1, p_2) \in E$ and then $f(p_1, p_2) < \kappa(p_1, p_2)$ or $(p_2, p_1) \in E$ and then $f(p_2, p_1) > 0$. Both cases contradict Theorem 28.

67. Let the set of edge $\overrightarrow{xy}$-cuts of minimal capacity in $\overrightarrow{G}$ be denoted by $\overrightarrow{\mathcal{W}}_0$, and the elements of $\overrightarrow{\mathcal{W}}_0$ by $\overrightarrow{W}_1, \overrightarrow{W}_2, \ldots, \overrightarrow{W}_k$. Let $\{X_i, Y_i\}$ be the partition of the set P induced by $\overrightarrow{W}_i$ ($i = 1, 2, \ldots, k$). Further, let $Y = P - X$ and let $\overrightarrow{W}$ be the set of $\overrightarrow{XY}$-edges. It has been verified in connection with the previous problem that $\overrightarrow{W} \in \overrightarrow{\mathcal{W}}_0$ and that $X \subseteq X_i$ ($i = 1, 2, \ldots, k$). Taking Problem 65 also into account,

$$X = \bigcap_{i=1}^{k} X_i$$

is obtained. This, however, proves the statement of the problem.

68. Let $\overrightarrow{W}_1$ and $\overrightarrow{W}_2$ be the edge $\overrightarrow{xy}$-cuts in $\overrightarrow{G}_1$ formed by the $\overrightarrow{X_1Y_1}$-edges and by the $\overrightarrow{X_2Y_2}$-edges, respectively. In connection with Problem 66, $\overrightarrow{W}_1$ has been shown to be an edge $\overrightarrow{xy}$-cut of minimal capacity in $\overrightarrow{G}_1$. Similarly, the set of $\overrightarrow{Y_2X_2}$-edges is an edge $\overrightarrow{yx}$-cut of minimal capacity in $\overrightarrow{G}_2$. Hence $\overrightarrow{W}_2$ is also an edge $\overrightarrow{xy}$-cut of minimal capacity in $\overrightarrow{G}_1$. Consequently, it suffices to verify that if $\overrightarrow{W}_1 = \overrightarrow{W}_2$ in $\overrightarrow{G}_1$ and $\overrightarrow{W}_3$ is an arbitrary edge $\overrightarrow{xy}$-cut of minimal capacity, then $\overrightarrow{W}_1 = \overrightarrow{W}_3$.

Assume that $\overrightarrow{W}_1 = \overrightarrow{W}_2$ in $\overrightarrow{G}_2$. Let us denote the set of $\overrightarrow{X_1Y_2}$-edges in $\overrightarrow{G}_1$ by $\overrightarrow{W}_{12}$. According to Problem 66, $X_1 \subseteq X_2$. Consequently, $\overrightarrow{W}_{12} \subseteq \overrightarrow{W}_2$. On the other hand, if $(p, q) \in \overrightarrow{W}_1 = \overrightarrow{W}_2$, then $p \in X_1$ and $q \in Y_2$, so $(p, q) \in \overrightarrow{W}_{12}$, i.e. $\overrightarrow{W}_2 \subseteq \overrightarrow{W}_{12}$. Therefore, $\overrightarrow{W}_1 = \overrightarrow{W}_2 = \overrightarrow{W}_{12}$.

Let now $\overrightarrow{W}_3$ be an arbitrary edge $\overrightarrow{xy}$-cut of minimal capacity in $\overrightarrow{G}_1$ and let $\overrightarrow{W}_3$ induce the partition $\{X_3, Y_3\}$ of the set P with $x \in X_3$. Finally let us denote the set of $\overrightarrow{X_3Y_2}$-edges by $\overrightarrow{W}_{32}$. According to Problem 66, $X_1 \subseteq X_3$

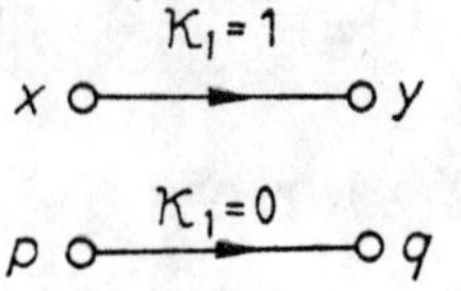

$$x \circ \xrightarrow{\kappa_1 = 1} \circ y$$

$$p \circ \xrightarrow{\kappa_1 = 0} \circ q$$

fig. 4.5

and $Y_2 \subseteq Y_3$. So $\overrightarrow{W}_1 = \overrightarrow{W}_{12} \subseteq \overrightarrow{W}_{32} \subseteq \overrightarrow{W}_3$. Since $\kappa_1(\overrightarrow{W}_1) = \kappa_1(\overrightarrow{W}_3)$, $\overrightarrow{W}_1 \subset \overrightarrow{W}_3$ cannot hold, because otherwise there would be an edge $e \in \overrightarrow{W}_3$ so that $\kappa_1(e) = 0$, although $\kappa_1(e) > 0$. Therefore, $\overrightarrow{W}_1 = \overrightarrow{W}_3$.

Remark. Figure 4.5 shows that the requirement $\kappa_1 > 0$ is essential. Indeed, here $X_1 = \{x\}$, $Y_2 = \{y\}$, both $\overrightarrow{W}_1$ and $\overrightarrow{W}_2$ contain nothing but the edge (x, y), still the two edges together constitute an edge $\overrightarrow{xy}$-cut of minimal capacity, too.

69. Algorithm 29 can be applied. Both the value of the requested flow and the capacity of the edge cut is 11.

70. For any pair i, j:

$$\min_{e \in E_i} \kappa(e) \leq \max_{e \in \overrightarrow{W}_j} \kappa(e)$$

because $\overrightarrow{W}_j$ includes at least one edge e' of the path $\overrightarrow{L}_i$, and this satisfies

$$\min_{e \in E_i} \kappa(e) \leq \kappa(e').$$

Consequently,

$$\max_i \min_{e \in E_i} \kappa(e) \leq \min_j \max_{e \in \overrightarrow{W}_j} \kappa(e).$$

Let us select at every i an edge e_i of the path $\overrightarrow{L}_i$ so that

$$\kappa(e_i) = \min_{e \in E_i} \kappa(e).$$

Let H be the set of the selected edges e_i. Since all $\overrightarrow{xy}$-paths include at least one edge of H, there is a number k so that

$$\overrightarrow{W}_k \subseteq H.$$

Consequently,

$$\max_i \min_{e \in E_i} \kappa(e) \geq \max_{e \in \overrightarrow{W}_k} \kappa(e),$$

and so,

$$\max_i \min_{e \in E_i} \kappa(e) \geq \min_j \max_{e \in \overrightarrow{W}_j} \kappa(e).$$

The desired equality follows from here and from the previous inequality of reversed direction.

71. The minimal time required by the concentration is 4. The requested plan can be seen in figure 4.6. The dotted lines indicate the time instants 0, 1, 2, 3 and 4. For simplicity, only the edges along which the concentration is executed have been drawn. The numbers written over the edges indicate the quantities to be transported.

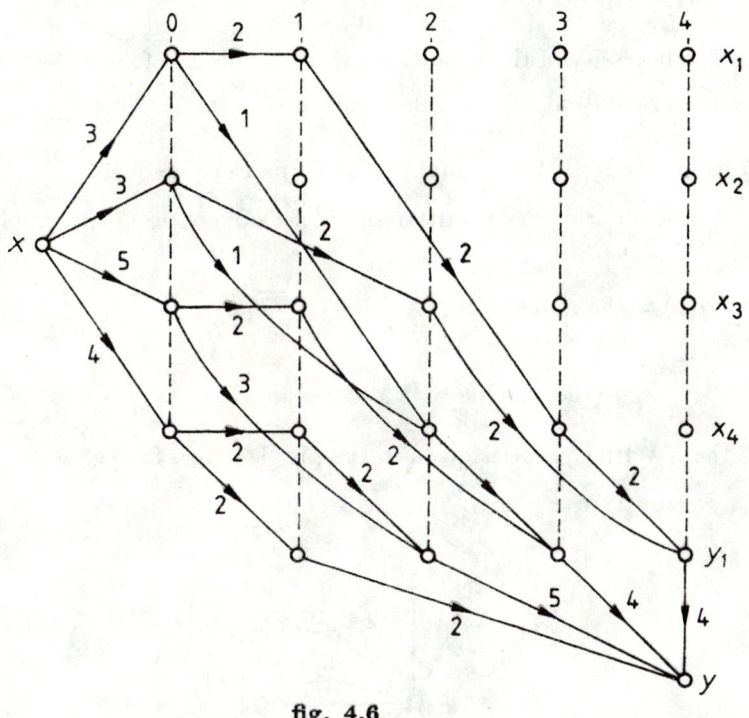

fig. 4.6

72. The fact that the condition given for feasibility is sufficient is a direct consequence of Theorem 32. In order to verify its necessity, assume the given requirements to be feasible by f. Let $\alpha^*(x) = f(x, B)$ for any vertex $x \in A$. Then Theorem 32 implies that for any sets $A_1 \subseteq A$ and $B_1 \subseteq B$:

$$\alpha^*(A_1) \leq \beta(B_1) + \kappa(A_1, B_2).$$

Since

$$\alpha(x) \leq \alpha^*(x)$$

for any $x \in A$, the inequality

$$\alpha(A_1) \leq \alpha^*(A_1)$$

holds, so the given condition is also necessary.

for any $x \in A$, the inequality
$$\alpha(A_1) \leq \alpha^*(A_1)$$
holds, so the given condition is also necessary.

73. By summing the equalities $f(p, P) - f(P, p) = 0$ over the elements of X:
$$0 = f(X, P) - f(P, X) = f(X, X) + f(X, Y) - f(Y, X) - f(X, X)$$
$$= f(X, Y) - f(Y, X) \geq \kappa_1(X, Y) - \kappa_2(Y, X).$$

74. The necessity of the condition can be verified as follows. Assume that the desired orientation is realisable yielding $\overrightarrow{G}$ from G. Then $\sum_{i=1}^{n} \alpha_i$ provides the number of edges of the graph G, and this clearly equals $\binom{n}{2}$. Further, let $\{X, Y\}$ be an arbitrary partition of the set of vertices of the graph G, and let $|X| = k$. Then:

$$\sum_{p \in X} \alpha(p) = \text{(the number of } \overrightarrow{XX}\text{-edges in } \overrightarrow{G}) \qquad (*)$$
$$+ \text{(the number of } \overrightarrow{XY}\text{-edges in } \overrightarrow{G}) \leq \binom{k}{2} + k(n - k).$$

Hence the $n - 1$ inequalities given in the problem also follow.

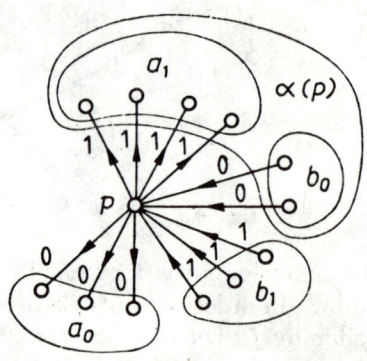

fig. 4.7

In order to verify the sufficiency of the condition, observe that if these $n - 1$ inequalities are satisfied then $(*)$ also holds for any subset X of the set of vertices of G, because with $|X| = k$:

$$\sum_{p \in X} \alpha(p) \leq \sum_{i=1}^{k} \alpha_i.$$

Now, consider the graph $\vec{G}_1 = (P, E, \vec{\mathcal{G}})$ obtained by orienting the edges of the graph G in an arbitrary way. Naturally, the desired orientation can be attained if and only if it is attainable by reversing the direction of certain edges of $\vec{G}_1$. Let a function f express this change of direction which is defined over the set of edges of $\vec{G}_1$ and its value is 0 or 1, depending on whether the direction of the relevant edge is reversed or not. Consider an arbitrary vertex p of $\vec{G}_1$ and the edges incident to P. These edges are assigned the values 0 or 1. Let a_0 and a_1 be the number of edges with value 0 and 1, respectively, with p being their tail, while b_0 and b_1 the number of edges with value 0 and 1, respectively, with p being their head. A possible case has been illustrated in figure 4.7 with $n = 13$. Observe that

$$f(p, P) - f(P, p) = a_1 - b_1.$$

It is clear from the diagram that the requested function f satisfies

$$a_1 + b_0 = \alpha(p).$$

Comparing with the former equality we obtain

$$f(p, P) - f(P, p) = \alpha(p) - (b_0 + b_1).$$

But

$$\varphi_{in}(p) = b_0 + b_1$$

in $\vec{G}_1$, so

$$f(p, P) - f(P, p) = \alpha(p) - \varphi_{in}(p).$$

Now, let the value of an edge-capacity function κ defined in $\vec{G}_1$ be 1 on every edge and let the integer function γ, defined over the set of vertices of $\vec{G}_1$, satisfy

$$\gamma(p) = \alpha(p) - \varphi_{in}(p).$$

We obtained that the desired orientation of the graph G is possible if the requirements

$$0 \le f(p, q) \le \kappa(p, q) \quad \text{if } (p, q) \in E,$$
$$f(p, P) - f(P, p) = \gamma(p) \quad \text{if } p \in P$$

are feasible in $\vec{G}_1$ by an integer function f defined over the set E. The sufficient (and necessary) condition can be seen from Theorem 35: (12) and (13) must be satisfied. Consider the former one:

$$\gamma(P) = \sum_{p \in P} \alpha(p) - \sum_{p \in P} \varphi_{in}(p) = \binom{n}{2} - |E| = 0.$$

For the latter one, let $\{X, Y\}$ be an arbitrary partition of the set P and $|X| = k$. Then, in view of the previous considerations:

$$\gamma(X) = \sum_{p \in X} \alpha(p) - \sum_{p \in X} \varphi_{\text{in}}(p) \leq \binom{n}{2} + k(n-k) - \sum_{p \in X} \varphi_{\text{in}}(p)$$

$$= \binom{k}{2} + k(n-k) - \text{(the number of } \overrightarrow{XX}\text{-edges in } \overrightarrow{G}_1)$$

$$- \text{(the number of } \overrightarrow{YX}\text{-edges in } \overrightarrow{G}_1)$$

$$= \text{(the number of } \overrightarrow{XY}\text{-edges in } \overrightarrow{G}_1) = \kappa(X, Y).$$

This completes the solution of the problem.

75. The desired arrangement is feasible, a solution can even be found by trial and error. To facilitate the solution, let us associate the complete graph $G = G(A, B)$ with the problem so that each vertex in A corresponds to a car and each vertex in B to a family. In case $p \in A$ let $\alpha(p)$ be the capacity of the car corresponding to the vertex p and in case $p \in B$ let $\beta(p)$ be the number of persons in the family corresponding to the vertex p. The desired arrangement is feasible if and only if G has a subgraph in which

$$\varphi(p) = \begin{cases} \alpha(p) & \text{if } p \in A, \\ \beta(p) & \text{if } p \in B. \end{cases}$$

However, according to Theorem 39, this is satisfied if and only if

$$\sum_{i=1}^{k} \alpha_i \leq \sum_{i=1}^{k} \beta_i^*$$

at $k = 1, 2, \ldots, 10$. Here, the values $\alpha_1, \alpha_2, \ldots, \alpha_{10}$ are the capacities of the cars in decreasing order, the values β_i^* can be computed on the basis of figure 2.40, and therein, the values $\beta_1, \beta_2, \ldots, \beta_5$ are the numbers of the persons in the families in decreasing order. So:

$$\beta_1^* = \beta_2^* = \beta_3^* = \beta_4^* = 5, \quad \beta_5^* = 4, \quad \beta_6^* = \beta_7^* = \beta_8^* = 2, \quad \beta_9^* = \beta_{10}^* = 0.$$

The ten inequalities above can be checked to be valid.

A desired arrangement can be found as follows. Let the complete graph $\overrightarrow{G}(A, B)$ obtained by assigning an orientation to G, be augmented by the source x, by the sink y and by an edge (x, a) and (b, y) for each $a \in A$ and $b \in B$. Let

$$\kappa(p, q) = \begin{cases} 1 & \text{if } p \in A \text{ and } q \in B, \\ \alpha(q) & \text{if } p = x \text{ and } q \in A, \\ \beta(p) & \text{if } p \in B \text{ and } q = y. \end{cases}$$

Any xy-flow of integer value $\alpha(A)$ in $\overrightarrow{G}$ determines a desired arrangement simply by those $\overrightarrow{AB}$-edges where the value of the flow is 1.

76. Let ϱ be a real function defined on the set E of edges of the graph $\overrightarrow{G} = (P, E, \overrightarrow{\mathcal{G}})$. It can verified by means of the considerations in §2.7 that if $\overrightarrow{G}$ contains no directed circuit of positive value then (and only then) the following algorithm provides an $\overrightarrow{xy}$-path of maximal value:

(1) If there is no $\overrightarrow{xy}$-path in $\overrightarrow{G}$ then the procedure has terminated.
(2) Let us define the real function λ over the set P as follows:
$$\lambda(p) = \begin{cases} 0 & \text{if } p = x, \\ -1 - \sum_{e \in E} |\varrho(e)| & \text{if } p \in P \text{ and } p \neq x. \end{cases}$$

(3) Let us form the number $\lambda(q) - \lambda(p)$ for each edge $(p, q) \in E$.
If there is an edge $(p, q) \in E$ with $\lambda(q) - \lambda(p) < \varrho(p, q)$,

then $\begin{cases} \text{select such an edge,} \\ \text{let } \lambda_1(r) = \begin{cases} \lambda(p) + \varrho(p, q) & \text{if } r = q, \\ \lambda(r) & \text{if } r \in P \text{ and } r \neq q; \end{cases} \\ \text{let } \lambda_1 \text{ take the role of the function } \lambda \text{ and} \\ \text{return to (3)} \end{cases}$

(4) Let us mark an edge (p, y) of $\overrightarrow{G}$ with $\lambda(y) - \lambda(p) = \varrho(p, y)$.

If $p \neq x$, then $\begin{cases} \text{let } p \text{ take the role of } y \text{ and} \\ \text{return to (4).} \end{cases}$

(5) Record the result: the marked edges constitute the edges of an $\overrightarrow{xy}$-path of maximal value in $\overrightarrow{G}$ and this maximal value is $\lambda(y)$.

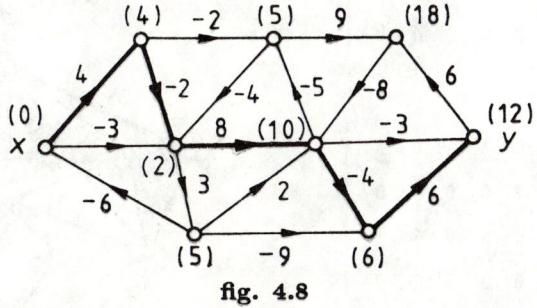

fig. 4.8

Applying this algorithm to the graph shown in figure 2.54, $\lambda(y) = 12$ is finally obtained. The edges of the requested $\overrightarrow{xy}$-path are formed by the heavy edges in figure 4.8.

77. Let $\overrightarrow{G}^*(A^*, B^*)$ be the complete bipartite graph governing the transportation, let α be the supply function and β be the demand function. Since

244   *Graph Theory: Flows, Matrices*

$$M_{01} = \begin{bmatrix} 3 & 0 & 4 & 0 & 1 & 3 & 0 \\ 3 & 3 & 9 & 2 & 0 & 9 & 0 \\ 0 & 5 & 0 & 1 & 1 & 0 & 0 \\ 7 & 3 & 7 & 2 & 3 & 7 & 0 \end{bmatrix}$$

$$M_{02} = \begin{bmatrix} 3 & 0 & 4 & 0 & 3 & 3 & 2 \\ 1 & 1 & 7 & 0 & 0 & 7 & 0 \\ 0 & 5 & 0 & 1 & 3 & 0 & 2 \\ 5 & 1 & 5 & 0 & 3 & 5 & 0 \end{bmatrix}$$

$$M_{03} = \begin{bmatrix} 0 & 1 & 1 & 2 & 2 & 0 & 4 \\ 0 & 4 & 6 & 4 & 1 & 6 & 4 \\ 0 & 9 & 0 & 6 & 5 & 0 & 7 \\ 0 & 0 & 0 & 0 & 0 & 0 & 0 \end{bmatrix}$$

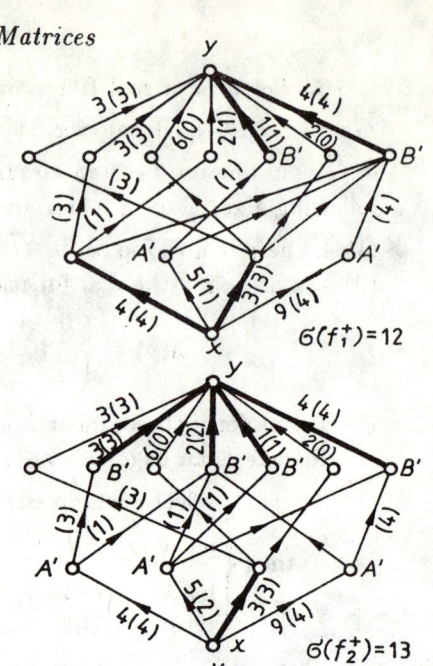

$\sigma(f_1^+) = 12$

$\sigma(f_2^+) = 13$

$$M_{04} = \begin{bmatrix} 0 & 0 & 0 & 1 & 1 & 0 & 3 \\ 0 & 3 & 5 & 3 & 0 & 6 & 3 \\ 1 & 9 & 0 & 6 & 5 & 1 & 7 \\ 1 & 0 & 0 & 0 & 0 & 1 & 0 \end{bmatrix}$$

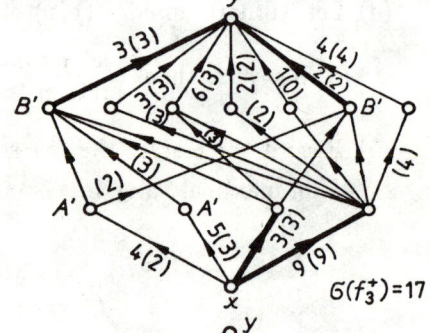

$\sigma(f_3^+) = 17$

$$M_{05} = \begin{bmatrix} 3 & 0 & 0 & 1 & 4 & 0 & 3 \\ 0 & 0 & 2 & 0 & 0 & 3 & 0 \\ 4 & 9 & 0 & 6 & 8 & 1 & 7 \\ 4 & 0 & 0 & 0 & 3 & 1 & 0 \end{bmatrix}$$

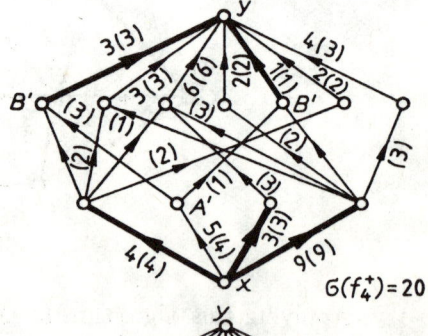

$\sigma(f_4^+) = 20$

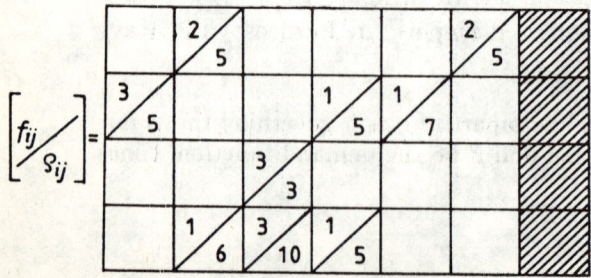

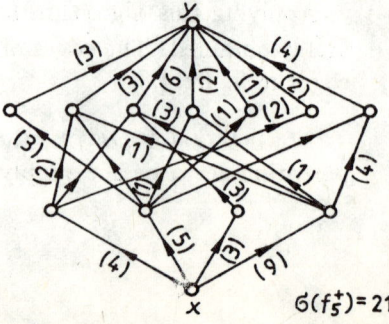

$\sigma(f_5^+) = 21$

fig. 4.9

$\left[f_{ij} \Big/ \varrho_{ij} \right] = $

$\alpha(A^*) - \beta(B^*) = 4 \neq 0$, a virtual demand of value 4 is introduced. Let the augmented graph be denoted by $\overrightarrow{G}(A, B)$. The corresponding cost matrix is:

$$M = [\varrho_{ij}] = \begin{bmatrix} 5 & 3 & 7 & 3 & 8 & 5 & 0 \\ 5 & 6 & 12 & 5 & 7 & 11 & 0 \\ 2 & 8 & 3 & 4 & 8 & 2 & 0 \\ 9 & 6 & 10 & 5 & 10 & 9 & 0 \end{bmatrix}.$$

Algorithm 44 is applied somewhat 'accelerated'. The steps of the solution have been illustrated in figure 4.9 similarly to figure 2.45. Subtractions have been applied to the columns of M, this yielded M_{01}. 2 has been subtracted from its second and fourth row and added to its fifth and seventh column. The numbers to be subtracted from the first, second and fourth row and to be added to the second, fourth, fifth and seventh column of M_{02} are $3, 1, 5$ and $4, 5, 2, 5$, respectively. For M_{03} and M_{04}, these numbers are 1 and 3, respectively. In the last chart, the values and costs of the required transportation appear together. Only the non-zero entries are indicated here. The minimal transportation cost is 93.

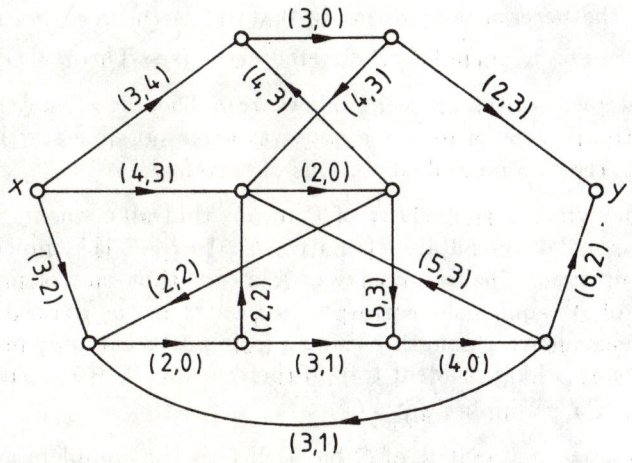

fig. 4.10

78. The problem is first reformulated to finding a flow limited by an edge-capacity function in a directed graph. The modified graph is shown in figure 4.10. It is not necessary to expand x and y into edges since their capacity is ∞. Now, Algorithm 47 can be applied. The result can be achieved in three steps, since the choice $\varepsilon_0 = 2$ can be made at each step. Figure 4.11(a) illustrates a situation when the algorithm gets stuck (this occurs in a unique way in our example) since an $\overrightarrow{xy}$-path should be found in figure 4.11(b) in order to further augment the flow. The numbers in figure 4.11(a) are the edge-capacity values and those written in rectangles are the values of the maximal xy-flow f on the edges, $\sigma(f) = 6$. The solution for the original graph is shown in figure 4.12. The cost of f can be seen here to be 40. An xy-flow of cost 44 and value 6 also exists which can be obtained by selecting the value of the flow to be 3 on the edge connecting the source x with the vertex of capacity 3 and by accordingly modifying the rest of the values.

79. According to the condition, there is a non-trivial flow of zero cost and zero value in $\vec{G}(f)$ (circulation). This can be decomposed into circuit circulations. Since, according to 47, the cost of all circuits in $\vec{G}(f)$ is non-negative, the statement is true.

Chapter 3

85. The statement of the first part is evident in view of Theorem 4, since in a loopless graph or directed graph any edge-sequence or directed edge-sequence of length 2 is also a path or directed path of length 2. For the generalisation, the necessary condition is that A^n is the zero matrix if n is sufficiently large, i.e. $\vec{G}$ includes no directed circle (see Theorem 5).

86. The statement of the problem follows from Theorem 4: any triangle is counted six times in view of its three nodes as three possible starting points and considering the two possible directions of traversal.

87. The reflexivity or irreflexivity of $\mathcal{R}$ means that all elements are 1 or 0 in the main diagonal of the adjacency matrix $A = [a_{ij}]$. $\mathcal{R}$ is symmetric if and only if A is symmetric. The asymmetry of $\mathcal{R}$ means that each element in the main diagonal of A^2 equals the corresponding entry in the main diagonal of A since an edge-sequence of length 2 leading from p to p can only be obtained by twice traversing a loop incident to p, if there is any. $\mathcal{R}$ is transitive if and only if $a_{ij}^{(2)} > 0$ $(i \neq j)$ implies $a_{ij} > 0$.

88. Let the adjacency matrix of G be A. If G is the complete graph with c vertices then let us subtract the first row of A from all the other rows and then let us add all of its columns to the first one. Then:

$$\det A = \det \begin{bmatrix} c-1 & 1 & 1 & \ldots & 1 \\ 0 & -1 & 0 & \ldots & 0 \\ 0 & 0 & -1 & \ldots & 0 \\ \ldots & \ldots & \ldots & \ldots & \ldots \\ 0 & 0 & 0 & \ldots & -1 \end{bmatrix} = (c-1)(-1)^{c-1}.$$

If G is not a complete graph then it must have two non-adjacent pairs of vertices having identical neighbours. The rows in A corresponding to these two rows are equal, so $\det A = 0$.

Solution of Problems

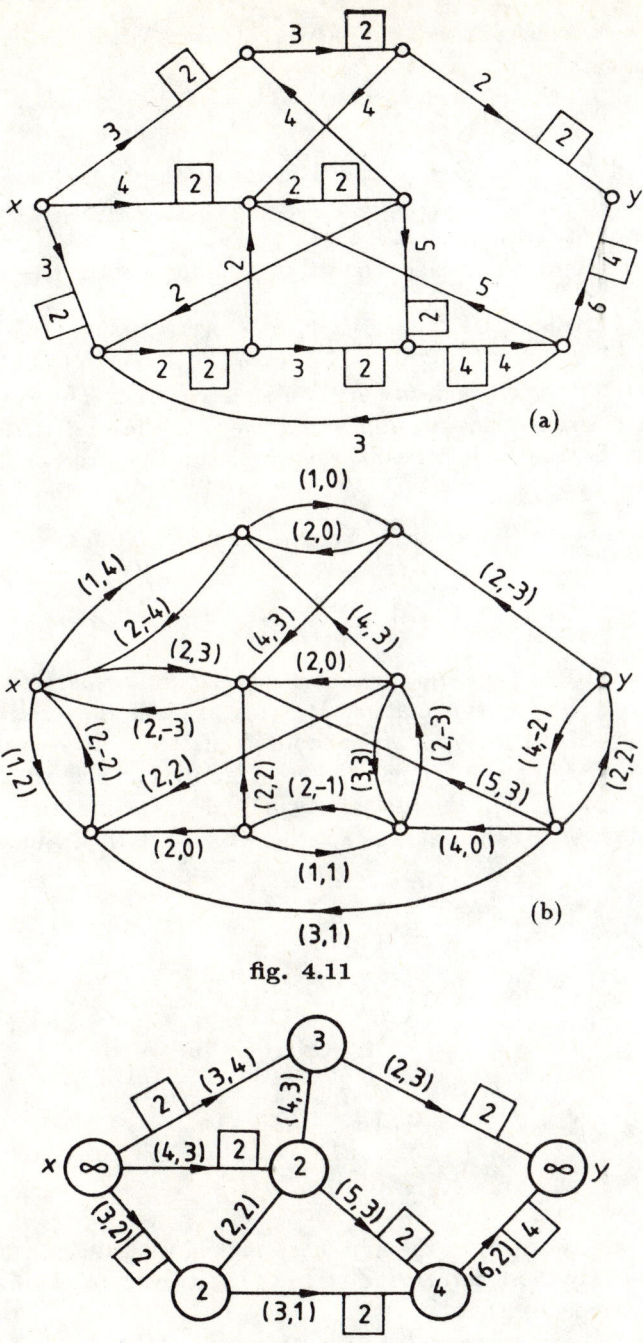

fig. 4.11

fig. 4.12

89. If G is disconnected then, in view of Theorem 2, the vertices of G can be ordered so that

$$A = \begin{bmatrix} A_{11} & N_{12} \\ N_{21} & A_{22} \end{bmatrix}$$

where N_{12} and N_{21} are zero matrices. The matrix $A + E$ is similar: N_{12} and N_{21} are unchanged. Now, N_{12} and N_{21} are present in their above positions in any power of matrices of this type.

If G is connected then consider the polynomial form of the $(c-1)$th power:

$$(A + E)^{c-1} = A^{c-1} + \binom{c-1}{1} A^{c-2} + \binom{c-1}{2} A^{c-3} + \ldots + \binom{c-1}{c-2} A + E.$$

In any of these terms, all elements are non-negative. Now Theorem 4 implies that, to any position, one can find a positive entry in one of these terms. This fact and the diagonal entries of E ensure that all entries of the sum are positive.

90. According to Theorem 2, we can assume that for a disconnected graph G:

$$D - A = \begin{bmatrix} A_{11} & N_{12} \\ N_{21} & A_{22} \end{bmatrix}$$

where N_{12} and N_{21} are zero matrices. Let us take into account that the sum of any column is zero in this matrix. Deleting any one row or column from $D - A$, the rows of either A_{22} or A_{11} remain unaffected in the obtained minor M, say those of A_{11} are unchanged. Let us apply Laplace expansion with respect to the rows of A_{11}, in order to calculate $\det M$. Since the rows of A_{11} are linearly dependent (their sum is a zero vector), $\det M = 0$ indeed holds.

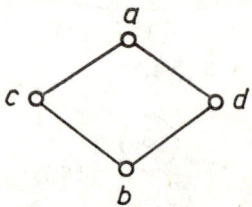

fig. 4.13

91. There are two rows in the adjacency matrix with their scalar product greater than 1 (say those corresponding to the vertices a and b) if and only if it contains a minor like this.

$$\begin{array}{c} \\ a \\ b \end{array} \begin{array}{cc} c & d \\ \begin{bmatrix} 1 & 1 \\ 1 & 1 \end{bmatrix} \end{array}.$$

The graph being simple, a, b, c and d are all distinct vertices, and this minor indicates a circuit of length four in the graph (see figure 4.13).

92. Since the graphs containing no circuit of odd length are the bipartite graphs, the statement of the problem follows from Theorem 4.

93. The number of triangles in T is evidently

$$\binom{n}{3}.$$

Now, at any orientation of the edges of a triangle H, either a directed circuit is obtained or exactly one node of H is the tail of the two edges incident with it. So it is easy to enumerate the triangles in T not yielding a directed circuit in $\vec{T}$, since s_i ($i = 1, 2, \ldots, n$) yields the number of edges in $\vec{T}$ having the ith vertex as their tail. Therefore, among the triangles in T containing an edge starting from the ith vertex, there are

$$\binom{s_i}{2}$$

not yielding a directed circuit in $\vec{T}$. So, the number S of the directed triangles in $\vec{T}$ is:

$$S = \binom{n}{3} - \sum_{i=1}^{n}\binom{s_i}{2} = \binom{n}{3} - \frac{1}{2}\sum_{i=1}^{n}(s_i^2 - s_i).$$

However, the number of edges of $\vec{T}$, i.e. of T is:

$$\sum_{i=1}^{n} s_i = \binom{n}{2}.$$

So:

$$S = \binom{n}{3} + \frac{1}{2}\binom{n}{2} - \frac{1}{2}\sum_{i=1}^{n} s_i^2.$$

In order to find the maximum of this number, the minimum of

$$S_1 = \sum_{i=1}^{n} s_i^2$$

is to be found. Since

$$\sum_{i=1}^{n} s_i$$

is constant, by standard results of calculus S_1 is minimal if the numbers s_i are all equal, i.e.

$$s_i = \frac{n-1}{2} \quad (i = 1, 2, \ldots, n).$$

If n is odd, this can be realised, so that cyclically ordering the vertices of T, edges are directed from each vertex towards the $(n-1)/2$ successively following vertices.

If n is even then, since the numbers s_i are integers, the condition of the minimality of S_1 is the satisfaction of the inequalities

$$|s_i - s_j| \leq 1$$

for any i and j. This can be realised as follows: for each of $n/2$ successively following vertices orient the adjacent edges of T towards the $(n-2)/2$ successively following vertices, and for the rest of the vertices do the same towards the $n/2$ successively following vertices. Simple arithmetic verifies that

$$\max S = \begin{cases} \dfrac{n^3 - n}{24} & \text{if } n \text{ is odd,} \\ \dfrac{n^3 - 4n}{24} & \text{if } n \text{ is even.} \end{cases}$$

Figure 4.14 illustrates solutions for the cases $n = 5$ and $n = 6$.

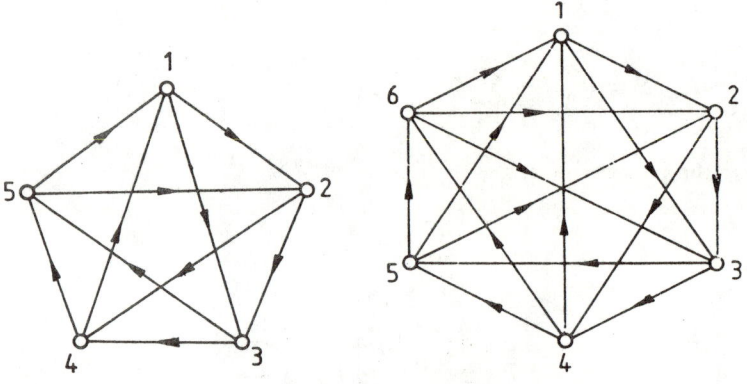

fig. 4.14

Observe that finding the minimum of S_1 can be reduced to the solution of Problem 6.36 of [2] since, obviously,

$$\left(\binom{n}{2}\right)^2 = \left(\sum_{i=1}^{n} s_i\right)^2 = \sum_{i=1}^{n} s_i^2 + 2\sum_{i<j} s_i s_j.$$

94. Let M, C and K denote a missionary, a cannibal unable to row and one able to row, respectively. Among the theoretically possible 24 states on the left bank, only 16 are allowed. These are shown in the table below; 0 indicates that there is no one on the left bank:

$MMMCCK$	$MMMK$	CCK	CK
$MMMCC$	$MMCC$	MC	C
$MMMCK$	$MMCK$	MK	K
$MMMC$	MMM	CC	O

The graph illustrating the changes in the states resulting from directly rowing from the left bank to the right one is shown in figure 4.15. We require the adjacency matrix of this graph for the calculations. The crossing will turn out to be possible. The steps of one solution, expressed by the states on the left bank are:

$MMMCCK$, $MMCK$, $MMMCK$, MMM, $MMMK$, MK, $MMCK$, MC, $MMCC$, CC, CCK, C, MC, O.

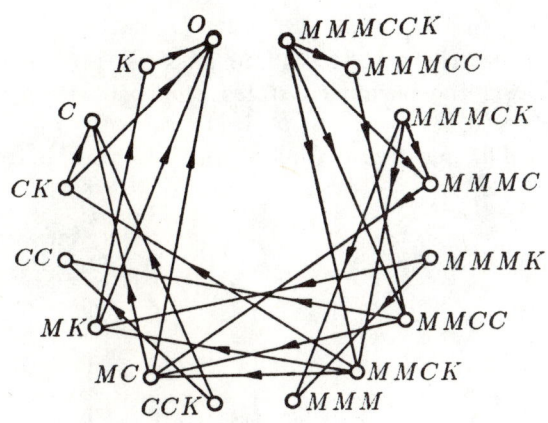

fig. 4.15

95. Yes, fewer crossovers are required. Indeed, if the difference between C and K is eliminated by writing C instead of K, too, then the partial sequence

$$MMCK, \quad MC, \quad MMCC, \quad CC$$

can be replaced in the solution by the crossovers

$$MMCC, \quad CC.$$

96. Assume that the solution requires us to reach the state p_k from the state p_1. The solution exists if and only if there is a positive integer m with non-zero entry in the position $(1, k)$ of the matrix

$$(AA^*)^m A.$$

This requires the presence of at least one non-zero element in the first row of $(AA^*)^m$ so that the corresponding entry in the kth column of the matrix

A is also non-zero. Therefore it must be decided, which elements in the first row of the matrix $(AA^*)^m$ will be non-zero at some exponent m. But this is easy even in the case of extremely large matrices. Indeed, assume that the elements of the set $\{p_a, p_b, \ldots, p_z\}$ are the vertices corresponding to the entries 1 in the first row of AA^*. These are the vertices accessible from p_1 by rowing once back and forth. If these vertices are augmented by those corresponding to the entries 1 in the ath, bth, ..., zth row of the matrix AA^* then those vertices are obtained which can be attained from p_1 by rowing twice back and forth. Let us repeat this procedure of augmenting the vertices, now starting from the rows corresponding to the subscripts of the vertices included in the enlarged set, and then let us continue in a similar fashion as long as the set keeps being augmented. If the set of the vertices obtained includes a vertex and in the position of its subscript the kth column of the matrix A is non-zero, then the problem has a solution, otherwise not.

Thereupon, consider the problem of four missionaries and four cannibals. The vertices denoting the permitted states are: $p_1(4,4)$, $p_2(4,3)$, $p_3(4,2)$, $p_4(4,1)$, $p_5(4,0)$, $p_6(3,3)$, $p_7(2,2)$, $p_8(1,1)$, $p_9(0,4)$, $p_{10}(0,3)$, $p_{11}(0,2)$, $p_{12}(0,1)$, $p_{13}(0,0)$. The question is whether p_{13} is accessible from p_1.

	p_1	p_2	p_3	p_4	p_5	p_6	p_7	p_8	p_9	p_{10}	p_{11}	p_{12}	p_{13}
p_1	0	1	1	0	0	1	0	0	0	0	0	0	0
p_2	0	0	1	1	0	1	0	0	0	0	0	0	0
p_3	0	0	0	1	1	0	1	0	0	0	0	0	0
p_4	0	0	0	0	1	0	0	0	0	0	0	0	0
p_5	0	0	0	0	0	0	0	0	0	0	0	0	0
p_6	0	0	0	0	0	0	1	0	0	0	0	0	0
p_7	0	0	0	0	0	0	0	1	0	0	1	0	0
p_8	0	0	0	0	0	0	0	0	0	0	0	1	1
p_9	0	0	0	0	0	0	0	0	0	1	1	0	0
p_{10}	0	0	0	0	0	0	0	0	0	0	1	1	0
p_{11}	0	0	0	0	0	0	0	0	0	0	0	1	1
p_{12}	0	0	0	0	0	0	0	0	0	0	0	0	1
p_{13}	0	0	0	0	0	0	0	0	0	0	0	0	0

$A =$ (the matrix above)

$$AA^* = \begin{array}{c} \\ p_1 \\ p_2 \\ p_3 \\ p_4 \\ p_5 \\ p_6 \\ p_7 \\ p_8 \\ p_9 \\ p_{10} \\ p_{11} \\ p_{12} \\ p_{13} \end{array} \begin{array}{c} \begin{array}{ccccccccccccc} p_1 & p_2 & p_3 & p_4 & p_5 & p_6 & p_7 & p_8 & p_9 & p_{10} & p_{11} & p_{12} & p_{13} \end{array} \\ \left[\begin{array}{ccccccccccccc} 1 & 1 & 0 & 0 & 0 & 0 & 0 & 0 & 0 & 0 & 0 & 0 & 0 \\ 1 & 1 & 1 & 0 & 0 & 0 & 0 & 0 & 0 & 0 & 0 & 0 & 0 \\ 0 & 1 & 1 & 1 & 0 & 1 & 0 & 0 & 0 & 0 & 0 & 0 & 0 \\ 0 & 0 & 1 & 1 & 0 & 0 & 0 & 0 & 0 & 0 & 0 & 0 & 0 \\ 0 & 0 & 0 & 0 & 0 & 0 & 0 & 0 & 0 & 0 & 0 & 0 & 0 \\ 0 & 0 & 1 & 0 & 0 & 1 & 0 & 0 & 0 & 0 & 0 & 0 & 0 \\ 0 & 0 & 0 & 0 & 0 & 0 & 1 & 0 & 1 & 1 & 0 & 0 & 0 \\ 0 & 0 & 0 & 0 & 0 & 0 & 0 & 1 & 0 & 1 & 1 & 1 & 0 \\ 0 & 0 & 0 & 0 & 0 & 0 & 1 & 0 & 1 & 1 & 0 & 0 & 0 \\ 0 & 0 & 0 & 0 & 0 & 0 & 1 & 1 & 1 & 1 & 1 & 0 & 0 \\ 0 & 0 & 0 & 0 & 0 & 0 & 0 & 1 & 0 & 1 & 1 & 1 & 0 \\ 0 & 0 & 0 & 0 & 0 & 0 & 0 & 1 & 0 & 0 & 1 & 1 & 0 \\ 0 & 0 & 0 & 0 & 0 & 0 & 0 & 0 & 0 & 0 & 0 & 0 & 0 \end{array} \right] \end{array}$$

The augmented set of vertices develops as:

$$\{p_1, p_2\}$$
$$\{p_1, p_2, p_3\}$$
$$\{p_1, p_2, p_3, p_4, p_6\}.$$

Since all elements in the positions 1, 2, 3, 4 and 6 in the 13th column of A are zero, the problem has no solution.

97. Yes, the circuit of three edges and this is the only graph with this property. Indeed, assume that the adjacency matrix of a connected graph G is A, its incidence matrix is B and $A = B$. G contains no loop, since, otherwise there would be a column in B consisting of zeros only which would indicate an isolated vertex in A. Now, it can be assumed that the edge 1 is $e_1 = \{p_2, p_3\}$, i.e. the entries at the positions $(2,1)$ and $(3,1)$ are 1 and, in view of the symmetry of A, are in the positions $(1,2)$ and $(1,3)$, too. Hence p_1, p_2 and p_3 are the nodes of a triangle. But no further edge can be incident with these three vertices since all columns in B contain two entries 1, so G is a triangle.

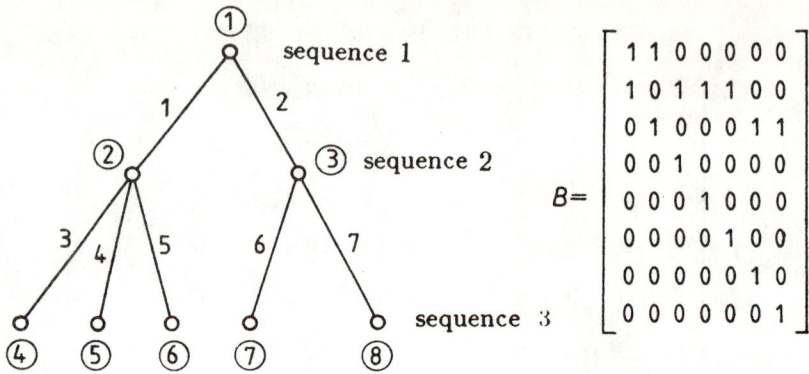

fig. 4.16

98. Assign numbers to the vertices and edges of F as follows: let some vertex p_1 of F obtain the number 1. p_1 is said to constitute the first vertex-sequence (the numbering of the vertex-sequences will play a role in the following problem). Let us now assign the following numbers to the edges of F incident to p_1. Thereupon, let us continue by assigning the following numbers to the other endpoints of the edges incident to p_1: these constitute the second vertex-sequence. Then the edges incident to these vertices are given successive numbers and thereafter, successively, the endpoints of these edges, i.e. the third vertex-sequence, and so on. All this is shown in figure 4.16 where the numbers of the vertices have been encircled. Now, if the first row of B is omitted, then a $(c-1)$-order square matrix B_1 is obtained with entries 1 in its main diagonal and zeros everywhere below, so $\det B_1 = 1$. This proves the statement of the problem.

99. As described in the solution of the previous example, let us first assign numbers to the vertices and edges of a spanning tree F of G and then to the chords of G with respect to F. In the obtained incidence matrix, let us add to the first row the rows with the corresponding vertices belonging to sequences of odd numbers, and then let us multiply by (-1) the rows with the corresponding vertices belonging to sequences with even numbers and let these also be added to the first row (in figure 4.16 there are three sequences in all).

If there is no circuit of odd length in G then both the chords of G with respect to F and its edges in F connect vertices belonging to sequences with their numbers of different parity. So, the first row of B is zero after the additions, i.e. with the last problem also taken into account,

$$r(B) = \varrho(G).$$

If, however, there is a circuit of odd length in G then the two entries 1 in the cth column of B can belong to sequences of the same parity, and so, the square minor B_1 constituting the first c columns of B after the additions satisfies
$$|\det B_1| = 2,$$
i.e.
$$c = r(B) > \varrho(G) = c - 1.$$

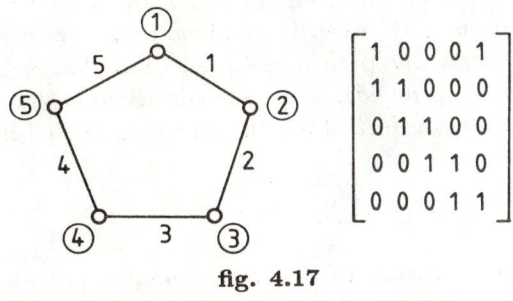

fig. 4.17

100. If the graph in question is not bipartite, consider one of its circuits of odd length, say K. Let us select the minor M of the incidence matrix of G with its rows and columns corresponding to the vertices and edges of K, respectively. Its rows and columns can be rearranged so that the entries in the main diagonal, in the positions below that and in the upper right corner are 1, and all other entries are 0. For a pentagon, such a renumbering is shown in figure 4.17. Taking into account that the order of M is odd, expansion with respect to the first row yields:
$$|\det M| = 2.$$

If G is bipartite, i.e. $G = G(P, Q)$, let us multiply the rows of the incidence matrix B corresponding to Q by (-1). All edges of G being PQ-edges, there will be exactly one entry 1 and one entry (-1) in each column of the modified B. Let us select an arbitrary square minor M_1 of this matrix. If there is a column in M_1 containing zeros only then
$$\det M_1 = 0.$$
The same is obtained if exactly two entries in all columns of M_1 are non-zero, because adding all the rows, a zero vector is obtained, i.e. the rows of M_1 are linearly dependent. If there is a column in M_1 with exactly one non-zero element then expanding $\det M_1$ with respect to this column and continuing the expansion in this way we obtain
$$|\det M_1| = 1$$

in the case of non-singular M_1. This was also true before B was modified.

101.
$$B^{-1} = \frac{\text{adj } B}{\det B},$$
and according to Theorem 18:
$$|\det B| = 1.$$
So, the entry of the matrix adj B at the position (i, j) is to be examined. This, however, equals the signed subdeterminant corresponding to the entry of the matrix B at the position (j, i). Keep in mind that any two vertices of T are connected in T by exactly one path (see [2], 2.9) and that T is decomposed into the trees T_1 and T_2 if the edge e_i is deleted. Assume that p_c belongs to T_2. The matrix remaining from B if its ith column is deleted is

$$\begin{bmatrix} B_1 & N_1 \\ N_2 & B_2' \end{bmatrix}$$

where B_1 is the incidence matrix of $\vec{T}_1$, B_2' is a reduced incidence matrix of T_2 and N_1, N_2 are zero matrices. Let us delete its jth row. Then, obviously, a singular matrix is obtained if the jth row belongs to B_2', and a matrix with determinant ± 1 if the jth row belongs to B_1. However, the latter is the case if and only if p_j belongs to T_1 which is equivalent to the statement that e_i is an edge of the $p_j p_c$-path.

102. The statement to be proved is dual to Theorem 27. To prove it, utilise the properties of the spanning forest and the fact that if any edge of a circuit is deleted then the remaining graph will be free of circuits.

103. The problem is the dual of [2] 2.17. The statement to be proved is easily verified using the properties of the spanning forest and of the cutset.

104. The requested determinants are 0 and (-2), and these are simultaneously the constant terms in the characteristic polynomials. Using Theorem 43 and the remark made in connection with figure 3.8, the characteristic polynomials can immediately be written:

$$\lambda^5 - 5\lambda^3 + 2\lambda \quad \text{and} \quad \lambda^5 - 5\lambda^3 + 5\lambda - 2.$$

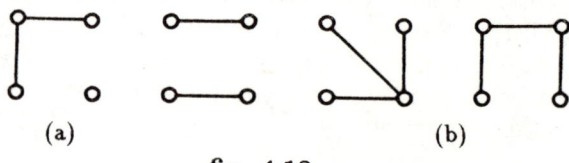

(a)   (b)

fig. 4.18

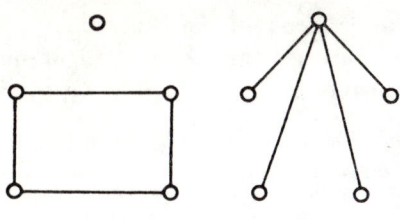

fig. 4.19

105. No. Figure 194 of [2], containing all simple graphs with four vertices, can be helpful for the answer, and so can Theorem 43 which excludes any pair different from the two shown in figure 4.18. However, in both pairs, the determinant of the adjacency matrix — which equals the constant term in the characteristic polynomial — of the first graph is zero, but that of the second one is not.

106. Only the two graphs shown in figure 4.19 are appropriate whose characteristic polynomial is $\lambda^5 - 4\lambda^3$. It is not very difficult to sketch the 34 different simple graphs with five vertices (see [74], pages 216–217). Taking Theorem 43 and the considerations in connection with figure 3.8 into account, the answer can be checked easily.

107. Let us add to the first row of the characteristic matrix of F, all the other rows multiplied by $1/\lambda$. Then the characteristic equation is

$$\lambda^{n-1}\left(n - \lambda^2\right) = 0.$$

Thus, the requested spectrum is:

$$\begin{bmatrix} 0 & -\sqrt{n} & \sqrt{n} \\ n-1 & 1 & 1 \end{bmatrix}.$$

On the basis of Theorem 44, the heuristic solution could be obtained as follows: consider an eigenvector whose coordinate associated with p_1 is 1. If λ is the corresponding eigenvalue then, on the one hand, the coordinate $1/\lambda$ is added to the rest of the vertices, and, on the other hand, the formula

$$n\frac{1}{\lambda} = \lambda$$

must also be satisfied. This yields the non-zero eigenvalues. To discover the rest, let us associate to p_1 a coordinate 0 of an eigenvector. This gives an eigenvalue 0. The coordinates corresponding to the vertices different from p_1 may be arbitrary as long as their sum is 0. This restriction results in $n-1$ linearly independent vectors, and this yields the spectrum.

108. Let us add the first row of the characteristic matrix multiplied by (-1) to the other rows, and then the rest of the columns to the first column. Then the characteristic equation

$$(\lambda + 1)^{n-1}((n-1) - \lambda) = 0$$

can easily be obtained. Therefore, the requested spectrum is:

$$\begin{bmatrix} -1 & n-1 \\ n-1 & 1 \end{bmatrix}.$$

For the heuristic solution using Theorem 44, the vector with all of its coordinates 1 is an eigenvector, and hence the eigenvalue $n-1$ is immediately obtained. The vector with one of its coordinates 1, another -1 and the rest 0 is also an eigenvector. There are $n-1$ linearly independent ones among these vectors. Hence, the complete spectrum can be obtained.

109. Let $|P| = p$, $|Q| = q$. The adjacency matrix of this graph can be written in the following form:

$$A = \begin{bmatrix} N_1 & J \\ J^* & N_2 \end{bmatrix}$$

where N_1 and N_2 are zero matrices of order p and q, respectively, and all elements of the $p \times q$ matrix J are 1. By rearranging the characteristic matrix, one can easily find that zero is a root of the characteristic equation with multiplicity $(p + q - 2)$: for example, add the last column multiplied by λ to the first one and the same multiplied by (-1) to the $(p+1)$th, $(p+2)$th, ..., $(p+q-1)$th columns. Let us expand the determinant with respect to the first row. Thereafter, let us add the columns $2, 3, \ldots, p$ to the first column and let us employ Laplace expansion with respect to the first $p - 1$ rows. Then zero is already seen to be a root of the characteristic polynomial with multiplicity at least $p - 1$. Removing the factor $(-\lambda)^{p-1}$ and adding the rows of the determinant obtained to the last row, the above assertion can be verified. Thus, the characteristic equation can be written as

$$\lambda^{p+q-2}(\lambda^2 + b\lambda + c) = 0$$

where b and c are appropriate constants. However, in view of Theorem 43, $b = 0$ and the number of edges of the graph $= p \cdot q = -c$. Consequently, the spectrum is:

$$\begin{bmatrix} 0 & -\sqrt{pq} & \sqrt{pq} \\ p+q-2 & 1 & 1 \end{bmatrix}.$$

110. In view of the corollary of Theorem 50, the spectrum can immediately be written:

$$\begin{bmatrix} 2n-4 & n-4 & -2 \\ 1 & n-1 & \frac{1}{2}n(n-3) \end{bmatrix}.$$

111. According to Problem 110, the spectrum of the edge graph of the complete graph with five vertices is:

$$\begin{bmatrix} -2 & 1 & 6 \\ 5 & 4 & 1 \end{bmatrix}.$$

So, according to Theorem 46, the requested spectrum is

$$\begin{bmatrix} 3 & 1 & -2 \\ 1 & 5 & 4 \end{bmatrix}.$$

Observe that the graph in question is the so-called *Petersen graph* shown in figure 4.20. The pairs of numbers at its nodes indicate how the vertices can correspond to the edges of the complete graph with five vertices.

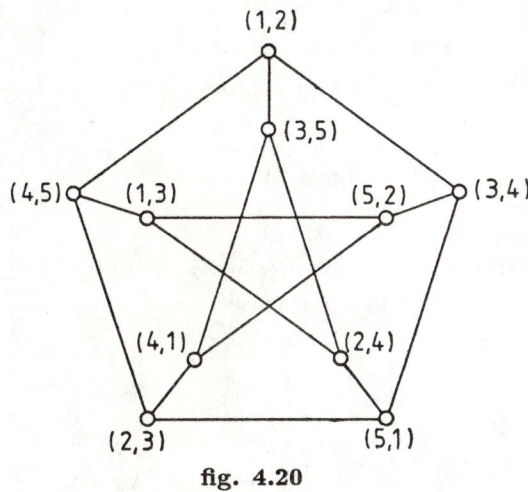

fig. 4.20

112. Let us number the vertices of the circuit along its traversal. Let A be the adjacency matrix of the circuit with n vertices and let ε denote the unit n-roots, i.e. one of the complex numbers $\sqrt[n]{1}$ and:

$$\mathbf{x}^* = [\varepsilon \quad \varepsilon^2 \quad \ldots \quad \varepsilon^n].$$

Seeking an eigenvector in this form, let us write the kth coordinate of $A\mathbf{x}$ as

$$\varepsilon^{k-1} + \varepsilon^{k+1} = \left(\varepsilon + \frac{1}{\varepsilon}\right)\varepsilon^k = (\varepsilon + \bar{\varepsilon})\varepsilon^k.$$

Hence $\varepsilon + \bar{\varepsilon}$ is an eigenvalue of A. So, the eigenvalues of A are:

$$2\cos\frac{2\pi k}{n} \quad (k = 0, 1, 2, \ldots, n-1).$$

Since there are coinciding ones among these, the requested spectrum is

$$\begin{bmatrix} 2 & 2\cos\dfrac{2\pi}{n} & 2\cos\dfrac{4\pi}{n} & \cdots & 2\cos\dfrac{(n-1)\pi}{n} \\ 1 & 2 & 2 & \cdots & 2 \end{bmatrix},$$

if n is odd, and

$$\begin{bmatrix} 2 & 2\cos\dfrac{2\pi}{n} & 2\cos\dfrac{4\pi}{n} & \cdots & 2\cos\dfrac{(n-2)\pi}{n} & -2 \\ 1 & 2 & 2 & \cdots & 2 & 1 \end{bmatrix},$$

if n is even.

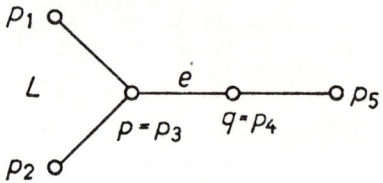

fig. 4.21

113. The following calculations, presented to the graph L in figure 4.21, indicate the line of the proof in the general case, too:

$$p(L, \lambda) = \left| \begin{array}{ccc|cc} \lambda & 0 & -1 & 0 & 0 \\ 0 & \lambda & -1 & 0 & 0 \\ -1 & -1 & \lambda & -1 & 0 \\ \hline 0 & 0 & -1 & \lambda & -1 \\ 0 & 0 & 0 & -1 & \lambda \end{array} \right|$$

$$= \begin{vmatrix} \lambda & 0 & -1 \\ 0 & \lambda & -1 \\ -1 & -1 & \lambda \end{vmatrix} \cdot \begin{vmatrix} \lambda & -1 \\ -1 & \lambda \end{vmatrix} - \begin{vmatrix} \lambda & 0 & 0 \\ 0 & \lambda & 0 \\ -1 & -1 & -1 \end{vmatrix} \cdot \begin{vmatrix} -1 & -1 \\ 0 & \lambda \end{vmatrix}$$

$$= \begin{vmatrix} \lambda & 0 & -1 \\ 0 & \lambda & -1 \\ -1 & -1 & \lambda \end{vmatrix} \cdot \begin{vmatrix} \lambda & -1 \\ -1 & \lambda \end{vmatrix} - \begin{vmatrix} \lambda & 0 \\ 0 & \lambda \end{vmatrix} \cdot \lambda$$

$$= \left| \begin{array}{ccc|cc} \lambda & 0 & -1 & 0 & 0 \\ 0 & \lambda & -1 & 0 & 0 \\ -1 & -1 & \lambda & 0 & 0 \\ \hline 0 & 0 & 0 & \lambda & -1 \\ 0 & 0 & 0 & -1 & \lambda \end{array} \right| - \left| \begin{array}{cc|c} \lambda & 0 & 0 \\ 0 & \lambda & 0 \\ \hline 0 & 0 & \lambda \end{array} \right| = p(L_1, \lambda) - p(L_2, \lambda).$$

114. Consider the adjacency matrix of the bipartite graph with vertices $p_1, p_2, \ldots, p_n$ in the form presented in Theorem 3:

$$A = \begin{bmatrix} N_{11} & A_{12} \\ A_{21}^* & N_{22} \end{bmatrix}.$$

Further, the rows of A_{12} can be assumed to correspond to the vertices $p_1, p_2, \ldots, p_k$ and $k \geq n/2$. Let us assume that zero is not an eigenvalue. Then $\det A \neq 0$, and so $\det A$ has a non-zero expansion term, i.e. the numbers $1, 2, \ldots, n$ have a permutation $i_1, i_2, \ldots, i_n$ so that

$$a_{1i_1} a_{2i_2} \ldots a_{ni_n} \neq 0.$$

Therefore, there exist edges $\{p_j, p_{i_j}\}$ in G with $j = 1, 2, \ldots, k$. Since the vertices p_{i_j} are all distinct and different from the vertices p_j, these edges constitute a one-factor of G and, of course, $k = n/2$.

115. Let G_k be the graph shown in figure 3.33 with its set of vertices P_k. Let the vertices of degree one and three in the graphs G_k constitute the sets Q_1 and Q_2, respectively, among the vertices of G. Then, $|Q_1| = 2|Q_2|$. Observe that every vertex of G has two neighbours in Q_1 and one neighbour in Q_2. Consequently, the choice

$$x_i = \begin{cases} -1 & \text{if } p_i \in Q_1, \\ 2 & \text{if } p_i \in Q_2, \end{cases}$$

$$\mathbf{x}^* = [x_1, x_2, \ldots, x_{6m}]$$

yields

$$A\mathbf{x} = \mathbf{0}$$

for the adjacency matrix A of the graph G and this implies that 0 is an eigenvalue of A.

116. If the incidence matrix of G is B and the adjacency matrix of $L(G)$ is A_L then, according to Statement 48,

$$A_L = B^* B - 2E_1,$$

where E_1 is a unit matrix. Since B^*B is positive semidefinite, it has no negative eigenvalue. Hence A_L has no eigenvalue less than (-2).

Consider the following estimate of ranks, taking into account that B has $|P|$ rows:

$$r(B^*B) \leq r(B) \leq |P| < |E|.$$

Consequently, $\det(B^*B) = 0$ and so 0 is an eigenvalue of the matrix B^*B, i.e. (-2) is an eigenvalue of A_L.

117. The spectra of an octahedron and of a cube are

$$\begin{bmatrix} 4 & 0 & -2 \\ 1 & 3 & 2 \end{bmatrix} \quad \text{and} \quad \begin{bmatrix} 3 & 1 & -1 & -3 \\ 1 & 3 & 3 & 1 \end{bmatrix},$$

respectively. The formula given in Theorem 61 can be used for the calculation. $\kappa = 384$ holds for both graphs.

118. M and N are $n/2$-order square matrices with the entries everywhere except in the main diagonal. In M, all entries in the main diagonal are $-\lambda$, while in N these are 0. The characteristic matrix of the graph in question is easily seen to be
$$\begin{bmatrix} M & N \\ N & M \end{bmatrix}.$$
If the second row of this hyper-matrix is subtracted from the first one and then its first column is added to the second one then the characteristic determinant is:
$$\det \begin{bmatrix} M & N \\ N & M \end{bmatrix} = \det \begin{bmatrix} M-N & 0 \\ N & M+N \end{bmatrix} = \det[M-N] \cdot \det[M+N].$$
Hence, the spectrum can be obtained by some arithmetic:
$$\begin{bmatrix} -2 & 0 & n-2 \\ \dfrac{n}{2}-1 & \dfrac{n}{2} & 1 \end{bmatrix}.$$
Accordingly, using the formula given in Theorem 61, the requested complexity is:
$$\frac{1}{n}(n-2)^{\frac{n}{2}} n^{\frac{n}{2}-1} = n^{n-2}\left(1 - \frac{2}{n}\right)^{\frac{n}{2}}.$$

119. A graph has as many spanning trees as sets of chords. Taking the statement of Theorem 65 into account, the statement of the problem is immediate.

120. Assume that the graph G_2 is the Whitney-dual of the graph G_1 and the one-to-one correspondence between their edges required by the definition is given. The fact that G_1 is a Whitney-dual of G_2 will be verified using this correspondence. Let G_2' be an arbitrary subgraph of G_2 and G_1' a subgraph of G_1 so that its edges are the correspondents of the edges of G_2' and let the number of edges in G_1 be e and that in G_1' be e_1. Then the number of edges in G_2 is e and that in G_2' is e_1. We have to prove that
$$\varrho(G_1 - G_1') = \varrho(G_1) - \mu(G_2').$$
According to (41):
$$\varrho(G_1 - G_1') = e - e_1 - \mu(G_1 - G_1').$$
Since G_2 is the Whitney-dual of G_1, the formula in the definition of Whitney-duals holds for the subgraph $G_1 - G_1'$ of G_1, i.e.
$$\varrho(G_2') = \varrho(G_2) - \mu(G_1 - G_1').$$

From the last two formulae:
$$\varrho(G_1 - G_1') = e - e_1 + \varrho(G_2') - \varrho(G_2).$$
According to (41):
$$\varrho(G_2') = e_1 - \mu(G_2'),$$
and according to Theorem 63:
$$\varrho(G_2) = \mu(G_1).$$
Now, the last three formulae imply
$$\varrho(G_1 - G_1') = e - \mu(G_2') - \mu(G_1),$$
but according to (41):
$$e - \mu(G_1) = \varrho(G_1),$$
which can be used to obtain the formula to be proved.

121. Let H_1' be an arbitrary subgraph of G_1' and H_2' a subgraph of G_2' so that its edges are the correspondents of the edges of H_1'. We have to prove that
$$\varrho(G_2' - H_2') = \varrho(G_2') - \mu(H_1').$$
Let H_1 and H_2 be the correspondents of H_1' and H_2' in G_1 and G_2, respectively. The graphs H_1 and H_1' are isomorphic which implies that
$$\mu(H_1) = \mu(H_1').$$
Since G_1 and G_2 are Whitney-duals:
$$\varrho(G_2 - H_2) = \varrho(G_2) - \mu(H_1).$$
If $\{a_2, b_2\}$ is a loop then its deletion does not affect the rank, so
$$\varrho(G_2') = \varrho(G_2) \quad \text{and} \quad \varrho(G_2' - H_2') = \varrho(G_2 - H_2).$$
If $\{a_2, b_2\}$ is not a loop then G_2' has one vertex less than G_2, but the number of the components of G_2' and of G_2 is the same, since the deletion of the edge $\{a_2, b_2\}$ is compensated by the unification of the vertices a_2 and b_2 as far as connectedness is concerned. Therefore,
$$\varrho(G_2') = \varrho(G_2) - 1 \quad \text{and} \quad \varrho(G_2' - H_2') = \varrho(G_2 - H_2) - 1.$$
The requested equality follows from these considerations.

122. For the indirect reasoning, let us assume the complete graph with five vertices to be planar and let G denote a planar diagram of the graph in the plane S with no intersecting edges. Further, let the vertices of G be p_1, p_2, p_3, p_4 and p_5. Let K be the circuit of length 3 in G with vertices p_1, p_2 and p_3. According to Jordan's curve theorem, K divides the points of S excluding the points of K into two disjoint regions: the interior and the exterior of K. Let us assume the vertex p_4 to be in the interior of K (if p_4 is in the exterior of K, the reasoning is similar). Since the edges of G do not intersect, all the edges $\{p_4, p_i\}$ ($i = 1, 2, 3$) run in the interior of K. These three edges divide the interior of K into three regions, these are T_1, T_2, T_3 and the exterior of K is T_4 (see figure 4.22). Now, if p_5 is in T_i then it cannot be adjacent with p_i which is a contradiction.

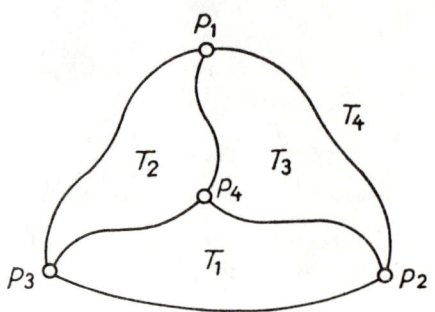

fig. 4.22

123. No, because it has a subgraph topologically equivalent to the three houses — three wells graph shown in figure 3.13: see the heavy lines in figure 4.23.

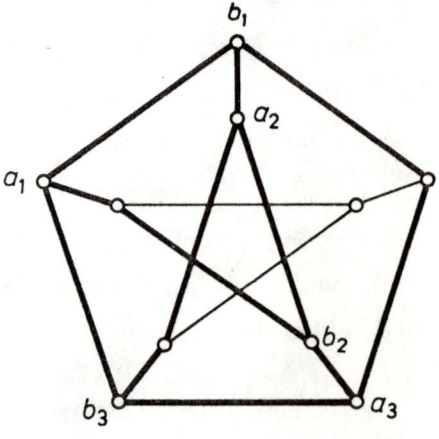

fig. 4.23

124. Figure 3.34 corresponds to figure 4.24 once the switch is closed. $c - 1 = 2$ and $e - c + 1 = 3$, it is advantageous to apply node transformation.

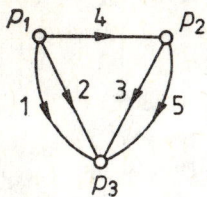

fig. 4.24

$$B_0 = \begin{bmatrix} 1 & 1 & 0 & 1 & 0 \\ 0 & 0 & 1 & -1 & 1 \end{bmatrix}, \quad Z = \begin{bmatrix} R_1 & 0 & 0 & 0 & 0 \\ 0 & sL_2 & sL_{23} & 0 & 0 \\ 0 & sL_{23} & sL_3 & 0 & 0 \\ 0 & 0 & 0 & R_4 & 0 \\ 0 & 0 & 0 & 0 & R_5 \end{bmatrix},$$

$$Y = Z^{-1} = \begin{bmatrix} \dfrac{1}{R_1} & 0 & 0 & 0 & 0 \\ 0 & \dfrac{L_3}{s(L_2 L_3 - L_{23}^2)} & \dfrac{-L_{23}}{s(L_2 L_3 - L_{23}^2)} & 0 & 0 \\ 0 & \dfrac{-L_{23}}{s(L_2 L_3 - L_{23}^2)} & \dfrac{L_2}{s(L_2 L_3 - L_{23}^2)} & 0 & 0 \\ 0 & 0 & 0 & \dfrac{1}{R_4} & 0 \\ 0 & 0 & 0 & 0 & \dfrac{1}{R_5} \end{bmatrix}.$$

With our parameters that yield time in milliseconds and the current in amperes:

$$Y = \begin{bmatrix} 4 & 0 & 0 & 0 & 0 \\ 0 & \dfrac{1}{2s} & \dfrac{-1}{20s} & 0 & 0 \\ 0 & \dfrac{-1}{20s} & \dfrac{1}{20s} & 0 & 0 \\ 0 & 0 & 0 & 1 & 0 \\ 0 & 0 & 0 & 0 & 2 \end{bmatrix}, \quad \mathbf{E} = \begin{bmatrix} \dfrac{100}{s} \\ 0 \\ 0 \\ 0 \\ 0 \end{bmatrix},$$

$$B_0 Y B_0^* = \begin{bmatrix} 4 & \frac{1}{4s} & \frac{-1}{20s} & 1 & 0 \\ 0 & \frac{-1}{20s} & \frac{1}{20s} & -1 & 2 \end{bmatrix} \begin{bmatrix} 1 & 0 \\ 1 & 0 \\ 0 & 1 \\ 1 & -1 \\ 0 & 1 \end{bmatrix} = \begin{bmatrix} 5 + \frac{1}{4s} & -1 - \frac{1}{20s} \\ -1 - \frac{1}{20s} & 3 + \frac{1}{20s} \end{bmatrix}$$

$$= \frac{1}{20s} \begin{bmatrix} 100s + 5 & -(20s+1) \\ -(20s+1) & 60s+1 \end{bmatrix}, \quad B_0 Y E = \begin{bmatrix} \frac{400}{s} \\ 0 \end{bmatrix},$$

$$(B_0 Y B_0^*)^{-1} = \frac{20s}{5(20s+1)(60s+1) - (20s+1)^2} \begin{bmatrix} 60s+1 & 20s+1 \\ 20s+1 & 100s+5 \end{bmatrix}$$

$$= \frac{20s}{(20s+1)(280s+4)} \begin{bmatrix} 60s+1 & 20s+1 \\ 20s+1 & 100s+5 \end{bmatrix},$$

$$\mathbf{V}_n = (B_0 Y B_0^*)^{-1} B_0 Y E$$

$$= \begin{bmatrix} \dfrac{8000(60s+1)}{(20s+1)(280s+4)} \\ \dfrac{8000}{280s+4} \end{bmatrix} = \begin{bmatrix} \dfrac{600s+10}{\left(s+\dfrac{1}{20}\right)\left(7s+\dfrac{1}{10}\right)} \\ \dfrac{200}{7s+\dfrac{1}{10}} \end{bmatrix} = \begin{bmatrix} v_{n1} \\ v_{n2} \end{bmatrix}.$$

Therefore, the Laplace transforms of the voltages and currents are:

$$\mathbf{V} = B_0^* \mathbf{V}_n = \begin{bmatrix} v_{n1} \\ v_{n1} \\ v_{n2} \\ v_{n1} - v_{n2} \\ v_{n2} \end{bmatrix} = \begin{bmatrix} \dfrac{600s+10}{\left(s+\dfrac{1}{20}\right)\left(7s+\dfrac{1}{10}\right)} \\ \dfrac{600s+10}{\left(s+\dfrac{1}{20}\right)\left(7s+\dfrac{1}{10}\right)} \\ \dfrac{200}{7s+\dfrac{1}{10}} \\ \dfrac{400s}{\left(s+\dfrac{1}{20}\right)\left(7s+\dfrac{1}{10}\right)} \\ \dfrac{200}{7s+\dfrac{1}{10}} \end{bmatrix},$$

$$\mathbf{I} = Y\mathbf{V} - Y\mathbf{E} = \begin{bmatrix} \dfrac{2400s+40}{\left(s+\dfrac{1}{20}\right)\left(7s+\dfrac{1}{10}\right)} - \dfrac{400}{s} \\ \dfrac{20}{s\left(s+\dfrac{1}{20}\right)} \\ \dfrac{-20}{\left(s+\dfrac{1}{20}\right)\left(7s+\dfrac{1}{10}\right)} \\ \dfrac{400s}{\left(s+\dfrac{1}{20}\right)\left(7s+\dfrac{1}{10}\right)} \\ \dfrac{400}{7s+\dfrac{1}{10}} \end{bmatrix}$$

With inverse Laplace transformation, the voltage and current vectors at $t \geq 0$ are:

$$\mathbf{v}(t) = \begin{bmatrix} 80e^{-\frac{t}{20}} + \dfrac{40}{7}e^{-\frac{t}{70}} \\ 80e^{-\frac{t}{20}} + \dfrac{40}{7}e^{-\frac{t}{70}} \\ \dfrac{200}{7}e^{-\frac{t}{70}} \\ 80e^{-\frac{t}{20}} - \dfrac{160}{7}e^{-\frac{t}{70}} \\ \dfrac{200}{7}e^{-\frac{t}{70}} \end{bmatrix}, \quad \mathbf{i}(t) = \begin{bmatrix} -400 + 320e^{-\frac{t}{20}} + \dfrac{160}{7}e^{-\frac{t}{70}} \\ 400 - 400e^{-\frac{t}{20}} \\ 80e^{-\frac{t}{20}} - 80e^{-\frac{t}{70}} \\ 80e^{-\frac{t}{20}} - \dfrac{160}{7}e^{-\frac{t}{70}} \\ \dfrac{400}{7}e^{-\frac{t}{70}} \end{bmatrix}$$

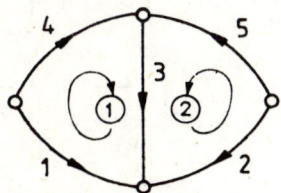

fig. 4.25

125. Figure 3.35 corresponds to figure 4.25 once the switch is closed. $c-1 = 3$ and $e-c+1 = 2$, it is advantageous to employ loop transformation.

$$K_0 = \begin{bmatrix} -1 & 0 & 1 & 1 & 0 \\ 0 & -1 & 1 & 0 & 1 \end{bmatrix}, \quad Z = \begin{bmatrix} R_1 & 0 & 0 & 0 & 0 \\ 0 & R_2 & 0 & 0 & 0 \\ 0 & 0 & R_3 & 0 & 0 \\ 0 & 0 & 0 & sL_4 & 0 \\ 0 & 0 & 0 & 0 & \dfrac{1}{sC_5} \end{bmatrix}.$$

With our parameters that yield time in milliseconds and the current in amperes:

$$Z = \begin{bmatrix} 0.35 & 0 & 0 & 0 & 0 \\ 0 & 0.35 & 0 & 0 & 0 \\ 0 & 0 & 0.65 & 0 & 0 \\ 0 & 0 & 0 & 0.21s & 0 \\ 0 & 0 & 0 & 0 & \dfrac{5.25}{s} \end{bmatrix}, \quad \mathbf{E} = \begin{bmatrix} \dfrac{10}{s} \\ \dfrac{20}{s} \\ 0 \\ 0 \\ 0 \end{bmatrix},$$

$$K_0 Z K_0^* = \begin{bmatrix} -0.35 & 0 & 0.65 & 0.21s & 0 \\ 0 & -0.35 & 0.65 & 0 & \dfrac{5.25}{s} \end{bmatrix} \begin{bmatrix} -1 & 0 \\ 0 & -1 \\ 1 & 1 \\ 1 & 0 \\ 0 & 1 \end{bmatrix}$$

$$= \begin{bmatrix} 1 + 0.21s & 0.65 \\ 0.65 & 1 + \dfrac{5.25}{s} \end{bmatrix}, \quad -K_0 \mathbf{E} = \begin{bmatrix} \dfrac{10}{s} \\ \dfrac{20}{s} \end{bmatrix},$$

$$(K_0 Z K_0^*)^{-1} = \dfrac{1}{(1+0.21s)\left(1+\dfrac{5.25}{s}\right) - 0.65^2} \begin{bmatrix} 1 + \dfrac{5.25}{s} & -0.65 \\ -0.65 & 1 + 0.21s \end{bmatrix}$$

$$= \dfrac{1}{0.21s^2 + 1.68s + 5.25} \begin{bmatrix} s + 5.25 & -0.65s \\ -0.65s & 0.21s^2 + s \end{bmatrix}$$

$$= \dfrac{1}{0.21\left((s+4)^2 + 9\right)} \begin{bmatrix} s + 5.25 & -0.65s \\ -0.65s & 0.21s^2 + s \end{bmatrix},$$

$$\mathbf{I}_m = (K_0 Z K_0^*)^{-1}(-K_0 \mathbf{E}) = \begin{bmatrix} \dfrac{-3s + 52.5}{0.21s\left((s+4)^2 + 9\right)} \\ \dfrac{4.2s^2 + 13.5s}{0.21s\left((s+4)^2 + 9\right)} \end{bmatrix} = \begin{bmatrix} \dfrac{-100s + 1750}{7s\left((s+4)^2 + 9\right)} \\ \dfrac{20s + \dfrac{450}{7}}{(s+4)^2 + 9} \end{bmatrix}.$$

Therefore, the Laplace transforms of the currents and voltages are:

$$\mathbf{I} = K_0^* \mathbf{I}_m = \begin{bmatrix} \dfrac{100s - 1750}{7s\left((s+4)^2 + 9\right)} \\ \dfrac{-20s - \dfrac{450}{7}}{(s+4)^2 + 9} \\ \dfrac{20s^2 + 50s + 250}{s\left((s+4)^2 + 9\right)} \\ \dfrac{100s - 1750}{7s\left((s+4)^2 + 9\right)} \\ \dfrac{20s + \dfrac{450}{7}}{(s+4)^2 + 9} \end{bmatrix},$$

$$\mathbf{V} = \mathbf{E} + Z\mathbf{I} = \begin{bmatrix} \dfrac{10}{s} + 5\dfrac{s - 17.5}{s\left((s+4)^2 + 9\right)} \\ \dfrac{20}{s} - \dfrac{7s + \dfrac{45}{2}}{(s+4)^2 + 9} \\ 13\dfrac{s^2 + 2.5s + 12.5}{s\left((s+4)^2 + 9\right)} \\ -3\dfrac{s - 17.5}{(s+4)^2 + 9} \\ 105\dfrac{s + \dfrac{45}{14}}{s\left((s+4)^2 + 9\right)} \end{bmatrix}.$$

With inverse Laplace transformation, the current and voltage vectors are:

$$\mathbf{i}(t) = \begin{bmatrix} -10 + 10e^{-4t}\cos 3t + \dfrac{380}{21}e^{-4t}\sin 3t \\ -20e^{-4t}\cos 3t + \dfrac{110}{21}e^{-4t}\sin 3t \\ 10 + 10e^{-4t}\cos 3t - \dfrac{70}{3}e^{-4t}\sin 3t \\ 10 - 10e^{-4t}\cos 3t - \dfrac{380}{21}e^{-4t}\sin 3t \\ 20e^{-4t}\cos 3t - \dfrac{110}{21}e^{-4t}\sin 3t \end{bmatrix},$$

$$\mathbf{v}(t) = \begin{bmatrix} 6.5 + 3.5e^{-4t}\cos 3t + \dfrac{19}{3}e^{-4t}\sin 3t \\ 20 - 7e^{-4t}\cos 3t + \dfrac{11}{6}e^{-4t}\sin 3t \\ 6.5 + 6.5e^{-4t}\cos 3t - \dfrac{91}{6}e^{-4t}\sin 3t \\ -3e^{-4t}\cos 3t + 21.5e^{-4t}\sin 3t \\ 13.5 - 13.5e^{-4t}\cos 3t + 17e^{-4t}\sin 3t \end{bmatrix}.$$

126. Let us construct the matrix below indicating the directed Hamiltonian paths:

$$M * \widehat{M}^4 = \begin{bmatrix} 0 & 0 & 126543 & 0 & 0 & 154326 \\ 0 & 0 & 0 & 0 & 263415 & 0 \\ 0 & 0 & 0 & 0 & \begin{bmatrix} 326415 \\ 341265 \end{bmatrix} & 0 \\ 0 & 0 & 0 & 0 & 0 & 0 \\ 0 & 0 & 541263 & 0 & 0 & 0 \\ 0 & 0 & 0 & 0 & 0 & 0 \end{bmatrix}.$$

If the values rather than the directed Hamiltonian paths are substituted, the requested Hamiltonian paths of values -2 and 9 can be selected:

$$\begin{bmatrix} 0 & 0 & 5 & 0 & 0 & 8 \\ 0 & 0 & 0 & 0 & 5 & 0 \\ 0 & 0 & 0 & 0 & \begin{bmatrix} 3 \\ -2 \end{bmatrix} & 0 \\ 0 & 0 & 0 & 0 & 0 & 0 \\ 0 & 0 & 9 & 0 & 0 & 0 \\ 0 & 0 & 0 & 0 & 0 & 0 \end{bmatrix}.$$

127. If D is a distance matrix, then it clearly has the given properties. Conversely, if D is given with the requested properties then let us define the graph $G = (P, E, \mathcal{G})$ as follows: $P = \{p_1, p_2, \ldots, p_c\}$ and $\{p_i, p_j\} \in E$ if and only if $d_{ij} = 1$. We must verify that the distance of the vertices p_i and p_j in G is d_{ij}. If $i = j$ or $\{p_i, p_j\} \in E$ then 0 or 1 yield the appropriate distances indeed. Assume that $i \neq j$ and $\{p_i, p_j\} \notin E$ so that $d_{ij} \geq 2$. The repeated application of (5) implies the existence of the positive integers $i_1, i_2, \ldots, i_k$ with

$$d_{ij} = d_{ii_1} + d_{i_1 i_2} + \ldots + d_{i_k j}$$

where all terms on the right-hand side equal 1. Hence $\{p_i, p_{i_1}\}, \{p_{i_1}, p_{i_2}\}, \ldots, \{p_{i_k}, p_j\}$ constitute an edge-sequence in G. Consequently there is a $p_i p_j$-path in G with its length not exceeding d_{ij}. If this distance were $< d_{ij}$ then there would be a $p_i p_j$-path in G shorter than d_{ij}. Let the vertices of such a path be $p_i, p_{j_1}, p_{j_2}, \ldots, p_{j_m}, p_j$ in the order following the traversal of the path. Then $d_{ij_1} = d_{j_1 j_2} = \ldots = d_{j_m j} = 1$. Now, according to (4):

$$d_{ij} \leq d_{ij_1} + d_{j_1 j_2} + \ldots + d_{j_m j} < d_{ij},$$

and this is a contradiction. Therefore, the distance of the vertices p_i and p_j is d_{ij}, and so the problem has been solved.

128. Let the columns of the incidence matrix B of the graph G be ordered in the same way as the edges corresponding to the columns of L. Let us delete from B the rows corresponding to the vertices x and y. This yields the matrix B_1. According to Theorem 84, $B_1 L^*$ is a zero matrix. Applying Sylvester's theorem as well as Theorem 11, we obtain

$$c - 2 + r(L^*) \leq e$$

which provides the desired inequality.

References

[1] Christofides, N: *Graph Theory, an Algorithmic Approach*, London, Academic Press, 1975.
[2] Andrásfai, B: *Introductory Graph Theory*, Budapest, Akadémiai Kiadó, Bristol, Adam Hilger, and New York, Pergamon Press, 1977.
[3] Aho, A V, Hopcroft, J E and Ullman, J D: *The Design and Analysis of Computer Algorithms*, Reading, MA, Addison–Wesley, 1975.
[4] Menger, K: *Kurventheorie*, Leipzig, B. G. Teubner, 1932.
[5] Grünwald, T (=Gallai): Ein neuer Beweis eines Mengerschen Sätzes, *J. London Math. Soc.*, **13**, 1938, 188–192
[6] Dirac, G A: Généralisations du théorème de Menger, *C. R. Acad. Sci. Paris*, **250**, 1960, 4252–4253.
[7] Menger, K: Zur allgemeinen Kurventheorie, *Fundamenta Mathematicae*, **10**, 1927, 96–105.
[8] Kőnig, D: *Theorie der endlichen und unendlichen Graphen*, Leipzig, Acad. Verlag. M. B. H., 1936.
[9] Egerváry, J: On Combinatorial Properties of Matrices, *George Washington University Logistics Papers*, 11, 1955.
[10] Kuhn, H W: The Hungarian method for the assignment problem, *Naval Res. Logist. Quart.*, **2**, 1955, 83–97.
[11] Ford, L R Jr and Fulkerson, D R: *Flows in Networks*, Princeton, Princeton University Press, 1962.
[12] Edmonds, J and Karp, P M: Theoretical improvements in algorithmic efficiency for network flow problems, *J. Assoc. Computing Machinery*, **19**, 1972, 2. 248–264.
[13] Dinic, E A: Algorithm for solution of a problem of maximum flow in a network with power estimation, *Sov. Mat. Dokl.*, **11**, 1970, 1277–1280.
[14] Karzanov, A V: Determining the maximal flow in a network by the method of preflows, *Sov. Mat. Dokl.*, **15**, 1974, 434–437.
[15] Ford, L R Jr and Fulkerson, D R: Constructing maximal dynamic flows from static flow, *Op. Res.*, **6**, 1958, 419–433.
[16] Ore, O: *Theory of Graphs*, Amer. Math. Soc. Coll. Publ., Vol. XXXVIII, 1962.
[17] Hall, P: On representations of subsets, *J. London Math. Soc.*, **10**, 1935, 26–30.
[18] Dantzig, G B: Application of the simplex method to a transportation problem, *Activity Analysis of Production and Allocation, Cowles Commission Monograph* 13, New York, Wiley, 1951, pp. 359–373.

[19] Kantorovich, L and Gavurin, M K: The application of mathematical methods in problems of freight flow analysis, *Collection of Problems Concerned with Increasing the Effectiveness of Transports, Publications of the Akademiya Nauk SSSR*, Moscow, 1949, pp. 110–138.
[20] Gale, D: A theorem on flows in networks, *Pacific J. Math.*, **7**, 1957, 1073–1082.
[21] Hoffman, A J: Some recent applications of the theory of linear inequalities to extremal combinatorial analysis, *Proc. Symposia on Applied Math.*, 10, 1960.
[22] Ryser, H J: Combinatorial properties of matrices of zeros and ones, *Canad. J. Math.*, **9**, 1957, 371–377.
[23] Ford L R: Network flow theory, *The RAND Corporation*, July 14, 1956, p. 923.
[24] Bellman, R E and Dreyfus, S E: *Applied Dynamic Programming*, Princeton University Press, 1962.
[25] Pollack, M and Wiebenson, W: Solutions of the shortest-route problem. A Review, *Op. Res.*, **8**, 1960, 224–230.
[26] Minty, G J: A comment on the shortest route problem, *Operations Res.*, **5**, 1957, 724.
[27] *P.E.R.T. Phase I and II*, Summary Report, Special Projects Office, Dept. of Navy, Washington D.C., 1958.
[28] Kaufmann, A and Desbazeille, G: *The Critical Path Method. Application of the PERT Method and its Variants to Production and Study Programs*, New York, Gordon & Breach, 1969.
[29] Hitchcock, F L: The distribution of a product from several sources to numerous localities, *J. Math. Phys.*, **20**, 1941, 224–230.
[30] Jewell, W S: Warehousing and distribution of a seasonal product, *Naval Res. Logist. Quart.*, **4**, 1957, 29–34.
[31] Busacker, R O and Gowen, P J: A procedure for determining a family of minimal-cost network patterns, *O.R.O. Technical Paper* 15, 1961.
[32] Busacker, R G and Saaty, T L: *Finite Graphs and Networks: An Introduction with Applications*, New York, McGraw-Hill, 1965.
[33] Fulkerson, D R: An out-of-kilter method for minimal cost flow problems, *J. Soc. Indust. Appl. Math.*, **9**, 1961, 18–27.
[34] Gallai, T: Maximum-minimum Sätze über Graphen, *Acta Math. Acad. Sci. Hung.*, **9**, 1958, 395–434.
[35] Gallai, T: Maximum-minimum Sätze und verallgemeinerte Faktoren von Graphen, *Acta Math. Acad. Sci. Hung.*, **12**, 1961, 131–173.
[36] Roy, B: *Algèbre moderne et théorie des graphes*, Vol. 2, Paris, Dunod, 1970.
[37] Hu, T C: *Integer Programming and Network Flows*, Reading, MA, Addison-Wesley, 1969.
[38] Lawler, E: *Combinatorial Optimization: Networks and Matroids*, New York, Holt, Rinehart & Winston, 1976.
[39] Lovász, L: *Combinatorial Problems and Exercises*, Budapest, Akadémiai Kiadó and Amsterdam, North-Holland, 1979.
[40] Edmonds, J and Fulkerson, D R: Bottleneck extrema, *J. Comb. Theory*, **8**, 1970, 299–306.
[41] Sholander, M C: The linear graphs, *Am. Math. Monthly*, **43**, 1942, 543–545.
[42] Veblen, O: *Analysis Situs, Am. Math. Soc.* Cambridge Colloquium Publications, Vol. 5, 1st edn, 1922, 2nd edn, 1931.

[43] Greub, W H: *Linear Algebra*, Berlin, Springer-Verlag, 1963.
[44] Poincaré, H: Second complément à l'analysis situs, *Proc. London Math. Soc.*, **32**, 1901, 277–308.
[45] Rózsa, P: *Lineáris algebra és alkalmazásai (Linear algebra and its applications)*, Budapest, Műszaki Könyvkiadó, 1974 (in Hungarian).
[46] Gantmacher, F R: *Applications of the Theory of Matrices*, New York, Interscience, 1959.
[47] Sachs, H: Über Teiler, Faktoren und characteristische Polynome von Graphen II., *Wiss. Z. Techn. Hochsch. Ilmenau*, **13**, 1967, 405–412.
[48] Lovász, L and Pelikán, J: On the eigenvalues of trees, *Periodica Math. Hung.*, **3**, (1–2), 1973, 175–182.
[49] Wilf, H S: The eigenvalues of a graph and its chromatic number, *J. London Math. Soc.*, **42**, 1967, 330–332.
[50] Erdős, P, Rényi, A, Rényi, T and Sós, V: On a problem of graph theory, *Studia Sci. Math. Hung.*, **1**, 1966, 215–235.
[51] Wilf, H S: The friendship theorem, in *Combinatorial Mathematics and its Applications*, ed. D J A Welsh, New York, Academic Press, 1971, pp. 307–309.
[52] Kirchhoff, G: Über die Auflösung der Gleichungen auf welche man bei der Untersuchung der linearen Verteilung galvanischer Ströme geführt wird, *Ann. Phys. Chem.*, **72**, 1847, 497–508; i.e. Leipzig, *Gesammelte Abhandlungen*, 1882, pp. 22–23.
[53] Cayley, A: A theorem on trees, *Quart. J. Math.*, **23**, 1889, 376–379. Mathematical papers, Cambridge, **13**, 1897, 26–28.
[54] Temperley, H N V: On the mutual cancellation of cluster integrals in Mayer's fugacity series, *Proc. Phys. Soc.*, **83**, 1964, 3–16.
[55] Hutschenreuther, H: Einfacher Beweis des Matrix-Gerüst-Sätzes der Netzwerktheorie, *Wiss. Z. Th. Ilmenau*, **13**, 1967, 403–404.
[56] Cvetković, D M, Doob, M and Sachs, H: *Spectra of graphs (Theory and Application)*, Berlin VEB Deutscher Verlag, 1980.
[57] Biggs, N: *Algebraic Graph Theory*, London, Cambridge University Press, 1974.
[58] Mayeda, W: *Graph Theory*, New York, Wiley–Interscience, 1972.
[59] Seshu, S and Reed, M B: *Linear Graphs and Electrical Networks*, Reading, MA, Addison-Wesley, 1961.
[60] Kuratowski, K: Sur le problème des courbes gauches en topologie, *Fund. Math.*, **15**, 1930, 271–283.
[61] Hopcroft, J and Tarjan, R: Efficient planarity testing, *J. Assoc. Computing Machinery*, **21**, 1974, 549–568.
[62] Tutte, W T: How to draw a graph, *Proc. London Math. Soc.*, **13**, 1963, 743–767.
[63] Whitney, H.: Congruent graphs and connectivity of graphs, *Amer. J. Math.*, **54**, 1932, 150–168.
[64] Whitney, H: Non-separable and planar graphs, *Trans. Amer. Math. Soc.*, **34**, 1932, 339–362.
[65] Whitney, H: Planar graphs, *Fund. Math.*, **21**, 1933, 78–84.
[66] Doetsch, G: *Introduction to the Theory and Application of the Laplace Transformation*, New York, Springer-Verlag, 1974.

[67] Bryant, P R: The order of complexity of electrical networks, *Proc. IEE*, **106**, 1959, 174.
[68] Kuh, E S and Rohrer, R A: The state variable approach to network analysis, *Proc. IEE*, **53**, 1965, 672.
[69] Rohrer, R A: *Circuit Theory: An Introduction to the State Variable Approach*, New York, McGraw-Hill, 1970.
[70] Welsh, D J A: *Matroid Theory*, London, Academic Press, 1976.
[71] Recski, A: *Matroid Theory and its Applications*, Budapest, Akadémiai Kiadó, and Berlin, Springer–Verlag, 1988.
[72] Wing, O and Kim, W H: The path matrix and its realizability, *IRE Trans. Circuit Theory*, **CT-6**, (3), 1959, 267–272.
[73] Harary, F and Manvel, B: On the number of cycles in a graph, *Math. Časopis Sloven. Akad. Vied.*, **21**, 1971, 55–63.
[74] Harary, F: *Graph Theory*, Reading, MA, Addison-Wesley, 1969.
[75] Wilson, R J: *Introduction to Graph Theory*, Edinburgh, Oliver & Boyd, 1972.
[76] Branin, F H Jr: The inverse of the incidence matrix of a tree and the formulation of the algebraic-first-order differential equations of an RLC network, *IEEE Trans. Circuit Theory*, **CT-10**, 1963, 543–544.
[77] Mowshowitz, A: The characteristic polynomial of a graph, *J. Combinatorial Theory*, **B12**, 1972, 177–193.
[78] Finck, H J and Sachs H: Über Beziehungen zwischen Struktur und Spektrum regulärer Graphen, *Wiss. Z. Th. Ilmenau*, **19**, 1973, 83–99.
[79] Hoffman, A J: Some recent results on spectral properties of graphs, *Beiträge zur Graphentheorie (Kolloquium Manebach 1967)*, Leipzig, 1968, pp. 75–80.

Subject index

Abstract graph, 1
Accessible vertex along a bypass path relative to a subgraph, 34
Accessible vertex along a flow augmenting path relative to a flow, 62
Accessible vertex along an alternating path, 44
Active network, 205
Acyclic graph, 22
Adjacency matrix, 133
Adjacent vertices, 1
Admittance matrix, 213
Algorithm, 36
Alternating path, 44
Articulation, 13
Attainability matrix, 216
Average degree, 180

Back-capacity, 60
Band, 47
Band covering one-entries in a matrix, 47
Basis to a matroid, 215
Binary relation, 2
Bipartite graph, 44
Block, 13
Block-chain, 15
Block-graph, 16–17
Bounded variation for a function, 203
Branch, 201
Bridge, 6
Bridge connects leaves, 7
Bypass path, 43
Bypass path relative to a subgraph, 34

Capacity, 28
Capacity of a vertex, 71
Capacity of an edge, 29
Capacity of an edge set, 56
Characteristic equation of a graph, 168
Characteristic matrix of a graph, 168
Characteristic polynomial of a graph, 168
Chromatic number, 180
Circuit matrix, 146
Circulation, 58
Circulation problem, 93
Complete bipartite directed graph, 81
Complete bipartite graph, 49
Complexity of a graph, 183
Component, 4
Connect, 1
Connected graph, 19
Construction of the dual graph, 196
Cost function, 108
Cost matrix, 108
Cost of a flow, 117
Covered edge, 49
Covering system of weights, 48
Covering system of weights covers the edges of a graph, 49
Critical job, 105
Critical path, 105
Current equations, 202
Cutset, 153, 157
Cutset matrix, 157
Cutset-transformation, 208
Cyclomatic number, 148

Decomposed into path flows, 58
Demand, 77
Detour matrix, 216
Diagram in space, 3
Directed edge, 1
Directed edge-sequence of length m, 135
Directed graph, 1
Directed leaf, 21
Directed leaf-graph, 22
Distance matrix, 216
Distance of vertices, 135
Dual, 193
Duals, 192
Dynamic problem, 28

Edge, 1
Edge-capacity function, 28, 56
Edge connects vertices, 1
Edge cut, 31, 43
Edge-disjoint directed paths, 20
Edge-disjoint paths, 4
Edge graph, 174
Edge matrix, 139
Edge-sequence of a lenght m, 135
Eigenvalue of a graph, 168
Eigenvector of a graph, 168
Electrical network, 204
Endpoint of an edge, 1
End-vertex of a directed graph, 22
Equivalence class, 2
Equivalence relation, 2
Event, 105
Exactly covered edge, 49

f-circuit, 148
f-circuit matrix, 150
f-cutset, 156
f-cutset matrix, 156, 159
Feasibility, 78
Flow, 29
Flow augmenting path relative to a flow, 60
Flow function, 56

Flow, limited by an edge-capacity function, 56
Flow, limited by capacity, 57, 72
f-system of circuits, 148
f-system of cutsets, 156
Fundamental circuit, 148
Fundamental circuit matrix, 148, 150
Fundamental cutset, 156
Fundamental cutset matrix, 159
Fundamental system of circuits, 148
Fundamental system of cutsets, 156

Geometrical realisation of a graph, 3
Graph, 1

Head of an edge, 1
Hungarian method, 52

Impedance matrix, 213
Incidence matrix, 139
Incident, 1
Independent edges, 44
Independent one-entries in a matrix, 47
Independent paths, 42
Independent sets to matroids, 215
Infinite graph, 1
Isomorphic graphs, 2

Job, 104

Kirchhoff's current law, 204
Kirchhoff's voltage law, 204

Latin matrix, 217
Leaf–bridge sequence, 8
Leaf-chain, 10
Leaf-graph, 10
Leaf of a directed graph, 21
Leaf of a graph, 6
Length of a directed path or circuit, 20
Linear network, 204
Linear, time-invariant network, 204

Linear, time-variant network, 204
Loop, 202
Loop-current, 208
Loop equation, 214
Loop transformation, 206

Matroid, 215
Maximal number of edge-disjoint paths, 31, 43
Maximal number of independent paths, 42, 43
Maximal number of vertex-disjoint paths, 38, 43
Method of alternating paths, 44
Minimal edge cut, 31, 43
Minimal number of covering vertices, 44
Minimal number of cut edges, 31, 43
Minimal number of cut vertices, 28, 43
Minimal number of vertices covering paths, 42, 43
Minimal vertex cut, 28, 43
Möbius ladder, 171
Multiple edges, 88

n-connected graph, 127
Node, 1
Node equation, 214
Node transformation, 207
Node-voltage, 208
Nullity, 148

Optimal assignment, 53
Optimal matching, 52
Over-covered edge, 49

Passive network, 205
Partition of a vertex set, 4, 11
Partition of a vertex set induced by a relation, 4
Path, 4, 42
Path flow, 56
Path matrix, 217
Petersen graph, 259
Planar diagram, 3
Planar graph, 3
Point, 1

Positive definite matrix, 205
Positive semidefinite matrix, 205
Program graph, 105

Rank of a directed graph, 144
Reciprocal network, 204
Reduced circuit matrix, 150
Reduced cutset matrix, 159
Reduced edge matrix, 141, 143, 145
Reduced incidence matrix, 141, 143, 145
Reflexive relation, 2
Relation $\mathcal{E}$, 20
Relation $\mathcal{L}$, 24
Relation $\mathcal{P}$, 24
Relation $\mathcal{R}$, 1
Relation $\mathcal{R}_i$, 31
Relation $\mathcal{R}_0$, 2
Relation $\mathcal{R}_1$, 3
Relation $\mathcal{R}_2$, 4
Relation $\mathcal{T}$, 12
Relation $\mathcal{V}_i$, 31
Relation $\mathcal{V}'$, 25
Relation $\mathcal{V}_1$, 24
Relation $\mathcal{V}_2$, 11
Residual capacity relative to a flow, 60

Separation into blocks, 196
Set of covering vertices, 44
Set of independent vertices, 20
Set of vertices covering paths, 42
Simple arc, 3
Sink, 28
Source, 28
Spanning forest, 144
Spanning tree, 144
Spectrum of a graph, 168
Starting vertex of a directed graph, 22
Static problem, 28
Strongly connected component, 21
Strongly connected graph, 19
Subgraph, 19
Subgraph, induced, 19
Subtraction of graphs, 199
Sum of weights of a covering system of weights, 48
Supply, 77

Support of path flow, 56
Symmetric relation, 2

Tail of an edge, 1
Topological image, 2
Topologically equivalent, 194
Transitive relation, 2
Transportation, 108
Transportation cost, 108
Turning a graph about two vertices, 197

Valency problem, 95
Value of a flow, 29, 72
Value of a function on an edge, 55
Value of a subgraph, 98

Value of a system of edges, 49
Value on an edge, 55
Value to an edge, 49
Vertex, 1
Vertex-capacity function, 71
Vertex cut, 28, 43
Vertex-disjoint directed paths, 20
Vertex-disjoint paths, 14
Virtual job, 105
Voltage equations, 202

Warehousing problem, 78
Weakly isomorphic graphs, 197
Weight of a vertex, 49
Whitney-dual, 199
Windmill, 181